# Tolerance and Response of Ornamental Plants to Abiotic Stress

# Tolerance and Response of Ornamental Plants to Abiotic Stress

Guest Editors

**Zhouli Liu**

**Yi Zhao**

MDPI

Basel • Beijing • Wuhan • Barcelona • Belgrade • Novi Sad • Cluj • Manchester

*Guest Editors*
Zhouli Liu
College of Life Science and Engineering
Shenyang University
Shenyang
China

Yi Zhao
School of Chemistry and Environmental Engineering
Liaoning University of Technology
Jinzhou
China

*Editorial Office*
MDPI AG
Grosspeteranlage 5
4052 Basel, Switzerland

This is a reprint of the Special Issue, published open access by the journal *Horticulturae* (ISSN 2311-7524), freely accessible at: https://www.mdpi.com/journal/horticulturae/special_issues/WG6839073D.

For citation purposes, cite each article independently as indicated on the article page online and as indicated below:

| |
|---|
| Lastname, A.A.; Lastname, B.B. Article Title. *Journal Name* **Year**, *Volume Number*, Page Range. |

**ISBN 978-3-7258-5343-4 (Hbk)**
**ISBN 978-3-7258-5344-1 (PDF)**
**https://doi.org/10.3390/books978-3-7258-5344-1**

# Contents

# About the Editors

**Zhouli Liu**

Zhouli Liu is currently associated with the College of Life Science and Engineering, Shenyang University. By virtue of the Key Laboratory of Regional Pollution Environment and Ecological Restoration (Ministry of Education), her research is mainly focused on stress responses to adverse environments and stress-resistant physiological ecology. From 2001 to 2005, she pursued her studies at Shenyang Agricultural University, where her research emphasis was placed on the stress resistance responses of horticultural plants, and she was awarded a Bachelor of Agronomy degree upon graduation. Between 2005 and 2010, she carried out research related to Ecology at the University of Chinese Academy of Sciences (formerly the Graduate University of Chinese Academy of Sciences) and obtained a Doctor of Science degree. During the ten-year period from 2010 to 2020, she engaged in scientific research on urban ecology at the Shenyang Institute of Applied Ecology, Chinese Academy of Sciences. Since 2020, she has been conducting research and teaching work in relevant fields at Shenyang University. Throughout her career, she has successively led or participated as a core member in more than 10 scientific research projects, accumulated rich academic achievements, and published over 60 academic papers. In addition, she also served as a Guest Editor for a Special Issue of *Horticulturae*, which further demonstrates her professional status in this field.

**Yi Zhao**

Yi Zhao is currently associated with the School of Chemistry and Environmental Engineering, Liaoning University of Technology. Her research mainly focuses on the carbon–nitrogen cycle, greenhouse gas emissions, and the physiological ecology of plant stress resistance. She has carried out extensive analytical research in the fields of plant greenhouse gas stress and soil carbon cycling. She is skilled at using the stable isotope 13C technology model to systematically analyze the changes in soil carbon pools in different production systems under various management measures. She has conducted systematic research on carbon turnover in three different systems: farmland, facility vegetable fields, and restored farmland in mining areas, achieving leading theoretical innovation and application promotion. At the same time, a large amount of analytical research work has been carried out in the study of the stress of greenhouse gases on plants. More than 20 academic achievements have been published in authoritative journals. She has presided over and participated in more than 10 other national and provincial projects, and served over 20 local research projects.

*horticulturae*

*Editorial*

# Tolerance and Response of Ornamental Plants to Abiotic Stress

**Zhouli Liu** [1,2,3]

[1] College of Life Science and Engineering, Shenyang University, Shenyang 110044, China; zlliu@syu.edu.cn
[2] Northeast Geological S&T Innovation Center of China Geological Survey, Shenyang 110000, China
[3] Key Laboratory of Black Soil Evolution and Ecological Effect, Ministry of Natural Resources, Shenyang 110000, China

## 1. Introduction

Ornamental plants play a pivotal role in environmental decoration, ecological balance, and air purification. However, they are facing escalating challenges from abiotic stresses, including heavy metal pollution, extreme temperatures, drought, salinity, and ozone exposure [1]. These stresses disrupt plant growth, metabolism, and survival, necessitating urgent research to unravel the tolerance mechanisms of plants and develop stress mitigation strategies. This Special Issue, "Tolerance and Response of Ornamental Plants to Abiotic Stress," compiles cutting-edge studies exploring how ornamental species adapt to diverse abiotic pressures, offering insights into their physiological, biochemical, and molecular responses. Through a rigorous peer review process, 10 research articles were selected, spanning topics ranging from stress physiology to innovative mitigation techniques, which collectively advance our understanding of ornamental plant resilience.

## 2. Overview of Published Articles

The articles in this Issue address various abiotic stressors and tolerance mechanisms associated with multiple ornamental species. They are categorized into four thematic areas:

### *2.1. Physiological and Biochemical Adaptations to Stress*

The studies in this category focus on how plants modulate physiological processes in order to cope with stress. For example, Kryuchkova et al. (2025) investigated winter hardiness in *Lavandula angustifolia* hybrids, identifying critical climatic factors (e.g., temperature fluctuations and snow cover) influencing their survival. The study revealed that hybrids with stable antioxidant enzyme activity exhibited superior cold tolerance, providing a foundation for breeding cold-resistant cultivars [2]. Wang et al. (2023) evaluated pH stress tolerance in Rhododendron genotypes through a 140-day greenhouse experiment, demonstrating that a neutral pH (6.3) significantly inhibited root development, plant height, and biomass accumulation, while causing chlorosis and reducing chlorophyll fluorescence (Fv/Fm). Notably, the genotype PB-T3-4 showed superior tolerance, maintaining root growth and photosynthetic efficiency under a neutral pH, highlighting how genetic variation can influence stress adaptation [3]. Shala et al. (2024) explored the role of gamma-aminobutyric acid (GABA) in alleviating the effects of salinity stress in *Lavandula dentata*. Foliar application of 40 mM GABA mitigated chlorophyll degradation, enhanced the accumulation of osmolytes (e.g., proline), and improved antioxidant enzyme activity, demonstrating its potential as a cost-effective stress mitigator [4].

 https://doi.org/10.3390/horticulturae11060704

### 2.2. Chemical and Molecular Interventions

Several studies highlight the efficacy of chemical compounds and gene regulation in enhancing stress tolerance. Liu et al. (2025) demonstrated that hydrogen sulfide ($H_2S$) alleviates manganese toxicity in *Malus hupehensis* by regulating the ascorbate–glutathione cycle and mineral homeostasis. $H_2S$ treatment reduced Mn accumulation in the roots and upregulated the activity of metal transporters (MTPs), emphasizing its dual role in detoxification and nutrient balance [5]. Elmongy and Abd El-Baset (2024) investigated melatonin's effects on heat-stressed carnations (*Dianthus caryophyllus*). Melatonin at a concentration of 5–10 mM enhanced the carnations' chlorophyll content, reduced the abundance of reactive oxygen species (ROS), and upregulated heat shock proteins (HSP70, HSP101), showcasing its potential to improve thermal resilience in ornamental crops [6].

### 2.3. Molecular and Genetic Mechanisms

Genomic and transcriptomic analyses dominate this theme. Yu et al. (2024) characterized the Isopentenyl Diphosphate Isomerase (IPI) gene in Fritillaria unibracteata, revealing its role in β-carotene synthesis and drought/salt tolerance. Transgenic Arabidopsis plants that overexpressed FuIPI exhibited higher abscisic acid levels and stress resistance, highlighting the utility of this gene for the genetic improvement of liliaceous plants [7]. Li et al. (contribution 5) used transcriptomics to dissect the effects of light and temperature on Hydrangea macrophylla. Key genes involved in photosynthesis (e.g., PHYB, PSBR) and hormone signaling (e.g., PIN3, EIN3) were identified, providing molecular markers for breeding shade-tolerant hydrangea cultivars [8].

### 2.4. Non-Destructive Monitoring and Stress Assessment

Innovative technologies for stress detection are showcased in two studies. Dmitriev et al. (2024) developed a hyperspectral imaging model to distinguish "Vegetation" and "Dormancy" states in conifers (Platycladus orientalis, Thuja occidentalis). Using linear discriminant analysis (LDA) and random forest algorithms, the model achieved >92% accuracy in the diagnosis of conifers' phenological states, enabling timely management of conifer plantations under climate change [9]. Qu et al. (contribution 4) proposed a comprehensive method to evaluate cold tolerance in Stenotaphrum accessions by integrating fall dormancy and spring green-up phenotypes. This approach outperformed traditional laboratory assays (e.g., LT50), identifying superior cold-tolerant genotypes (e.g., ST003, S13) for temperate regions [10].

## 3. Conclusions

The studies presented in this Special Issue collectively advance our understanding of ornamental plants' responses to abiotic stress, from physiological adaptations to molecular mechanisms. They highlight the potential of chemical priming (e.g., $H_2S$, GABA), genetic engineering, and non-destructive monitoring techniques to enhance stress resilience. These findings not only provide theoretical insights, but also offer practical tools that can be used by breeders and horticulturists to develop climate-smart ornamental varieties.

We extend our gratitude to all of the authors who made valuable contributions and to the reviewers who conducted rigorous evaluations. We hope that this issue will foster cross-disciplinary collaboration and inspire further research aiming to safeguard ornamental plant sustainability in an increasingly stressful environment.

**Funding:** This study was supported by the LiaoNing Revitalization Talents Program (XLYC2203070), the Science and Technology Plan Joint Project Natural Science Foundation–General Program of Liaoning Province (2024-MSLH-506), and the funding project of the Northeast Geological S&T Innovation Center of China Geological Survey (QCJJ2022-44).

**Acknowledgments:** Z.L. is grateful to Y.Z., from the School of Chemistry and Environmental Engineering, Liaoning University of Technology, for her valuable contributions to and support of this research.

**Conflicts of Interest:** The author declares no conflicts of interest.

## References


1. Liu, Z.; Hu, B.; Zhao, Y.; Zhang, S.; Duan, X.; Liu, H.; Meng, L. Visual Analysis of Research Progress on the Impact of Cadmium Stress on Horticultural Plants over 25 Years. *Horticulturae* **2025**, *11*, 28. [CrossRef]
2. Kryuchkova, V.; Evtyukhova, A.; Avdeev, S.; Donskih, V.; Shelepova, O.; Ladyzhenskaya, O.; Gorbunov, Y. Lavender Breeding for Winter Hardiness in a Temperate Climate. *Horticulturae* **2025**, *11*, 139. [CrossRef]
3. Wang, S.; Van Labeke, M.C.; Dhooghe, E.; Van Huylenbroeck, J.; Leus, L. Greenhouse Screening for pH Stress in *Rhododendron* Genotypes. *Horticulturae* **2023**, *9*, 1302. [CrossRef]
4. Shala, A.Y.; Aboukamar, A.N.; Gururani, M.A. Exogenous Application of Gamma Aminobutyric Acid Improves the Morpho-Physiological and Biochemical Attributes in *Lavandula dentata* L. under Salinity Stress. *Horticulturae* **2024**, *10*, 410. [CrossRef]
5. Liu, B.; Wang, B.; Chen, T.; Zhang, M. Hydrogen Sulfide Mitigates Manganese-Induced Toxicity in *Malus hupehensis* Plants by Regulating Osmoregulation, Antioxidant Defense, Mineral Homeostasis, and Glutathione Ascorbate Cycle. *Horticulturae* **2025**, *11*, 133. [CrossRef]
6. Elmongy, M.S.; Abd El-Baset, M.M. Melatonin Application Induced Physiological and Molecular Changes in Carnation (*Dianthus caryophyllus* L.) under Heat Stress. *Horticulturae* **2024**, *10*, 122. [CrossRef]
7. Yu, X.; Chen, J.; Yan, H.; Huang, X.; Chen, J.; Ma, Z.; Zhou, J.; Liao, H. Functional Identification of the Isopentenyl Diphosphate Isomerase Gene from *Fritillaria unibracteata*. *Horticulturae* **2024**, *10*, 887. [CrossRef]
8. Li, Z.; Lyu, T.; Lyu, Y. The Molecular Biology Analysis for the Growing and Development of *Hydrangea macrophylla* 'Endless Summer' under Different Light and Temperature Conditions. *Horticulturae* **2024**, *10*, 586. [CrossRef]
9. Dmitriev, P.A.; Kozlovsky, B.L.; Dmitrieva, A.A. Vegetation and Dormancy States Identification in Coniferous Plants Based on Hyperspectral Imaging Data. *Horticulturae* **2024**, *10*, 241. [CrossRef]
10. Qu, J.; Hao, D.L.; Zhou, J.Y.; Chen, J.B.; Sun, D.J.; Liu, J.X.; Zong, J.Q.; Wang, Z.Y. Evaluating the Cold Tolerance of *Stenotaphrum* Trin Plants by Integrating Their Performance at Both Fall Dormancy and Spring Green-Up. *Horticulturae* **2024**, *10*, 761. [CrossRef]

horticulturae

MDPI

*Article*

# Lavender Breeding for Winter Hardiness in a Temperate Climate

**Viktoriya Kryuchkova [1,*], Anastasia Evtyukhova [1], Sergey Avdeev [2], Vitaly Donskih [1], Olga Shelepova [1], Olga Ladyzhenskaya [1] and Yuri Gorbunov [1]**

[1] N.V. Tsitsin Main Botanical Garden of Russian Academy of Science, Botanicheskaya, 4, 127276 Moscow, Russia; anastasia_flowerdesign@bk.ru (A.E.); donskih.65@yandex.ru (V.D.); shov_gbsad@mail.ru (O.S.); o.ladyzhenskaya91@mail.ru (O.L.); gbsran@mail.ru (Y.G.)

[2] All-Russia Research Institute of Agricultural Biotechnology, Timiryazevskaya 42, 127550 Moscow, Russia; avdeevbio@yandex.ru

* Correspondence: vkruchkova@mail.ru

**Abstract:** *Lavandula angustifolia* is a promising essential oil and ornamental crop whose distribution in the temperate zone and northern regions is limited by its low winter hardiness. Analyzing the causes of low winter hardiness will facilitate the selection of the most winter-hardy hybrids. The study goal is to evaluate the climatic conditions and winter hardiness of narrow-leaved lavender and to determine critical conditions for the successful overwintering of plants in the conditions of Moscow. The studies were conducted in the laboratory of cultivated plants of MBG RAS from 2015 to 2022. The research objects were 72 lavender hybrids. The assessment of hybrids' winter hardiness was carried out after complete snow melt. Average daily temperature, snow cover height, and precipitation were considered daily. Data statistical processing was carried out using Microsoft Excel and PAST 4.5 software. Optimal lavender overwintering conditions were formed in 2018 and the greatest plant damage was observed in 2017. The research years were grouped by winter hardiness structured into clusters, which allowed us to identify common features in climatic conditions and to identify critical periods of the winter period leading to a decrease in winter hardiness. Temperature fluctuations in winter, frequent temperature transitions over the 0 °C mark, high levels of snow cover and the formation of ice deposits led to severe damage to some lavender hybrids. Severe frosts in the absence of snow cover can lead to the death of lavender plants in the temperate zone. Lavender hybrids were grouped by winter hardiness into two clusters and 11 subclusters. A group of hybrids with consistently high resistance has been selected throughout the years of the study; these hybrids are the most promising for further hybridization.

**Keywords:** *Lavandula angustifolia*; hybrids; climatic conditions; desiccation; winter hardiness

## 1. Introduction

Lavender is a plant of the Lamiaceae family cultivated as an essential oil, ornamental and medicinal plant [1]. The *Lavandula* genus includes more than 45 species. The genus representatives' growing region is the Mediterranean Sea, southern Europe, North and East Africa, the Middle East, and southwest Asia and southeast India [2,3]. Representatives of four lavender species are most often used in the industry—*Lavandula angustifolia* Mill., *Lavandula stoechas* L., *Lavandula latifolia* Medik., and *Lavandula x intermedia* Emeric ex Loisel. The last one is a hybrid between *L. latifolia* and *L. angustifolia* [4–6].

The main purpose of lavender cultivation in the world is to obtain essential oil, which is included in a wide range of products [7–9]. This is possible mainly in countries with a high sum of active temperatures and a warm, dry climate. The leading countries for


https://doi.org/10.3390/horticulturae11020139

lavender production are Bulgaria and France, followed by Russia, Ukraine, Moldova, Romania and others [10].

Among *Lavandula* species, the most important and most winter-hardy is the *Lavandula angustifolia*, which makes it increasingly important to move the cultivation area northward.

Lavender essential oils contain more than 100 components, including linalyl acetate (30–55%), linalool (20–35%), tannins (5–10%), and caryophyllene (8%) as well as other sesquiterpenoids, perillyl alcohols, esters, oxides, ketones, cineole, camphor, β-ocimene, limonene, caproic acid, and caryophyllene oxide [11]. Lavender essential oils are used as a medicinal agent [12], characterized by a variety of pharmacological properties—anti-inflammatory, antioxidant, antibacterial, antiseptic, antiviral, antidepressant, sedative, immunostimulant [13]. Lavender oil is used in cosmetic products and for insect repellent. Also, lavender oils are used in aromatherapy [14,15]. However, it is known that the accumulation of essential oils is higher at higher temperatures and decreases at lower temperatures [16–18]; therefore, industrial cultivation of lavender for the production of essential oils may not be the main goal in temperate climates.

There is currently a trend towards the development of zero-waste agricultural production technologies, which makes essential oils not the only lavender processing product. Secondary lavender processing products are particularly attractive, and studying these will allow for the best utilization of the crop [19].

Lavender is rich in phenolic compounds, with 8 anthocyanins and 19 flavonols identified in the plants [19]. Lavender leaves are widely used as flavorings for food and beverages and are a source of phenolic components and antioxidants. *Lavandula angustifolia* is a good honey grower and is also used as one of the ingredients for tea blends and ice cream flavoring [20]. Кроме того, лаванда популярна в качестве декоративного растения [21–23]. These uses will be available in regions with temperate and cold climates if plants are obtained that are resistant to the adverse conditions of the winter period—frost-resistant and winter-hardy.

The trends of *Lavandula angustifolia* breeding programs in the world are different depending on the climatic conditions of the growing region. In regions with a warm climate, breeding work is mainly aimed at increasing the content and quality of essential oils [24]. A separate breeding direction is obtaining interspecific hybrids with a complex of valuable traits, including ornamentality, resistance to frost and drought, and content and quality of essential oils [25], but their winter hardiness does not allow growing them in the temperate zone. In Romania, one of the main directions is breeding for drought and salinity tolerance [26].

In temperate regions, low temperatures are the main limiting factor in lavender propagation; therefore, the main breeding direction in these regions is now increasing the winter hardiness and frost resistance of lavender. Even though *Lavandula angustifolia* is one of the most frost-tolerant species, and its ecological optimum is in climates with warm sunny summers but cold winters [21], the region of its industrial use is concentrated in southern Europe, Mediterranean countries, southern Russia and Ukraine [27–29]

In one study, the most resistant varieties of *Lavandula angustifolia* for the conditions of the forest-steppe zone of Ukraine were identified [30]; however, their stability was insufficient for cultivation in the conditions of the Central region of Russia.

Therefore, for the temperate climatic zone, which includes a large part of Russia, наиболее, an important breeding task is to produce, in particular, winter-hardy hybrids. These hybrids should have maximum stability to enable them to be stable for many years since the climatic conditions of the Central region of Russia are very variable in different years. Fluctuations in climatic conditions have become even stronger in recent years due to the influence of global warming. Climate change-related factors are increasingly affecting plant growth and development worldwide [18].

Among these hardy hybrids, it will be necessary to further select plants with a complex of economically valuable characteristics (ornamentality, content of biologically active agents, not only essential oils).

The purpose of this work is to assess the climatic conditions and winter hardiness of *Lavandula angustifolia* and to determine the critical conditions for successful overwintering of plants in the conditions of Moscow, Russia.

## 2. Materials and Methods

### *2.1. Plant Materials*

The studies were conducted during 2015–2022 in the laboratory of cultivated plants of the N.V. Tsitsin Main Botanical Garden of the Russian Academy of Sciences (Moscow, Russia). The geographic coordinates of the plant collection location are [55.84360134560778 N; 37.6201296230011 E].

The objects of the study were 72 second-generation seedlings of lavender narrow-leaved from free pollination (Figure 1), parental forms of which were produced as a result of seed exchange between botanical gardens in the 1980s.

**Figure 1.** Lavender field.

The plantation grows only narrow-leaved lavender plants, so the origin of seedlings as a result of interspecific hybridization is excluded.

Each hybrid represents five plants produced by the rooting of semi-timbered cuttings. The age of plants in 2014 (when the experiment started) was 5 years.

The plants are planted in rows; the distance between rows is 80 cm, and the distance between plants in a row is 60 cm. One of the lavender cultivation problems is weed control [19]; to reduce the number of weeds, we used mulching agro-fabric (manufactured NPK "PROTECT" Pereslavl-Zalesskiy, Russia) in our cultivation technology, which is permeable to water and air but prevents weeds growth and development. During the growing season, feeding three times with complex mineral fertilizers was carried out—in spring, fertilizers with high nitrogen content were applied, and in autumn, phosphorus–potassium fertilizers were applied. Formative pruning (removal of peduncles and shortening of shoots) was carried out at the end of the vegetation period (September–early October); damaged, broken and frozen shoots were also removed in spring after snow melt.

### 2.2. Plants Winter Hardiness Assessment

Winter hardiness was assessed according to a 6-point system: 0 points—the plant died, 1 point—the plant froze to the soil level, 2 points—perennial shoots were damaged, 3 points—biennial and annual shoots were damaged, 4 points—annual shoots were damaged, 5 points—the plant survived the winter without damage. Winter hardiness tests were performed after the complete snow melt in April-May, before spring pruning. Figure 2 shows the growing plants for better visibility.

### 2.3. Climatic Conditions Analysis

During the winter period, the average daily temperature, snow cover height, and precipitation were considered. Climatic data were provided by the meteorological observatory named after V.A. Mikhelson by RSAU-MAA named after K.A. Timiryazev (Russia, Moscow) (license for activities in the field of hydrometeorology and related fields No. L039-00117-77/00654436). The main climatic data were obtained according to the methodology created Tudriy and Ismagilov [31].

### 2.4. Statistical Analysis

The volume of the necessary sample was determined according to the previously developed methodology [32]. Sample statistical analysis (basic statistics) [33] was used to assess the reliability of the obtained results, and cluster analysis was used to group objects. Cluster analysis was carried out according to the comprehensive scores of principal component evaluation using Ward's method. Analysis of the results and their visualization were performed using Microsoft Excel, Microsoft Visio, PAST 4.5 software.

(a) (b) (c) (d) (e) (f)

**Figure 2.** Post-winter damage assessment methodology of regrowing plants. Litera (**a**)—5 points, (**b**)—4 points, (**c**)—3 points, (**d**)—2 points, (**e**)—1 points, (**f**)—0 points.

## 3. Results

### 3.1. Analysis of Lavender Hybrids' Winter Hardiness in Research Years

The winter hardiness of narrow-leaved lavender seedlings varied significantly among the years of the study (Table 1 and Figure 3). The worst indicators in the sample, on average, were observed in 2017—all plants were damaged to different degrees after overwintering;

the mode and median were 2 points, and the average overwintering score for the sample was 1.9 points. Also, the first plant dropouts were noted.

**Table 1.** Statistical parameters of lavender hybrids for winter hardiness.

| Year | Mediana | Mode | Average Point | Overwintered Plants, % |
|---|---|---|---|---|
| 2015 | 4 | 4 | 3.5 | 100 |
| 2016 | 4 | 4 | 3.4 | 100 |
| 2017 | 2 | 2 | 1.9 | 97.2 |
| 2018 | 5 | 5 | 4.5 | 98.6 |
| 2019 | 4 | 5 | 3.9 | 97.1 |
| 2020 | 4 | 4 | 3.6 | 91.0 |
| 2021 | 4 | 5 | 4.2 | 95.1 |
| 2022 | 3 | 4 | 3.3 | 98.3 |

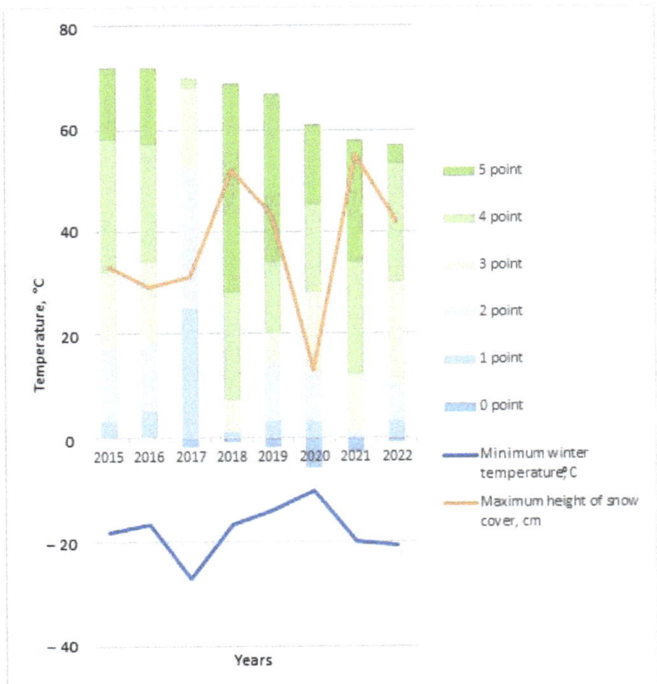


**Figure 3.** Structure of winter hardiness of *Lavandula angustifolia* hybrids in absolute numbers.

According to the set of statistical parameters of winter hardiness, the optimal year for overwintering was 2018; the mode and median were 5 points, and the average winter hardiness score for the sample was 4.5. This was the maximum average score for the whole time of observations, and only one seedling died.

The winter hardiness in 2019–2021 was quite leveled, but in 2020, the maximum death of plants during the observation period (6 pcs.) was noted. In general, more than 90% of plants came out of winter alive with different degrees of damage during the 8 years of the study.

### 3.2. Climatic Conditions Analysis

It was assumed that winter hardiness may be influenced by the following climatic features of the year: minimum winter temperature, sum of active temperatures, maximum snow cover height, and precipitation amount for the year (Table 2).

**Table 2.** Mean annual climatic parameters in the research years.

| Year | Minimum Winter Temperature, °C | The Sum of Active Temperatures Is Above 10 °C | Maximum Snow Cover Height, cm | Precipitation Amount for the Year, mm |
|---|---|---|---|---|
| 2015 [1b] | −18.3 | 2911 | 33 | 717 |
| 2016 [1b] | −16.8 | 2877 | 29 | 881 |
| 2017 [3] | −27.0 | 2555 | 31 | 870 |
| 2018 [1a] | −16.7 | 3355 | 52 | 652 |
| 2019 [1a] | −13.9 | 3093 | 43 | 556 |
| 2020 [2a] | −10.3 | 2966 | 13 | 901 |
| 2021 [2b] | −20.0 | 2897 | 55 | 817 |
| 2022 [2b] | −20.6 | 3065 | 42 | 648 |

[1a, 1b] years included in the first cluster, [2a, 2b] years included in the second cluster, [3] year included in the third cluster—are presented in Figure 3.

The lowest winter minimum temperatures were observed in 2017, 2021, and 2022 and the highest in 2019 and 2020.

In a more detailed analysis of the temperature regime, the minimum and average ten-day temperatures from December 1 to March 30 for 2014–2022 were determined (minimum temperatures are marked in Table 3 in red font, near-zero temperatures in green font).

**Table 3.** Minimum and average ten-day temperatures for the winter period. Minimum temperatures are marked in red font, near-zero temperatures in green font.

| Month | Ten-Day Period | 2014–2015 | 2015–2016 | 2016–2017 | 2017–2018 | 2018–2019 | 2019–2020 | 2020–2021 | 2021–2022 |
|---|---|---|---|---|---|---|---|---|---|
| | | | | Minimum temperature, °C | | | | | |
| December | 1 | −12 | −1 | −14 | −4 | −15 | −5 | −13 | −14 |
| | 2 | −3 | −10 | −19 | 0 | −16 | −2 | −11 | −14 |
| | 3 | −17 | −10 | −4 | −4 | −10 | −4 | −11 | −21 |
| January | 1 | −20 | −20 | −30 | −4 | −14 | −3 | −9 | −8 |
| | 2 | −12 | −20 | −12 | −9 | −11 | −6 | −20 | −17 |
| | 3 | −18 | −19 | −23 | −10 | −20 | −7 | −9 | −8 |
| February | 1 | −14 | −5 | −23 | −13 | −5 | −15 | −17 | −8 |
| | 2 | −12 | −8 | −11 | −10 | −10 | −4 | −18 | −3 |
| | 3 | −2 | −8 | −7 | −18 | −12 | −5 | −17 | −3 |
| March | 1 | −5 | −8 | −2 | −13 | −12 | −3 | −16 | −9 |
| | 2 | −4 | −9 | −5 | −13 | −11 | −9 | −11 | −4 |
| | 3 | −10 | −10 | −6 | −6 | −3 | −6 | −2 | −4 |
| | | | | Average temperature, °C | | | | | |
| December | 1 | −7.2 | 0.7 | −8.9 | −1.4 | −5.9 | −1.1 | −8.2 | −5.3 |
| | 2 | −0.5 | −3.5 | −10.8 | 1.8 | −8.1 | 0.0 | −5.6 | −4.1 |
| | 3 | −9.5 | −0.6 | −2.0 | −0.7 | −9.1 | −1.0 | −5.9 | −12.8 |
| January | 1 | −10.2 | −16.6 | −19.0 | 0.0 | −8.5 | −1.4 | −2.8 | −5.3 |
| | 2 | −2.8 | −13.7 | −6.3 | −6.9 | −7.3 | −1.2 | −13.4 | −5.8 |
| | 3 | −9.3 | −9.4 | −10.0 | −6.0 | −12.3 | −2.8 | −0.8 | −5.1 |
| February | 1 | −7.5 | −1.8 | −14.1 | −5.5 | −3.3 | −6.1 | −10.4 | −2.8 |
| | 2 | −6.0 | −3.3 | −4.9 | −6.4 | −3.6 | −0.9 | −13.3 | 0.1 |
| | 3 | −0.5 | −3.8 | −3.9 | −14.4 | −5.4 | −1.5 | −5.8 | −0.1 |
| March | 1 | −0.7 | −1.2 | 0.4 | −8.8 | −6.2 | 2.3 | −5.0 | −2.9 |
| | 2 | −2.0 | −4.0 | −1.1 | −5.4 | −2.3 | −0.4 | −1.7 | 0.0 |
| | 3 | −2.8 | −4.2 | −1.8 | −1.5 | −0.5 | −2.2 | 2.2 | 1.4 |

The minimum winter temperature was most often observed in January, but in some years, it was observed in late December (2022) and early February (2017). In some years, winter temperatures never dropped below −15–18 °C (2018 and 2020).

The ten-day average temperatures became close to critical for lavender starting in January, while December, February and March generally had average temperatures ranging from 0 to −5 °C, which is not a problem for lavender plants.

The temperatures in the winter months mainly range from 0 to minus 10–12 °C, but in recent years, due to global warming, there have been sharp temperature changes and days with extremely low temperatures, which can be clearly seen when analyzing the daily temperature using basic statistics and box-plot visualization (Figure 4). In winter 2017, there were 4 days with extreme temperature variations, and in 2022, there were three days.

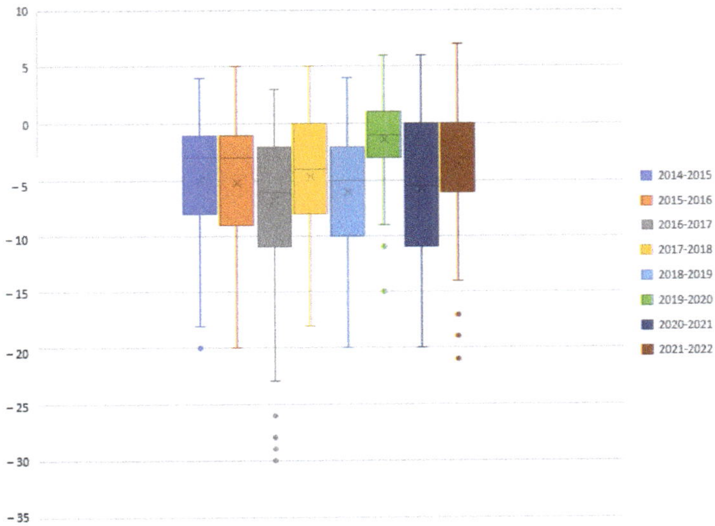


**Figure 4.** Box-plot of average daily temperatures for winter months.

Through a detailed analysis of climatic conditions in winter for the research years, possible critical points causing damage to plants in winter were determined based on winter hardiness indicators (Figure 5).

The following critical points were identified (shown in Figure 5):

(1) Absence of snow cover and temperature decrease below zero (blue vertical arrows);
(2) Sufficient snow cover and increase in temperatures to positive values (red vertical arrows);
(3) Formation of a dense layer of snow with prolonged preservation of positive temperatures and subsequent decrease in temperatures to negative values (red double-sided horizontal arrow).

Visualizing the values of temperature and snow cover height, differences between years became visible. The winter of 2014–2015 was characterized by a snow-free period followed by a sharp cold snap, numerous temperature increases to positive values, and two periods with the formation of a dense ice crust. In the winter of 2015–2016, there was only one episode of sharp (temperature difference from +6° to −18 °C) cold weather in the absence of snow and a thaw at the end of January, which led to the formation of ice crust. In the winter of 2016–2017, we did not observe sharp fluctuations with complete absence of snow, but three episodes with temperature increase to zero and formation of frost were observed. The first half of winter 2017–2018 was characterized by almost complete absence of snow and temperature fluctuations around zero, in mid-January temperatures dropped to −10 °C in the absence of snow, and in early February heavy snowfalls were observed; despite a one-time temperature increase to zero, the formation of snow crust was not observed. Conditions in the winter months of 2018–2019 were characterized by stable snow cover and temperature, with one critical point in early December with a short-term temperature increase to zero, which did not lead to the formation of ice, as well as temperature fluctuations near zero in February, leading to the formation and melting of snow crust. Winter conditions in 2019–2020 resembled climatic conditions in the regions of traditional cultivation of narrow-leaved lavender in all parameters—practically no snow cover and temperature fluctuations around zero, with a single cold snap to −5 °C in the absence of snow and to −10 °C in the presence of 10 cm of light cover. In the winter of 2020–2021, three critical points characterized by zero temperature in the presence of snow cover were observed, one of which was accompanied by the formation of ice crust; however, the height of snow under the crust was significantly higher than normal. In 2021–2022, the snow cover height was also higher than normal, and three thaws were noted during the winter period, two of which led to the formation of ice crust; however, in the first case, the ice crust was thin for a short time and then broke up.

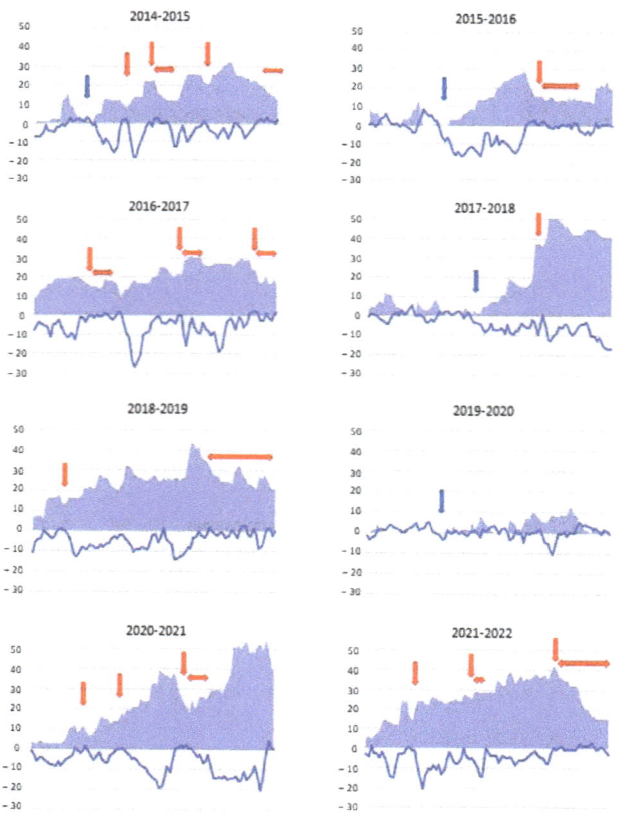


**Figure 5.** Fluctuations in temperature and snow cover height during the study period from early December to late February: (a) 2014–2015; (b) 2015–2016; (c) 2016–2017; (d) 2017–2018; (e) 2018–2019; (f) 2019–2020; (g) 2020–2021; (h) 2021–2022. The blue line shows the variation in minimum daily mean temperature, and the filled box shows the snow cover height. The blue vertical arrow is a Type 1 critical point, the red vertical arrow is a Type 2 critical point, and the red horizontal arrow is a Type 3 critical point.

*3.3. Relationship Between Winter Hardiness and Winter Climatic Conditions*

For further identification of climatic features affecting the winter hardiness of lavender hybrids, the years of the study were grouped according to the complex winter hardiness score (Figure 6). At an association distance of 19 Euclidean distances, 2017 was separated into a separate cluster, and at a distance of 17 Euclidean distances, the remaining years were divided into two clusters; the first was included in subgroup 1—2015 and 2016 and in subgroup 2—2018 and 2019, and the second cluster was included in 2020–2022, with 2021 and 2022 being close to each other in terms of complex winter hardiness, and 2020 differing from them to some extent.

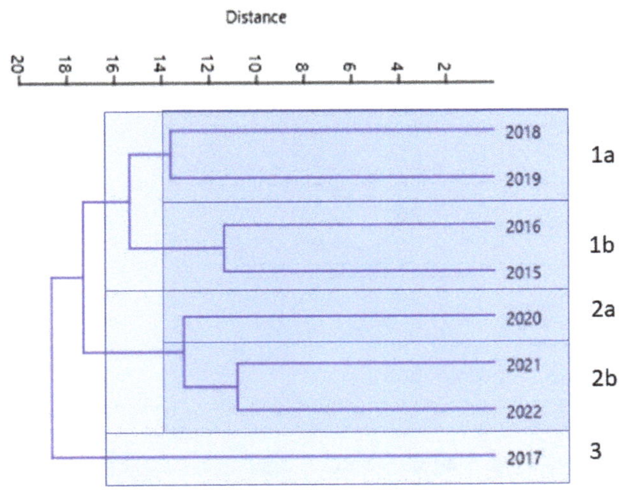


**Figure 6.** Grouping of research years based on the structure of winter hardiness of lavender hybrids. Numbers indicate clusters, and literals indicate subclusters.

Based on the data on winter hardiness obtained over 8 years of research, we grouped the seedlings. The grouping was carried out for all hybrids, except for those that completely fell out during the years of research (Figure 7).

The sample was divided into 2 large clusters (groups), including 11 smaller ones. The most interesting in this case is the separation of the second- and third-order clusters. The first and second subclusters of the third order, combined into a second-order cluster, united objects with relatively low winter hardiness in all years of research, except 2018, and the average winter hardiness for them was 3.2 points. The third, fourth, fifth and sixth subclusters were characterized by high winter hardiness, averaging 3.9–4.1 points in all years, except 2017, when all plants in the study showed low winter hardiness. This group of plants showed high winter hardiness compared to the others regardless of the conditions that year. The seventh and eighth subclusters included plants with average winter hardiness that reacted quite strongly to the climatic conditions of the year—in more favorable years their winter hardiness was high, while in years with a complex of unfavorable conditions it sharply decreased. Finally, the ninth, tenth and eleventh subclusters united objects characterized by a rather low average (2.5–3.4 points) and, at the same time, unstable winter hardiness.

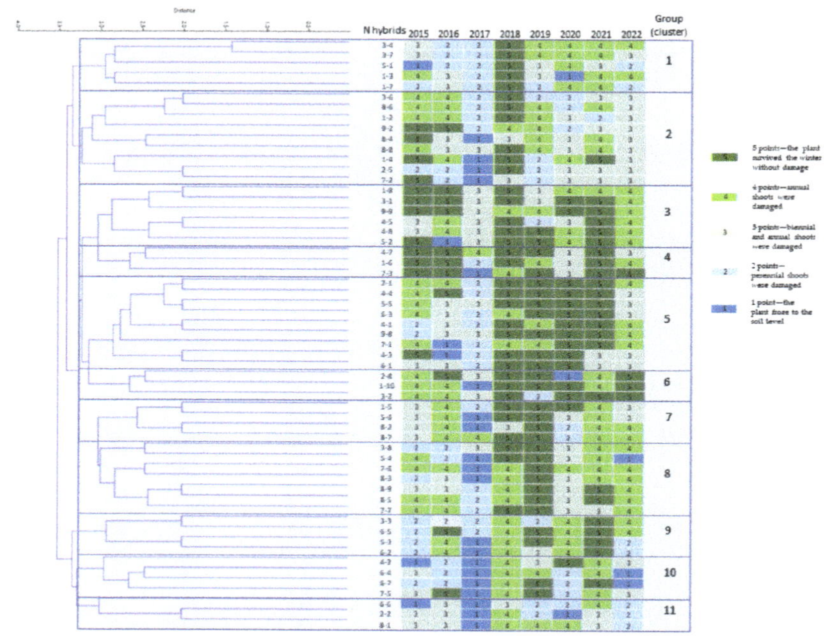


**Figure 7.** Lavender seedlings grouped by winter hardiness.

## 4. Discussion

Based on the analysis of climatic conditions and the degree of damage to lavender plants in different years of research, the years with the greatest damage to plants were identified. We noted the greatest degree of damage to lavender plants in 2017, which was characterized by the lowest temperatures, the lowest amount of active temperatures, (Table 2), strong temperature fluctuations (Figure 4, Table 3) and a high level of snow cover. Similar average conditions were observed in 2021 and 2022, characterized by significantly higher winter hardiness of plants. It is possible that plants' winter hardiness in 2017 was influenced by the conditions of the previous year of 2016, which was characterized by a low sum of active temperatures, resulting in insufficient shoot maturation.

In temperate climates, lavender behaves as a chamaephyte, a perennial evergreen semi-shrubby plant with perennial buds located on single-trunked shoots above the soil surface [19,34]. Young plants are most sensitive to adverse weather conditions as their resistance increases with their age [22]; also, young tissues are more susceptible to spring damage [35]. The risk of damage exists for plants buds in forced dormancy during warming in winter or in early spring [36].

Usually, low temperatures have been recognized as a major factor limiting the geographical distribution of plants [16,37], as cold stress leads to morphological changes, such as a decrease in biomass and leaf surface area [38]. However, in the case of lavender, as our data show, the effect of a complex of climatic factors should be considered.

Plant winter hardiness is a complex parameter, which depends on many factors both in the external environment and in the state and plant genotype. Winter hardiness includes resistance to low negative temperatures of the above-ground part and root system, as well as resistance to uprooting, soaking, and fungal diseases, which can progress in conditions of high humidity and positive temperature under the snow crust. There are many ways woody plants adapt to lower temperatures—nutrient outflow before the dormant period, biochemical acclimatization, and phenology peculiarities [39,40].

A comparative analysis of temperature and snow cover fluctuations revealed critical climatic conditions that are most likely to cause severe damage to plants (Figure 5). Critical conditions determining the winter hardiness of *Lavandula angustifolia* hybrids are as follows: lack of snow cover with temperature decreases below 0 °C; sufficient snow cover and temperature increase to positive values; and snow crust formation with prolonged preservation of positive temperatures and consequent temperature decrease to negative values.

The combination of such critical climatic conditions led to the greatest damage to plants in the winter of 2017. However, most of the plants, despite severe damage, remained alive (Table 1)

The highest plant death rate was recorded in 2020, which was characterized by relatively high temperatures and light snow cover.

Thus, according to the influence of climatic conditions on lavender plants in winter, it can be concluded that with strong temperature fluctuations, especially with frequent passage through 0 °C, high snow cover and the formation of vegetation, severe damage to plants is observed, but this does not lead to their death; also, with low negative temperatures and the absence of snow cover, the death of plants is seen.

Such climatic conditions in the temperate zone are repeated regularly and have recently been exacerbated by climate change due to global warming [18]. Hybrids that have shown severe or even moderate damage in such climatic conditions should not be used in further breeding work. As our research has shown, lavender hybrids vary quite significantly in the degree of damage, which allowed us to identify individual samples that are consistently more resistant to adverse climatic conditions in winter.

Plant biological characteristics play an important role, as the results of the research show significant differences between seedlings in terms of winter hardiness in the dynamics over the years of the study (Figure 7). As a result of cluster analysis of the long-term winter hardiness of seedlings, we identified groups of plants that consistently had the highest resistance to unfavorable conditions: cluster 3 (Figure 7), hybrids 1-8, 3-1, 9-9, 4-5, 4-8 and 5-2. These hybrids, even in the most unfavorable year of 2017, had the least damage at the end of winter compared to the rest. It is recommended to use these hybrids in further breeding work as sources of the complex trait "winter hardiness", since studies have shown that winter-hardy genotypes from the most northern regions (northern border of the cultivation area) transmit this trait to offspring better [41]. Also, hybrids with stable winter hardiness, which does not depend on a set of winter factors, are more important in breeding work due to global climate change, which creates additional risks in plant cultivation [42].

In the future, understanding which biological characteristics of plants are related to high winter hardiness will enable us to select seedlings at the earliest development stages, as happens in the directed selection of more-studied crops.

Some studies show that aromatic plants' essential oils components are involved in some mechanisms that protect against biotic and abiotic factors [19,43], and growing conditions affect the composition of lavender essential oils, which requires additional study in temperate climates [9]. Also morphological and anatomical features are considered as important indicators for cold tolerance studies [44].

## 5. Conclusions

Temperature fluctuations in winter, frequent temperature transitions over the 0 °C mark, high levels of snow cover and the formation of ice deposits lead to severe damage to some lavender hybrids. Severe frosts in the absence of snow cover can lead to the death of lavender plants in the temperate zone. Information about the reaction of lavender hybrids to various unfavorable winter conditions will make it possible to plan breeding work in other regions, as well as plan the industrial cultivation of the most winter-hardy hybrids. The lavender hybrids obtained on the basis of free pollination in our collection vary significantly in the degree of stability and damage during the winter period during the years of research. Hybrids that are moderately or severely damaged are eliminated from breeding work. As a result of our research, we have identified groups of the most winter-hardy hybrids, as well as hybrids showing stable winter hardiness regardless of unfavorable winter conditions. These hybrids will be used for further breeding work.

The next stage of our work, studying the winter hardiness of narrow-leaved lavender and breeding to increase winter hardiness, will be to analyze the relationship between morphological, anatomical and phenological parameters of hybrids from clusters selected by us and their winter hardiness.

**Author Contributions:** Conceptualization, V.K.; methodology, S.A. and A.E.; validation, V.K. and O.S.; formal analysis, O.S., S.A. and O.L.; investigation, A.E., V.D. and O.L.; resources, V.K. and O.S.; data curation, V.K.; writing—original draft preparation, V.K. and V.D.; writing—V.K. and S.A.; visualization, V.K. and V.D.; supervision, Y.G.; project administration, Y.G. All authors have read and agreed to the published version of the manuscript.

**Funding:** The work was carried out within the framework of the state task of the MBG RAS "Biological diversity of natural and cultural flora: fundamental and applied issues of study and conservation", state registration number 122042700002-6 and the framework of the state task of the All-Russia Research Institute of Agricultural Biotechnology "Investigation of the molecular genetic potential of agricultural plant genotypes to ensure the breeding process of creating highly productive and cultured plants resistant to adverse factors", state registration FGUM-2025-0004.

**Data Availability Statement:** The original contributions presented in this study are included in the article. Further inquiries can be directed to the corresponding author.

**Acknowledgments:** The authors thank the editors and reviewers for their contributions to the current form of this manuscript.

**Conflicts of Interest:** The authors declare no conflicts of interest.

## References

1. Blankespoor, J. *Lavender's Medicinal and Aromatherapyuses and Lavender Truffles*; Chestnut School of Herbal Medicine: Weaverville, NC, USA, 2012.
2. Mokhtarzadeh, S.; Hajyzadeh, M.; Ahmad, H.; Khawar, K.M. The problems in acclimatisation of in vitro multiplied plants of *Lavandula angustifolia* Miller under field conditions. *Acta Hortic.* **2013**, *988*, 71–76. [CrossRef]
3. Lis-Balchin, M. Lavender. In *Handbook of Herbs and Spices*; Elsevier: Amsterdam, The Netherlands, 2012; pp. 329–347.
4. Adaszyńska, M.; Swarcewicz, M.; Markowska-Szczupak, A. Comparison of the chemical composition and antimicrobial activity of essential oil obtained from various domestic varieties of narrow-leaved lavender (*Lavandula angustifolia* L.). *Postępy Fitoter.* **2013**, *2*, 90–96.
5. Bejar, E. Adulteration of english lavender (*Lavandula angustifolia*) essential oil. *J. Pharm. Biomed. Anal.* **2020**, *199*, 114050.
6. Détár, E.; Németh, É.Z.; Gosztola, B.; Demján, I.; Pluhár, Z. Effects of variety and growth year on the essential oil properties of lavender (*Lavandula angustifolia* Mill.) and lavandin (*Lavandula x intermedia* Emeric ex Loisel.). *Biochem. Syst. Ecol.* **2020**, *90*, 104020. [CrossRef]
7. Gallotte, P.; Fremondière, G.; Gallois, P.; Bernier, J.-P.B.; Buchwalder, A.; Walton, A.; Piasentin, J.; Fopa-Fomeju, B. Lavandula angustifolia Mill. and *Lavandula* x *Intermedia* Emeric Ex Loisel: Lavender and Lavandin. In *Medicinal, Aromatic and Stimulant Plants*; Novak, J., Blüthner, W.-D., Eds.; Handbook of Plant Breeding; Springer International Publishing: Cham, Switzerland, 2020; pp. 303–311. ISBN 978-3-030-38792-1.
8. Pokajewicz, K.; Białoń, M.; Svydenko, L.; Fedin, R.; Hudz, N. Chemical Composition of the Essential Oil of the New Cultivars of *Lavandula angustifolia* Mill Bred in Ukraine. *Molecules* **2021**, *26*, 5681. [CrossRef]
9. Miastkowska, M.; Kantyka, T.; Bielecka, E.; Kałucka, U.; Kamińska, M.; Kucharska, M.; Kilanowicz, A.; Cudzik, D.; Cudzik, K. Enhanced Biological Activity of a Novel Preparation of *Lavandula angustifolia* Essential Oil. *Molecules* **2021**, *26*, 2458. [CrossRef]
10. Giray, F.H. An Analysis of World Lavender Oil Markets and Lessons for Turkey. *J. Essent. Oil Bear. Plants* **2018**, *21*, 1612–1623. [CrossRef]

11. Ciesielska, K.; Ciesielski, W.; Girek, T.; Kołoczek, H.; Oszczęda, Z.; Tomasik, P. Reaction of *Lavandula angustifolia* Mill. to Water Treated with Low-Temperature, Low-Pressure Glow Plasma of Low Frequency. *Water* **2020**, *12*, 3168. [CrossRef]
12. Lis-Balchin, M.; Hart, S. Studies on the mode of action of the essential oil of Lavender (*Lavandula angustifolia* P. Miller). *Phytother. Res.* **1999**, *13*, 540–542. [CrossRef]
13. Dezici, S. Promising anticancer activity of lavender (*Lavandula angustifolia* Mill.) essential oil through induction of both apoptosis and necrosis. *Ann. Phytomed.* **2018**, *7*, 38–45.
14. Cavanagh, H.M.A.; Wilkinson, J. Biological activities of lavender essential oil. *Phytother. Res.* **2002**, *16*, 301–308. [CrossRef] [PubMed]
15. Kasper, S.; Gastpar, M.; Müller, W.E.; Volz, H.P.; Möller, H.J.; Dienel, A.; Schläfke, S. Silexan, an orally administered Lavandula oil preparation, is effective in the treatment of 'subsyndromal' anxiety disorder: A randomized, double-blind, placebo controlled trial. *Int. Clin. Psychopharm.* **2010**, *25*, 277–287. [CrossRef]
16. Rastogi, S.; Shah, S.; Kumar, R.; Vashisth, D.; Akhtar, M.Q.; Kumar, A.; Dwivedi, U.N.; Shasany, A.K. Ocimum metabolomics in response to abiotic stresses: Cold, flood, drought and salinity. *PLoS ONE* **2019**, *14*, e0210903. [CrossRef] [PubMed]
17. Manukyan, A. Effects of PAR and UV-B radiation on herbal yield, bioactive compounds and their antioxidant capacity of some medicinal plants under controlled environmental conditions. *Photochem. Photobiol.* **2013**, *89*, 406–414. [CrossRef] [PubMed]
18. Mansinhos, I.; Gonçalves, S.; Romano, A. How climate change-related abiotic factors affect the production of industrial valuable compounds in Lamiaceae plant species: A review. *Front. Plant Sci.* **2024**, *15*, 1370810. [CrossRef] [PubMed]
19. Crișan, I.; Ona, A.; Vârban, D.; Muntean, L.; Vârban, R.; Stoie, A.; Mihăiescu, T.; Morea, A. Current Trends for Lavender (*Lavandula angustifolia* Mill.) Crops and Products with Emphasis on Essential Oil Quality. *Plants* **2023**, *12*, 357. [CrossRef]
20. Bagget, N. *The Art of Cooking with Lavender*; Kitchenslane Productions: Columbia, MD, USA, 2016.
21. Lis-Balchin, M. *Lavender: The Genus Lavandula*; CRC Press: Boca Raton, FL, USA, 2003.
22. Mason, J. *Growing and Knowing Lavender*; ACS Distance Education: Nerang, QLD, Australia, 2014; ISBN 978-0-9925878-0-2.
23. Demasi, S.; Caser, M.; Lonati, M.; Gaino, W.; Scariot, V. Ornamental traits of *Lavandula angustifolia* Mill. are affected by geographical origin and cultivation substrate composition. *Acta Hortic.* **2021**, *1331*, 49–56. [CrossRef]
24. Urwin, N. Lavender Breeding for Commercial Yield. *Comb. Proc. Int. Plant Propagator Soc.* **2008**, *58*, 78–84.
25. Van Oost, E.; Leus, L.; De Rybel, B.; Van Laere, K. Determination of Genetic Distance, Genome Size and Chromosome Numbers to Support Breeding in Ornamental Lavandula Species. *Agronomy* **2021**, *11*, 2173. [CrossRef]
26. Szekely-Varga, Z.; González-Orenga, S.; Cantor, M.; Jucan, D.; Boscaiu, M.; Vicente, O. Effects of Drought and Salinity on Two Commercial Varieties of *Lavandula angustifolia* Mill. *Plants* **2020**, *9*, 637. [CrossRef]
27. Kremenchuk, P.I.; Kitaev, O.I. Estimation of lavender (*Lavandula angustifolia*) frost resistance. *Plant Var. Stud. Prot.* **2013**, *13*, 155–161. [CrossRef]
28. Zheljazkov, V.; Astatkie, T.; Hristov, A. Lavender and hyssop productivity, oil content, and bioactivity as a function of harvest time and drying. *Ind. Crop. Prod.* **2012**, *36*, 222–228. [CrossRef]
29. Stanev, S.; Zagorcheva, T.; Atanassov, I. Lavender cultivation in Bulgaria—21st century developments, breeding challenges and opportunities. *Bulg. J. Agric. Sci.* **2016**, *22*, 584–590.
30. Verlet, N. The world herbs and essential oils economy-analysis of a medium term development. *Acta Hortic.* **1992**, *306*, 474–481. [CrossRef]
31. Tudrij, V.D.; Ismagilov, N.V. *Metody i Sredstva Gidrometeorologicheskih Izmerenij*; Kazanskij Universitet: Kazan', Russia, 2011; p. 296. (In Russian)
32. Isachkin, A.V.; Kryuchkova, V.A. Algoritmy opredeleniya dostatochnyh ob"emov vyborok (na primere sadovyh rastenij). *Byulleten' Gl. Bot. Sada Bul. MBS RAS* **2020**, *4*, 68–78. (In Russian)
33. Isachkin, A.V. *Osnovy Nauchnyh Issledovanij v Sadovodstve: Uchebnik Dlya Bakalavrov i Magistrov po Napravleniyu «Sadovodstvo*; Isachkin, A.V., Kryuchkova, V.A., Eds.; Izdatel'stvo "Lan": Moskva, Russisa, 2019; (In Russian). ISBN 978-5-8114-5019-0.
34. Colțun, M. Step-by-Step Creation of a Lavender Plantation. *J. Bot.* **2016**, *8*, 76–80.
35. Kovaleski, A.P. Woody species do not differ in dormancy progression: Differences in time to budbreak due to forcing and cold hardiness. *Proc. Natl. Acad. Sci. USA* **2022**, *119*, e2112250119. [CrossRef]
36. Nole, A.; Rita, A.; Ferrara, A.M.S.; Borghetti, M. Effects of a large-scale late spring frost on a beech (*Fagus sylvatica* L.) dominated Mediterranean mountain forest derived from the spatio-temporal variations of NDVI. *Ann. For. Sci.* **2018**, *75*, 83. [CrossRef]
37. Niu, R.; Zhao, X.; Wang, C.; Wang, F. Physiochemical Responses and Ecological Adaptations of Peach to Low-Temperature Stress: Assessing the Cold Resistance of Local Peach Varieties from Gansu, China. *Plants* **2023**, *12*, 4183. [CrossRef]
38. Lianopoulou, V.; Bosabalidis, A.M.; Patakas, A.; Lazari, D.; Panteris, E. Effects of chilling stress on leaf morphology, anatomy, ultrastructure, gas exchange, and essential oils in the seasonally dimorphic plant *Teucrium polium* (Lamiaceae). *Acta Physiol. Plant* **2014**, *36*, 2271–2281. [CrossRef]
39. Vitasse, Y.; Lenz, A.; Körner, C. The interaction between freezing tolerance and phenology in temperate deciduous trees. *Front. Plant Sci.* **2014**, *5*, 541. [CrossRef] [PubMed]

40. Guo, X.; Khare, S.; Silvestro, R.; Huang, J.; Sylvain, J.D.; Delagrange, S.; Rossi, S. Minimum spring temperatures at the provenance origin drive leaf phenology in sugar maple populations. *Tree Physiol.* **2020**, *40*, 1639–1647. [CrossRef] [PubMed]
41. Butnor, J.R.; Wilson, C.P.; Bakır, M.; D'Amato, A.W.; Flower, C.E.; Hansen, C.F.; Keller, S.R.; Knight, K.S.; Murakami, P.F. Cold Tolerance Assay Reveals Evidence of Climate Adaptation Among American Elm (*Ulmus americana* L.) Genotypes. *Forests* **2024**, *15*, 1843. [CrossRef]
42. Li, X.; Liu, Y.; Yang, S.; Wang, J.; Xia, H.; Liu, X.; Chen, Q. Invisible Frost Stress on Introduced Dalbergia odorifera: A Bioassay on Foliar Parameters in Seedlings from Six Provenances. *Sustainability* **2023**, *15*, 14097. [CrossRef]
43. Demasi, S.; Caser, M.; Lonati, M.; Cioni, P.L.; Pistelli, L.; Najar, B.; Scariot, V. Latitude and Altitude Influence Secondary Metabolite Production in Peripheral Alpine Populations of the Mediterranean Species *Lavandula angustifolia* Mill. *Front. Plant Sci.* **2018**, *9*, 1–12. [CrossRef]
44. Wisniewski, M.; Fuller, M.; Glenn, D.; Gusta, L.; Duman, J.; Griffith, M. Extrinsic ice nucleation in plants: What are the factors and can they be manipulated. In *Plant Cold Hardiness: Gene Regulation and Genetic Engineering*; Kluwer Academic/Plenum Publishing: New York, NY, USA, 2002.

horticulturae

MDPI

*Article*

# Hydrogen Sulfide Mitigates Manganese-Induced Toxicity in *Malus hupehensis* Plants by Regulating Osmoregulation, Antioxidant Defense, Mineral Homeostasis, and Glutathione Ascorbate Cycle

**Bowen Liu, Baozhu Wang, Tianlnog Chen and Manrang Zhang ***

College of Horticulture, Northwest A&F University, Xianyang 712100, China; liubowen@nwafu.edu.cn (B.L.); wangbaozhu@nwafu.edu.cn (B.W.); chentianlong@nwafu.edu.cn (T.C.)
* Correspondence: mrz@nwsuaf.edu.cn; Tel.: +86-133-6390-3687

**Abstract:** Manganese (Mn) is a toxic metal element that adversely affects plant growth. Hydrogen sulfide ($H_2S$) is considered an important signaling molecule with significant potential in alleviating various abiotic stresses. However, there is limited information available on the role of $H_2S$ in alleviating manganese stress in plants. In this study, the effects of exogenous $H_2S$ and its scavenger, homocysteine thiolactone (HT), on the physiological and biochemical parameters of *Malus hupehensis* var. *pingyiensis* seedlings were evaluated. Our results show that $H_2S$ treatment significantly alleviates growth inhibition and oxidative damage induced by manganese stress in *Malus hupehensis* seedlings, primarily by enhancing antioxidant enzyme activity and up-regulating the ascorbate-glutathione (ASA-GSH) cycle. $H_2S$ treatment increased photosynthetic pigment content and helped maintain osmotic balance in leaves, thereby enhancing key gas exchange parameters and mitigating manganese-induced suppression of photosynthesis. $H_2S$ treatment enhanced the absorption of Ca, Mg, Fe and Zn under manganese stress, significantly reduced manganese accumulation in *Malus hupehensis* seedlings, and modulated the transcriptional expression of *MTPs*, facilitating the transfer of manganese to the leaves. Thus, $H_2S$ reduces oxidative damage and promotes growth under Mn stress, highlighting its important role in plant stress tolerance.

**Keywords:** antioxidants; ASA-GSH cycle; hydrogen sulfide; *Malus hupehensis*; manganese toxicity

## 1. Introduction

Manganese (Mn), the second most abundant metal widely available worldwide, is an essential trace element for all living organisms [1]. In plants, Mn plays an important role in growth, development, and metabolism mainly in the form of $Mn^{2+}$ [2]. Within chloroplasts, Mn supports chlorophyll formation and structural stability and acts as a cofactor in the redox reactions of the electron transport chain during photosynthesis [3,4]. However, excessive Mn levels in the environment can be detrimental to plants, a phenomenon that often occurs in acidic and flooded soils. Under low pH and reducing conditions, the concentration of soil $Mn^{2+}$ increases, making it more accessible to plants and leading to toxicity [5]. With the development of industry and long-term overuse of fertilizers and pesticides, more than half of the potentially arable soils around the world are acidic, and manganese stress has become the second most important plant growth limiting factor after aluminum toxicity [6,7]. Therefore, understanding Mn toxicity and plant tolerance mechanisms is important for agricultural development.

Many studies have shown that Mn toxicity disrupts a variety of physiological processes in plant cells, such as triggering oxidative stress, inhibiting enzyme activities, hindering chlorophyll biosynthesis and photosynthesis [8], and hindering the uptake and transport of other essential minerals [9]. This results in a reduction in the number of lateral roots of the plant, reduced root vigor, leaf greening and necrosis, and growth inhibition [10]. Prolonged exposure to metal stress leads to the production of reactive oxygen species (ROS), such as hydrogen peroxide ($H_2O_2$), superoxide ($O_2^-$), and singlet oxygen ($^1O_2$). Excessive accumulation of ROS leads to lipid peroxidation, which damages proteins involved in cellular activities and impairs their normal functions [11].

It has been found in past studies that plants can mitigate Mn toxicity by complexing ligands and chelates, regulating manganese transport proteins, regulating antioxidant systems, and altering biochemical pathways [12]. Organic acids such as citrate, malate, and oxalate form stable Mn complexes that reduce its bioavailability and toxicity [13,14]. Heavy metal transport proteins segregate Mn into subcellular compartments such as the cell wall, Golgi apparatus, and vesicles, which in turn reduces the concentration of $Mn^{2+}$ in the cytoplasm and other organelles, decreasing the effective concentration and toxicity [1]. The manganese tolerance protein (MTP) transporter plays a critical role in Mn transport, distribution, and the maintenance of Mn balance in plants. A total of 20 *MTP* family members have been identified in the apple genome. Based on the substrate specificity of its members, the apple *MTP* family is primarily classified into three subfamilies: Mn-MTP, Zn-MTP, and Zn/Fe-MTP. Among these, MTP8 to MTP11 belong to the Mn-MTP subfamily [15]. MTP8, MTP9, and MTP11 are localized to the vacuolar membrane, cell membrane, and Golgi apparatus, respectively, and play roles in the compartmentalization and efflux of manganese ions [15–17]. Plants also induce multiple antioxidant enzyme activities to interact with non-enzymatic antioxidants to scavenge excess ROS and reduce oxidative stress [18].

Hydrogen sulfide ($H_2S$) has emerged as an important signaling molecule involved in regulating various plant physiological processes, such as seed germination, root development, stomatal movement, photosynthesis, and stress responses [19,20]. Studies demonstrate that $H_2S$ acts as a signal transducer during abiotic stresses such as salinity, osmotic pressure, and extreme temperatures [21,22]. Under metal stress, endogenous $H_2S$ levels rise, while the exogenous application of $H_2S$ in appropriate concentrations can enhance plant tolerance and reduce stress-induced damage [23]. For instance, exogenous $H_2S$ reduces damage caused by Cd stress in *Brassica oleracea*, *Oryza sativa*, and *Triticum aestivum* L., alleviates Al toxicity in *Brassica napus* and *Glycine max*, and promotes wheat seed germination under Cr stress [23–25]. These protective effects are attributed to $H_2S$-mediated reductions in metal uptake and accumulation or by increasing the activity of antioxidant systems [26].

While the benefits of exogenous $H_2S$ under various metal stresses are well-documented, research on its role in mitigating plant manganese stress remains limited. *Malus hupehensis* var. *pingyiensis* is a unique germplasm resource in northern China, known for its well-developed root system, strong waterlogging resistance, and high disease tolerance. It is frequently used as rootstock and in research on resistance mechanisms [27]. Therefore, this study will evaluate how exogenous $H_2S$ affects the physiological and biochemical responses of *Malus hupehensis* under Mn stress. The results will provide insight into the potential use of $H_2S$ as a tool for mitigating Mn toxicity in agricultural settings.

## 2. Materials and Methods

### 2.1. Materials and Treatments

We selected approximately 300 healthy and uniform *Malus hupehensis* var. *pingyiensis* seeds (previously stored in the laboratory) for surface sterilization for 10 min using 5% sodium hypochlorite (NaClO) solution, and washed them thoroughly with distilled water, then randomly placed them in 15cm diameter Petri dishes containing gauze moistened with distilled water for one week. Sprouted seedlings were transplanted into cavity trays filled with media comprising nutrient soil (Xingyuxing, Wuhan, China), peat moss (Xinghe, Jinan, China), and vermiculite (Luqing, Nanning, China) at a 3:1:1 ratio (*v*:*v*:*v*), and placed in a light incubator (Yiheng, Shanghai, China) under a 16 h light/8 h dark photoperiod at 25 ± 2 °C. Following two weeks of normal growth, uniform seedlings were transferred to full Hoagland's nutrient solution (HB8870, Haibo, Qingdao, China) for hydroponic culture, and after one week, Mn stress was induced by adding an extra 1 mM $MnSO_4$ to the Hoagland's solution, while control plants continued to grow in Hoagland's solution. To assess the role of $H_2S$ under Mn stress, the $H_2S$ donor sodium hydrogen sulfide (NaHS) was introduced into the Hoagland's solution for both the control and Mn-stressed plants. Additionally, the $H_2S$ scavenger taurine (HT; 200 μM) was supplied under conditions with exogenous $H_2S$ to verify the positive effects of $H_2S$ under Mn stress. We added the reagent combinations specified in Table 1 to the complete Hoagland solution for plant treatment. The pH of the nutrient solution for all treatment groups was adjusted to 5.9 prior to use.

**Table 1.** Experimental treatments.

| Sample Number | Processing Reagents |
|---|---|
| CK | Mn 0 mmol/L + NaHS 0 mmol/L |
| HS0.2 | Mn 0 mmol/L + NaHS 0.2 mmol/L |
| Mn | Mn 1 mmol/L + NaHS 0 mmol/L |
| MHS0.05 | Mn 1 mmol/L + NaHS 0.05 mmol/L |
| MHS0.1 | Mn 1 mmol/L + NaHS 0.1 mmol/L |
| MHS0.2 | Mn 1 mmol/L + NaHS 0.2 mmol/L |
| MHS0.5 | Mn 1 mmol/L + NaHS 0.5 mmol/L |
| HT | Mn 1 mmol/L + NaHS 0.2 mmol/L + HT 0.2 mmol/L |

The nutrient solution was replaced every three days throughout the experimental period, and the environmental conditions were set to remain the same as before. After two weeks of treatment, samples were collected to measure the specified indicators. Each treatment was repeated three times under identical experimental conditions, with ten plants per replicate, using a completely randomized design to ensure reliability.

### 2.2. Estimation of Growth and Photosynthetic Pigment Parameters

Plants were harvested after 14 days of treatment and fresh weight (FW) and dry weight (DW) of the plants were determined using a standard weighing balance.

FW was measured using whole plants, while DW was measured by drying the plants at 70 °C for 48 h. Fresh leaf (100 mg) samples were impregnated in 80% acetone following a previously described method of Copper [28]. The absorbance of chlorophyll a and b was measured using a spectrophotometer (UV3600, Shimadzu, Kyoto, Japan) at 663 nm and 645 nm, respectively. The chlorophyll estimates were calculated using the following equation:

$$\text{Chlorophyll a (mg g}^{-1}\text{ FW)} = 100 \times [(A663 \times 0.0127 - A645 \times 0.00269)]/0.5$$

The net photosynthetic rate (Pn), transpiration rate (E), stomatal conductance (gsw), and intercellular $CO_2$ concentration (Ci) of *M. hupehensis* seedlings were measured using the LI-6800 Photosynthesis Measurement System (LI-COR, Tucson, AZ, USA) during the periods of 8:30–11:30 a.m. or 2:30–4:30 p.m. All measurements were performed on the third fully expanded mature leaf.

### 2.3. Estimation of Leaf RWC

The leaf relative water content (RWC) was measured as previously described by Lazcano-Ferrat [29].

The FW of the leaves was recorded before immersing the leaves in distilled water for 24 h at room temperature to measure the turgid weight (TW). The leaves were then dried at 70 °C for 48 h to measure the DW. RWC was calculated using the following equation:

$$RWC\% = (FW - DW)/(TW - DW) \times 100\%$$

### 2.4. Estimation of Soluble Sugar and Proline Contents

The soluble sugar content was determined using the plant soluble sugar assay kit (BC0030, Solarbio, Beijing, China), based on the anthrone colorimetric method. Leaves (0.1 g) from treated seedlings were ground in 1 mL of distilled water and incubated in a boiling water bath for 10 min. The supernatant was then collected and the final volume was adjusted to 10 mL with distilled water. The reaction system was prepared according to the kit instructions and incubated at 95 °C for 10 min. After cooling to room temperature, absorbance was measured at 620 nm, and soluble sugar content was calculated using a standard curve.

Proline content was determined using the proline assay kit (BC0290, Solarbio, Beijing, China), based on the ninhydrin method. Leaves (0.1 g) from treated seedlings were homogenized in 1 mL of extraction buffer on ice, followed by a 10-min boiling water bath and centrifugation at 10,000× *g* for 10 min. The reaction system was prepared according to the kit instructions and incubated at 100 °C for 30 min. After cooling to room temperature, absorbance was measured at 520 nm, and proline content was calculated using a standard curve.

### 2.5. Estimation of Metal Element Content

The plant tissue was processed using the ashing method. The treated plants were rinsed three times with 10mM ethylene diamine tetraacetic acid (EDTA) and then dried at 80 °C for 48 h and ground into powder. Then, we accurately weighed 0.1 g in a crucible and heated in a muffle furnace at 300 °C for 48 h or more. We dissolved the ash residue in 0.5 mol $L^{-1}$ nitric acid, transferred the solution to a centrifuge tube, and diluted it with the same concentration of nitric acid to a final volume of 10 mL. The metal ion concentrations were analyzed using inductively coupled plasma atomic emission spectrometry (ICP-AES; Prodigy, Leemanlabs Inc., Hudson City, NY, USA).

### 2.6. Evaluation of MDA and $H_2O_2$ Contents

Malondialdehyde (MDA) content was measured using the MDA assay kit (BC0020, Solarbio, Beijing, China), based on the thiobarbituric acid (TBA) method. MDA reacts with TBA under acidic and high-temperature conditions, forming a reddish-brown compound with maximum absorption at 532 nm. According to the kit instructions, leaves (0.1 g) from treated seedlings were homogenized in 1 mL of extraction buffer on ice, followed by centrifugation at 8000× *g* for 10 min at 4 °C. The supernatant was collected, and 200 μL of the supernatant was mixed with 800 μL of the reaction solution and incubated at 100 °C for 60 min. After cooling in ice and centrifugation, absorbance was measured at 520 nm to estimate MDA content.

Hydrogen peroxide ($H_2O_2$) content was measured using the $H_2O_2$ assay kit (BC3590, Solarbio, Beijing, China). $H_2O_2$ reacts with titanium sulfate to form a yellow titanium peroxide complex, which absorbs light at 415 nm. According to the kit instructions, leaves (0.1 g) from treated seedlings were ground in 1 mL of cold acetone and homogenized, followed by centrifugation at 8000× *g* for 10 min at 4 °C. The supernatant was collected, and the reaction solution was added. Absorbance was measured at 415 nm to estimate $H_2O_2$ content in the sample.

### 2.7. Determination of Electrolyte Leakage

Electrolyte leakage (EL) was assessed using the method of Dionisio-Sese and Tobita [30]. From the removed plants, 10 small discs were taken from 4 to 6 leaves with the same functional leaf position using a 0.8 cm hole punch and immersed in a medium of 10 mL of deionized distilled water. They were immersed at room temperature for 4 h, during which they were shaken several times and the initial conductance value (EC1) of the solution was measured after 4 h. The samples were then boiled for 30 min, cooled to room temperature, and the final conductivity (EC2) was measured. EL was calculated as follows:

$$EL\ (\%) = (EC1/EC2) \times 100$$

### 2.8. RNA Extraction and RT-qPCR Analysis

Total RNA was extracted using the FastPure Universal Plant Total RNA Isolation Kit (Vazyme Biotech Co., Nanjing, China) following the manufacturer's instructions. The cDNA was obtained by reverse transcription using the HiScript II 1st Strand cDNA Synthesis Kit (Vazyme Biotech Co., Nanjing, China). Quantitative real-time PCR (RT-qPCR) was performed using the ArtiCanATM SYBR qPCR Mix kit (Tsingke Biotech Co., Ltd., Beijing, China) on the QuantStudio 3 system (Life Technologies, Carlsbad, CA, USA), with actin serving as the internal reference. The qRT-PCR procedure included 95 °C for 1 min, 95 °C for 10 s, and 58 °C for 20 s for 40 cycles. Three replicates were run for each sample. RT-qPCR primers are shown in Table 2.

**Table 2.** Primer sequences of RT-qPCR.

| Gene Name | Primers' Sequences (5′-3′) |
|---|---|
| *MdMTP8.2F* | TCTTATTTGTTTCGCTGCGTGTAG |
| *MdMTP8.2R* | GCATTGTGACCGTTCGATTTTC |
| *MdMTP9.1F* | GACAGTGGACGCCGTCGTTT |
| *MdMTP9.1R* | ACTGCCATCCGCTCGCTCTT |
| *MdMTP9.3F* | GGCCTCATGTTGAATTAACCTGTA |
| *MdMTP9.3R* | GGCCTCATGTTGAATTAACCTGTA |
| *MdMTP11.1F* | TCCCACACACTCTCTCCTTTTACCT |
| *MdMTP11.1R* | CGTCGAAGTTCAACCGCCAC |

### 2.9. Estimation of Enzymatic and Non-Enzymatic Antioxidant Activity

The activities of superoxide dismutase (SOD), peroxidase (POD), catalase (CAT) and ascorbate peroxidase (APX) were measured using assay kits (BC5160, BC0090, BC0020 and BC0220, Solarbio, Beijing, China). Leaves (0.5 g) from treated seedlings were ground to a powder in liquid nitrogen. Then, 5 mL of 50 mM/L phosphate-buffered saline containing 1% β-mercaptoethanol and 1% polyvinylpyrrolidone (Ph = 7.8) was added to the sample and incubated for 60 min at 4 °C. After centrifugation at 8000× *g* for 20 min at 4 °C, the supernatant (enzyme extract) was used for further analysis. The reaction systems were prepared according to the kit instructions, and absorbance changes were measured at 450 nm, 470 nm, 240 nm, and 290 nm over a two-minute period using a spectrophotometer. Enzyme activities were calculated based on the rate of absorbance change.

MDHAR and DHAR activities were measured using activity kits (BC0650, BC0660, Solarbio, Beijing, China). According to the kit instructions, leaves (0.1 g) from treated seedlings were homogenized with 1 mL of 50 mM Tris-HCl buffer (pH = 7.8) on ice. After centrifugation at 10,000 rpm for 10 min at 4 °C, the supernatant (enzyme extract) was used for analysis. The reaction solution was added according to the manufacturer's instructions, and the absorbance changes at 340 nm and 412 nm were measured for two minutes using a spectrophotometer. The enzyme activities were calculated based on the rate of absorbance.

The ascorbate acid (ASA) content was determined using the assay kit (BC1230, Solarbio, Beijing, China), based on the oxidation of AsA by ascorbic acid oxidase (AAO), which converts AsA to dehydroascorbic Acid (DHA). The content of AsA was calculated by measuring the oxidation rate. According to the kit instructions, leaves (0.1 g) from treated seedlings were homogenized with 1 mL of extraction buffer on ice, following the kit instructions. After centrifugation at 8000× *g* for 10 min at 4 °C, the supernatant was collected for analysis. The reaction solution was prepared according to the manufacturer's instructions, and absorbance changes at 265 nm were measured over two minutes using a spectrophotometer. The ASA content was calculated from the rate of absorbance.

Glutathione (GSH) and oxidized glutathione (GSSG) contents were determined using the assay kits (BC1170, BC1180, Solarbio, Beijing, China). GSH reacts with 5,5′-dithiobis-2-nitrobenzoic acid (DTNB), producing 2-nitro-5-thiobenzoic acid (DNCB) and GSSG. The DNCB is a yellow product with maximum absorption at 412 nm. According to the kit instructions, leaves (0.1 g) from treated seedlings were homogenized with 1 mL of extraction buffer on ice, following the kit instructions. After centrifugation at 8000× *g* for 10 min at 4 °C, the supernatant was collected for analysis. The reaction solution was prepared according to the manufacturer's instructions, and absorbance changes at 412 nm were measured over two minutes using a spectrophotometer. The GSH and GSSG contents were calculated from the rate of absorbance.

### 2.10. Statistical Analysis

Statistical analysis of all experiments was performed using SPSS. Data were analyzed using a one-way ANOVA. Significant differences among the treatments were determined using the least significant difference (LSD) test ($p < 0.05$). All presented values are repeated means ± standard error of three independent experiments.

## 3. Results

### 3.1. Effect of $H_2S$ on Plant Biomass and Leaf RWC Under Manganese Stress

Mn severely affected plant biomass and leaf RWC (Table 3). The FW and DW of plants were reduced after exposure to Mn compared to control seedlings (CK). The reduction of leaf RWC verified the water deficit condition of *M. hupehensis* seedlings under Mn stress. Leaf RWC was reduced by 25% in Mn-stressed seedlings compared to the control. However, all four levels of $H_2S$ significantly improved water balance by increasing leaf RWC compared to the stress treatment without $H_2S$. In addition, the program HT treatments with $H_2S$ can be used in a similar representation to the Mn treatments.

**Table 3.** The effects of different treatments on plant biomass, and leaf relative water content (%) of the aerial parts and roots of *M. hupehensis* seedlings under manganese stress.

| Treatments | Aerial Parts | | Roots | | Leaf RWC (%) |
|---|---|---|---|---|---|
| | FW (mg) | DW (mg) | FW (mg) | DW (mg) | |
| control | 1.74 ± 0.08 a | 0.66 ± 0.02 ab | 0.7 ± 0.13 a | 0.18 ± 0.03 a | 91.52 ± 1.88 a |
| HS0.2 | 1.77 ± 0.05 a | 0.69 ± 0.03 a | 0.68 ± 0.08 a | 0.18 ± 0.02 a | 90.4 ± 3.77 a |
| Mn | 0.82 ± 0.07 e | 0.29 ± 0.02 e | 0.33 ± 0.07 e | 0.05 ± 0.01 d | 64.34 ± 5.16 de |
| MHS0.05 | 0.96 ± 0.1 d | 0.32 ± 0.04 e | 0.39 ± 0.04 de | 0.06 ± 0.01 cd | 65.88 ± 4.06 de |
| MHS0.1 | 1.16 ± 0.08 c | 0.47 ± 0.03 d | 0.42 ± 0.1d e | 0.08 ± 0.02 c | 67.66 ± 2.89 d |
| MHS0.2 | 1.35 ± 0.07 b | 0.54 ± 0.04 c | 0.49 ± 0.04 cd | 0.11 ± 0.01 b | 75.01 ± 3 c |
| MHS0.5 | 1.44 ± 0.1 b | 0.63 ± 0.05 b | 0.58 ± 0.07 bc | 0.12 ± 0.02 b | 81.81 ± 2.48 b |
| HT | 0.87 ± 0.04 de | 0.31 ± 0.02 e | 0.35 ± 0.09 e | 0.06 ± 0.02 cd | 60.78 ± 3.93 e |

Dates are expressed as average values ± standard errors. In the table, different letters indicate the significant difference at $p < 0.05$.

### 3.2. Effect of $H_2S$ on Proline and Soluble Sugar Contents Under Manganese Stress

Mn stress significantly increased proline levels of seedlings (Figure 1A). However, the application of 0.2 mM $H_2S$ (HS0.2) reduced proline content significantly, with the 0.5 mM $H_2S$ treatment (MHS0.5) leading to a 41% decrease compared to Mn-only stressed plants. Under non-stressed conditions, NaHS application had little effect on soluble sugars and proline content in plant leaves (Figure 1A,B). Mn stress also elevated soluble sugar levels by 24% compared to control plants (CK), and co-application of NaHS further increased this content compared to Mn alone. However, the inclusion of taurine (HT, an $H_2S$ scavenger) reversed the beneficial effects of $H_2S$.

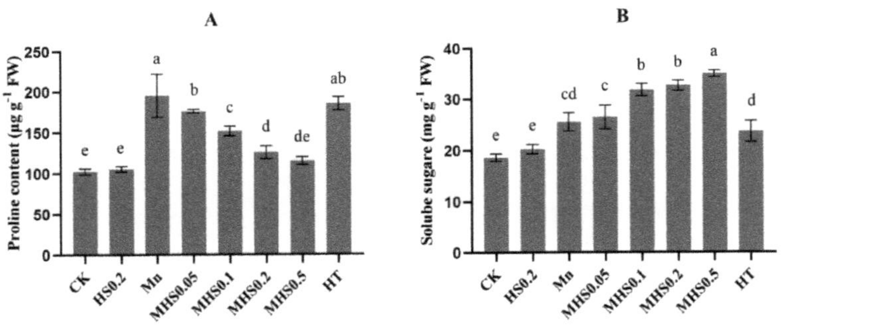


**Figure 1.** The effects of different treatments on proline (**A**) and soluble sugar (**B**) contents of *M. hupehensis* seedlings under manganese stress. Data are expressed as average values ± standard errors. In the charts, different letters indicate the significant difference at $p < 0.05$.

### 3.3. Effect of $H_2S$ on Chlorophyll Content and Gas Exchange Parameters Under Manganese Stress

Chlorophyll a and chlorophyll b content decreased by 21% in Mn-stressed plants compared to control plants (CK), reflecting the impact of Mn toxicity (Figure 2A,B). Supplementation with $H_2S$ significantly restored chlorophyll levels, with the 0.5 mM $H_2S$ treatment (MHS0.5) showing a more potent effect in restoring both chlorophyll a and chlorophyll b content. Mn stress negatively impacted gas exchange parameters, reducing Pn, E, gsw, and Ci by 52%, 80%, 79%, and 28%, respectively, compared to the control plants (Figure 2C–F). However, the addition of $H_2S$ improved these parameters under Mn stress, with the MHS0.5 treatment increasing Pn, E, gsw, and Ci by 41%, 75%, 74% and 39%, respectively, compared to Mn-only treated plants.

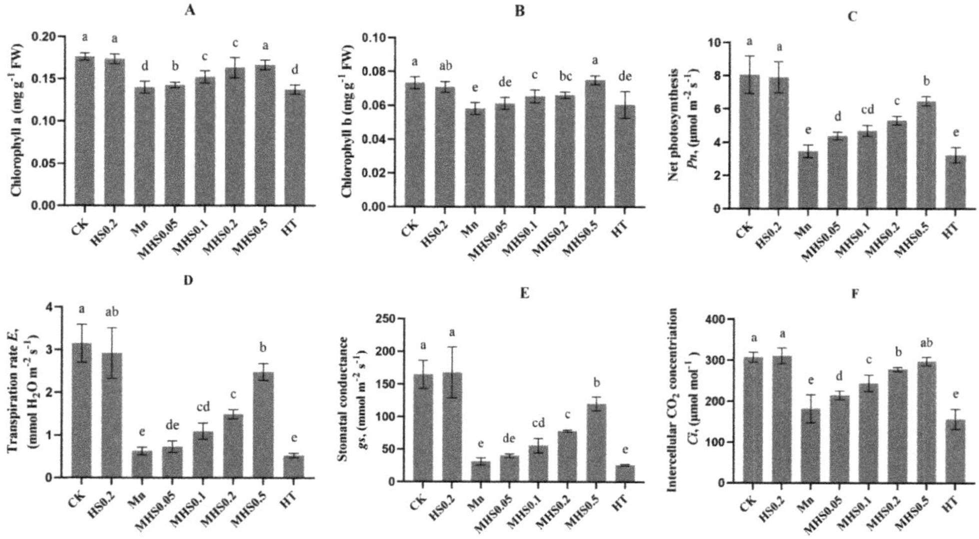


**Figure 2.** The effects of different treatments on chlorophyll a (**A**), chlorophyll b (**B**), Net photosynthesis (**C**), Transpiration rate (**D**), Stomatal conductance (**E**) and Intercellular $CO_2$ concentration (**F**) of *M. hupehensis* seedlings under manganese stress. Data are expressed as average values ± standard errors. In the charts, different letters indicate the significant difference at $p < 0.05$.

### 3.4. Effect of $H_2S$ on Mineral Homeostasis and Manganese Accumulation Under Mn Toxicity

Mn stress reduced the contents of Fe, Zn, Mg, and Ca by 41%, 44%, 53%, and 25%, respectively, compared to control plants (Table 4). Under non-stress conditions, 0.2 mM $H_2S$ application (HS0.2) had little effect on plant metal content. $H_2S$ supplementation alleviated this reduction, with MHS0.5 treatment reducing these losses to 8%, 12%, 49%, and 10%, respectively, compared to the control. There was almost no difference in the metal contents of the plants in the HT treatment group and the Mn alone treatment group.

**Table 4.** The effects of different treatments on nutrient contents (Fe, Zn, Mg and Ca) of *M. hupehensis* seedlings under manganese stress.

| Treatments | Fe (mg g$^{-1}$ DW) | Zn (mg g$^{-1}$ DW) | Mg (mg g$^{-1}$ DW) | Ca (mg g$^{-1}$ DW) |
|---|---|---|---|---|
| CK | 1.82 ± 0.46 a | 0.09 ± 0.01 a | 6.68 ± 0.23 a | 4.89 ± 0.49 ab |
| HS0.2 | 1.71 ± 0.13 ab | 0.09 ± 0.02 a | 6.58 ± 0.5 a | 5.44 ± 1 a |
| Mn | 1.07 ± 0.08 d | 0.05 ± 0 c | 3.16 ± 0.51 c | 3.65 ± 0.5 cd |
| MHS0.05 | 1.36 ± 0.42 bcd | 0.05 ± 0.01 c | 3.4 ± 0.6 bc | 3.89 ± 0.26 cd |
| MHS0.1 | 1.41 ± 0.03 abcd | 0.05 ± 0 c | 3.41 ± 0.61 bc | 4.05 ± 0.35 bcd |
| MHS0.2 | 1.63 ± 0.11 abc | 0.06 ± 0 b | 3.83 ± 0.42 bc | 3.82 ± 0.11 cd |
| MHS0.5 | 1.71 ± 0.14 ab | 0.08 ± 0.01 a | 4.1 ± 0.52 b | 4.43 ± 0.38 bc |
| HT | 1.15 ± 0.06 cd | 0.05 ± 0 c | 2.27 ± 0.43 d | 3.51 ± 0.22 d |

Dates are expressed as average values ± standard errors. In the table, different letters indicate the significant difference at $p < 0.05$.

Additionally, Mn content in the leaves, stems, and roots decreased by 41%, 76%, and 57%, respectively, under MHS0.5 treatment compared to Mn-alone treatment (Table 5). Mn accumulation in the roots showed the greatest reduction, followed by the stems and leaves. Interestingly, the proportion of Mn in leaves relative to the total plant Mn content increased with rising $H_2S$ concentrations, ranging from 24% to 39%.

**Table 5.** The effects of different treatments on the accumulation of manganese in different parts of *M. hupehensis* seedlings under manganese stress.

| Treatments | Mn (mg g$^{-1}$ DW) | | | | Percentage of Mn in Leaves (%) |
|---|---|---|---|---|---|
| | **Leaf** | **Root** | **Stem** | **Total** | **Leaf** |
| CK | 0.06 ± 0.01 d | 0.01 ± 0 f | 0.07 ± 0.02 f | 0.14 ± 0.03 e | 0.44 ± 0.07 a |
| HS0.2 | 0.07 ± 0.01 d | 0.01 ± 0.01 f | 0.06 ± 0 f | 0.15 ± 0.02 e | 0.65 ± 0.31 a |
| Mn | 0.79 ± 0.16 ab | 1.7 ± 0.03 a | 0.77 ± 0.03 b | 3.26 ± 0.17 a | 0.24 ± 0.05 c |
| MHS0.05 | 0.74 ± 0.22 ab | 1.2 ± 0.15 c | 0.57 ± 0.02 d | 2.51 ± 0.25 b | 0.3 ± 0.09 bc |
| MHS0.1 | 0.66 ± 0.15 bc | 1.24 ± 0.07 c | 0.69 ± 0.07 c | 2.59 ± 0.29 b | 0.25 ± 0.06 c |
| MHS0.2 | 0.52 ± 0.06 c | 0.59 ± 0.01 d | 0.63 ± 0.02 d | 1.74 ± 0.09 c | 0.3 ± 0.03 bc |
| MHS0.5 | 0.47 ± 0.11 c | 0.42 ± 0.04 e | 0.33 ± 0.01 e | 1.22 ± 0.14 d | 0.39 ± 0.09 ab |
| HT | 0.91 ± 0.05 a | 1.38 ± 0.02 b | 0.9 ± 0 a | 3.18 ± 0.07 a | 0.28 ± 0.02 bc |

Data are expressed as average values ± standard errors. In the table, different letters indicate the significant difference at $p < 0.05$.

### *3.5. Effect of $H_2S$ on the Expression of Manganese Tolerance Protein Genes Under Mn Stress*

Mn stress significantly up-regulated the expression of *MdMTP8.2* and *MdMTP11.1* by 4.1-fold and 5.5-fold, respectively, compared to the control plants (Figure 3A,D). The co-application of Mn and $H_2S$ further enhanced their expression. Mn treatment increased *MdMTP9.1* expression by 1.3-fold compared to the control, and MHS0.5 treatment elevated this by 2.3-fold relative to Mn alone (Figure 3B). While Mn stress did not alter *MdMTP9.3* expression, $H_2S$ application significantly up-regulated its expression under both stressed and non-stressed conditions (Figure 3C).

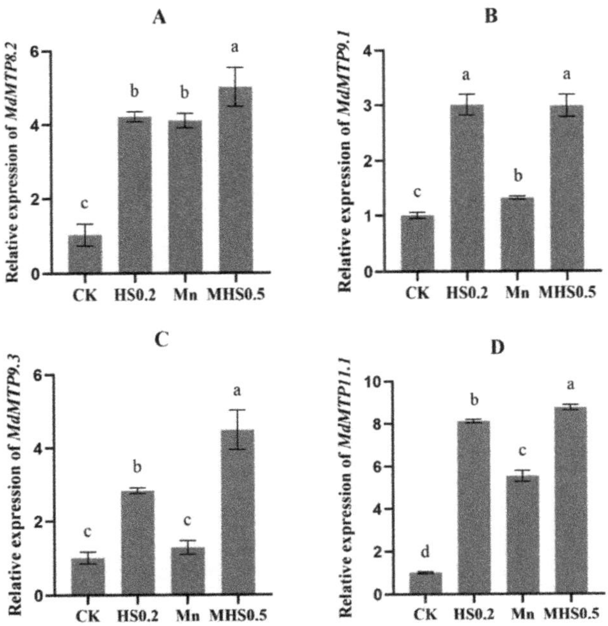


**Figure 3.** The effects of different treatments on the Expression of Manganese Tolerance Protein (MTP) family genes, *MdMTP8.2* (**A**), *MdMTP9.1* (**B**), *MdMTP9.3* (**C**) and *MdMTP11.1* (**D**), of *M. hupehensis* seedlings under manganese stress. Data are expressed as average values ± standard errors. In the charts, different letters indicate the significant difference at $p < 0.05$.

### 3.6. *Effect of $H_2S$ on ROS Accumulation and Electrolyte Leakage Under Mn Stress*

Mn stress elevated $H_2O_2$ and MDA levels by 65% and 71%, respectively, compared to the control plants (Figure 4A,B). However, $H_2S$ supplementation reduced these levels significantly, with the MHS0.5 treatment showing the greatest reduction. In addition, EL increased from 42% to 68% under Mn stress; however, 0.5 Mm $H_2S$ application (MHS0.5) decreased EL to 48% (Figure 4C). Similar results were observed in the HT-treated group of plants as Mn-alone stressed plants in terms of $H_2O_2$, MDA, and EL levels.

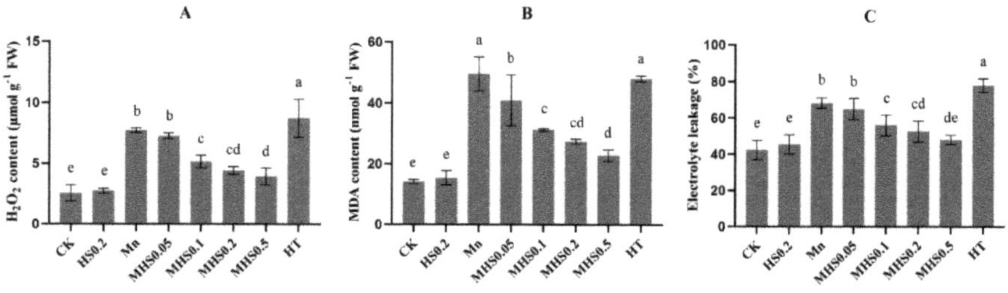


**Figure 4.** The effects of different treatments on hydrogen peroxide ($H_2O_2$) content (**A**), malondialdehyde (MDA) content (**B**) and electrolyte leakage (**C**) of *M. hupehensis* seedlings under manganese stress. Data are expressed as average values ± standard errors. In charts, different letters indicate the significant difference at $p < 0.05$.

### 3.7. *Effect of $H_2S$ on Antioxidant Enzyme Activities and Antioxidant Content Under Mn Stress*

Mn stress increased SOD, POD, and CAT by 48%, 48% and 37%, respectively, compared to the controls (Figure 5A–D). Supplementation with $H_2S$ further enhanced these enzyme activities under Mn stress, with increases of 36%, 27%, 74%, and 35% in SOD, CAT and POD activities, respectively, under MHS0.5 treatment. Mn stress reduced MDHAR and DHAR activities, but $H_2S$ supplementation increased these by 56% and 69%, respectively, under MHS0.5 treatment (Figure 5E,F). There was no significant difference in antioxidant enzyme activity between plants treated with HT and Mn alone.

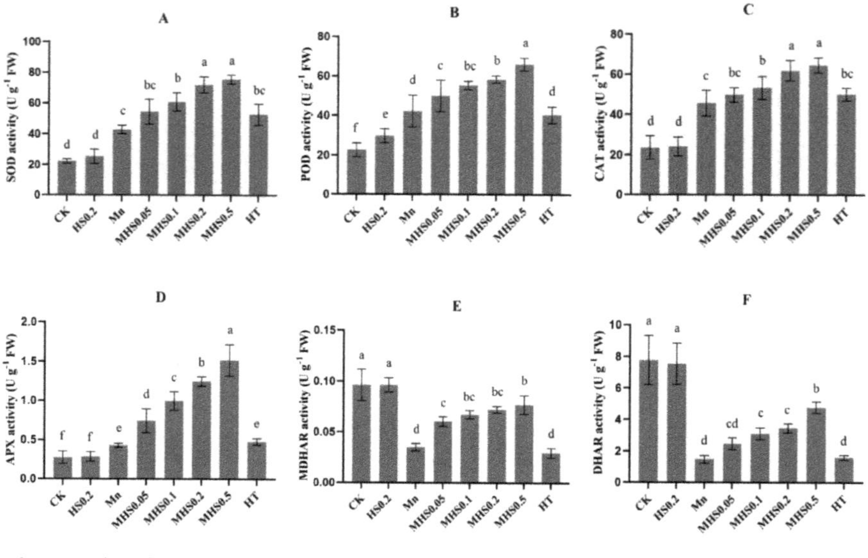


**Figure 5.** The effects of different treatments on SOD (**A**), POD (**B**), CAT (**C**), APX (**D**), MDHAR (**E**), and DHAR (**F**) activities of *M. hupehensis* seedlings under manganese stress. Data are expressed as average values ± standard errors. In the charts, different letters indicate the significant difference at $p < 0.05$.

GSH and GSSG contents increased by 33% and 44%, respectively, while ASA content decreased by 53% under Mn stress (Figure 6). Increasing the concentration of $H_2S$ applied to Mn-stressed plants further significantly increased the effect on antioxidant content compared to Mn-treated plants only, with the supply of 0.5 mM $H_2S$ increasing the levels of ASA and GSH by 52% and 26%, respectively, and decreasing the level of GSSG by 28%. The HT treatment reversed these improvements.

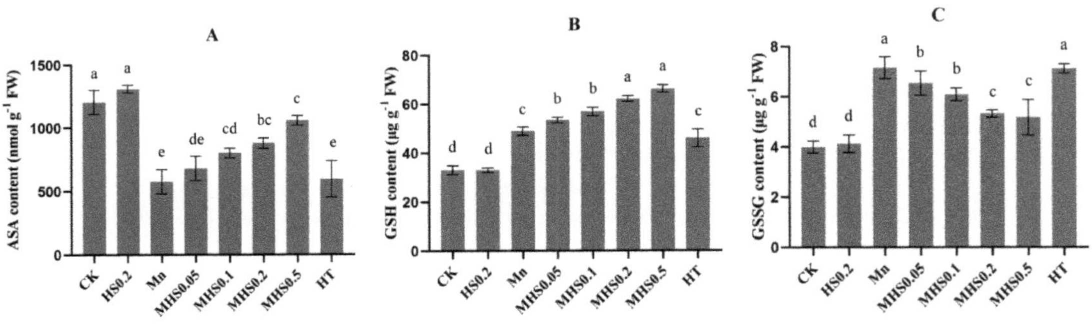


**Figure 6.** The effects of different treatments on ASA (**A**), GSH (**B**), and GSSG (**C**) contents of *M. hupehensis* seedlings under manganese stress. Data are expressed as average values ± standard errors. In the charts, different letters indicate the significant difference at $p < 0.05$.

## 4. Discussion

The literature reports that $H_2S$ is involved in a variety of growth and developmental processes and adversity responses in plants [31]. In this study, it was observed that there was significant potential for $H_2S$ to alleviate manganese (Mn) toxicity in *M. hupehensis*. Our results suggest that Mn toxicity reduces the biomass of *M. hupehensis* (Table 3), findings that are consistent with earlier reports on other species such as *Triticum polonicum* [32]. Manganese-induced plant growth arrest or damage may be due to causing oxidative stress, mineral element imbalance, inhibition of photosynthesis, and the destruction of the cellular structure [33,34]. In this study, we found that $H_2S$ treatment mitigated the adverse effects of Mn toxicity on the morphological and physiological properties of seedlings. Supplementation with $H_2S$ significantly improved the growth of Mn-stressed *M. hupehensis*, consistent with previous studies demonstrating that exogenous $H_2S$ enhances plant growth under toxic metal stress [23,24,35]. Furthermore, substantial evidence indicates that the application of $H_2S$ scavengers (HT) effectively reverses the positive effects of $H_2S$ on plant growth under heavy metal stress [25,36,37].

In the present study, $H_2S$ treatment significantly alleviated the negative effects of Mn toxicity, which may be due to the activation of complex mechanisms by $H_2S$ to resist metal toxicity. Manganese toxicity disturbs the water balance of plants. In response to this stress, plants usually accumulate osmoprotectants such as proline and soluble sugars [13]. In this study, $H_2S$ maintained soluble sugar and proline synthesis in leaves of seedlings under Mn stress to help maintain cellular osmotic pressure and protect organelles from damage (Figure 1). Notably, increased leaf RWC (Table 3) may have improved the stomatal conductance (Figure 2C) of *M. hupehensis* seedlings under Mn treatment with $H_2S$ application, thereby increasing photosynthetic activity and, thus, improving plant growth [38]. This is in agreement with findings in other plant systems, where $H_2S$ has been shown to mitigate drought and salinity stress by modulating osmotic balance and water relations [39,40].

Mn stress has been shown to adversely affect chloroplasts and inhibit chlorophyll biosynthesis [41]. Our study showed that Mn stress significantly reduced the chlorophyll content of *M. hupehensis* seedlings (Figure 2A,B), which is in line with previous findings on *Arabidopsis thaliana* [42] and *Marchantia polymorpha* [43]. Under Mn stress, Mn in chloroplasts generates ROS through photo-oxidation, and free photosynthetic electron transport chain electrons will react with the large amount of $O_2^-$ in the chloroplasts and reduce the photosynthetic rate [41]. Intercystic swelling and basal deformation in chloroplasts have also been associated with Mn toxicity [44]. In addition, Mn may replace Mg in the chlorophyll molecule or bind to ferredoxin in the stroma of the cysts, ultimately destroying the ultrastructure of the chloroplasts [45]. $H_2S$ treatment significantly reduced the adverse effects of Cd on photosynthesis in *Brassica rapa* [37] and As in *Pisum sativum* L. [46], and in agreement with these studies, we observed that under manganese toxicity, the chlorophyll content and the gas exchange parameters(Figure 2) were at their lowest levels. In contrast, elevated $H_2S$ concentration significantly increased chlorophyll content (Figure 2A,B).

It was found that a decrease in E and gsw was detected only in the more damaged young leaves under Mn stress, while no decrease in gsw was detected in the mature leaves with higher Mn accumulation, suggesting that the reduction in gsw by Mn may be a direct result of leaf damage [47]. Thus, $H_2S$ treatment alleviated the decrease in photosynthesis and gas exchange parameters due to Mn stress, which is likely due to the ability of $H_2S$ to reduce ROS levels, maintain chloroplast integrity, and enhance the activity of photosynthetic enzymes (Figure 4). Similar improvements in photosynthesis and chlorophyll content have been observed in plants exposed to other heavy metals, such as Cd-stressed *Triticum aestivum* L. [25] and Ni-stressed *Oryza sativa* L. [48]. Furthermore, $H_2S$ has been shown to promote chloroplast biogenesis and $CO_2$ fixation efficiency in *Spinacia oleracea* [49].

Our results showed that the exogenous application of $H_2S$ increased the plant uptake of Ca, Fe, Mg, and Zn (Table 4). $Ca^{2+}$ signaling was triggered in *Ziziphus jujuba* in response to environmental stimuli that triggered corresponding cellular responses in plants [50]. In addition, it has been shown that Ca can significantly reduce the cell membrane permeability of date palm under salt stress [51]. The increase in Ca content in $H_2S$-treated plants may have reduced the EL of *M. hupehensis* seedlings (Figure 4C). $H_2S$ increased the uptake of Mg and Fe in plants, which contributes to the synthesis of photosynthetic pigments under Mn stress. The increase in Zn content in $H_2S$-treated plants contributed to the increase in biomass [52]. In this study, $H_2S$ treatment improved the uptake of essential nutrients and promoted a more favorable distribution of Mn within the plant. Specifically, $H_2S$ reduced Mn accumulation in the roots and stems while increasing its translocation to leaves, where it can be stored in non-toxic forms.

The observed changes in Mn distribution are likely mediated by the up-regulation of manganese tolerance protein (MTP) genes, such as *MdMTP8.2*, *MdMTP9.1*, *MdMTP9.3*, and *MdMTP11.1*. These proteins are involved in Mn sequestration into vacuoles, the Golgi apparatus, and extracellular spaces, thereby reducing its cytoplasmic concentration and associated toxicity [15]. Previous studies have highlighted the role of MTP8.2 in conferring Mn tolerance in *Pyrus* spp. [16]. It has been shown that in *Arabidopsis thaliana*, the calcium-dependent signaling protein complex CBL2/3-CIPK3/9/26 and CPK5 regulate the vesicular transporter protein AtMTP8 through sequential phosphorylation, enhancing their activity [53]. $H_2S$ treatment significantly up-regulated the expression of the *MdMTP8.2* gene in *M. hupehensis* under Mn stress, suggesting that $H_2S$ may be involved in the maintenance of Mn homeostasis by regulating the $Ca^{2+}$ signaling pathway. $H_2S$ can affect the signaling pathway by modulating the sulfhydrylation modification of proteins. Exploring whether $H_2S$ regulates key signaling pathways in response to Mn stress through sulfhydrylation may be a worthwhile direction of research.

Heavy metal stress can lead to excessive accumulation of ROS, which will cause serious damage to plant physiological metabolism [54]. $H_2O_2$ and MDA are potential biomarkers of oxidative stress. Therefore, we examined the ROS content and antioxidant enzyme activities in plants after $Mn^{2+}$ stress, and the results showed that in addition to reducing ROS accumulation, $H_2S$ treatment significantly enhanced membrane stability, as evidenced by the decreased levels of $H_2O_2$ and MDA (Figure 4A,B), thereby protecting plant cells from oxidative damage induced by Mn stress.

ROS scavenging in plants protects against oxidative stress and this scavenging system is regulated by enzymes and non-enzymatic antioxidants [11]. Among these enzymes, SOD is considered to be the first line of defense for the diversification of $O_2^-$ to $H_2O_2$ [55]. $H_2O_2$ is highly destructive because of its ability to be transferred across the cell membrane to other parts of the cell in which it is produced. $H_2O_2$ and $O_2^-$ are then converted by other antioxidant enzymes into less harmful molecules, thereby protecting cellular components. In our study, the activities of SOD, POD, and CAT were significantly increased under Mn stress, which is similar to previously observed responses in other species, such as *Pisum sativum* L. [46] and *Brassica rapa* [37]. The plants treated with Mn + NaHS (0.5 mM) showed a further increase in the three enzyme activities (Figure 5A–C) and a corresponding further decrease in ROS content. The observed increase in antioxidant enzyme activities aligns with previous studies on $H_2S$-mediated stress tolerance in other species, such as *Brassica napus* and *Spinacia oleracea* [37,49]. While the exact molecular mechanisms are not fully elucidated, it is likely that $H_2S$ modulates the activity of key enzymes involved in antioxidant defense (e.g., SOD, CAT, POD) through post-translational modifications such as sulfhydrylation, which may influence the efficiency of ROS scavenging

Additionally, the AsA-GSH cycle is a critical pathway for ROS detoxification and redox homeostasis [56]. MDHAR, DHAR and APX, as well as two nonenzymatic antioxidants, AsA and GSH, are important components of the AsA-GSH cycle. In this study, Mn stress significantly suppressed MDHAR and DHAR activities while increasing APX activity (Figure 5D–F). This imbalance led to reduced levels of AsA and GSH, compromising the plant's ability to scavenge ROS. However, $H_2S$ supply attenuated the inhibition of MDHAR and DHAR activities and further increased APX activity in Mn-treated plants, and these findings suggest that $H_2S$ may regulate the redox state of AsA and GSH by inducing enzyme activities in the AsA-GSH cycle (Figure 5D–F). AsA is a non-enzymatic water-soluble antioxidant that directly interacts with ROS, thereby reducing the ROS content in cells [56]. GSH is another non-enzymatic antioxidant that can act as a cofactor in the glyoxalase system, reducing ROS levels and alleviating oxidative stress [57]. Our results clearly showed that Mn stress disrupted the redox balance in *M. hupehensis* by depleting AsA and GSH levels (Figure 6A,B). The addition of $H_2S$ restored redox homeostasis, as evidenced by increased AsA and GSH levels, which enhanced the redox buffering capacity of the plants. The restoration of the AsA-GSH cycle by $H_2S$ is indicative of a broader redox-regulatory role. The increased activities of MDHAR, DHAR, and APX under $H_2S$ treatment suggest that $H_2S$ not only enhances the enzymatic scavenging of ROS but also helps restore the levels of AsA and GSH, which is critical for cellular detoxification mechanisms. Consistent with our findings, $H_2S$ application has been shown to attenuate arsenate toxicity in *Pisum sativum* L. by up-regulating the AsA-GSH cycle [58].

However, under Mn stress, the positive regulatory effects of $H_2S$ on both ROS levels and the antioxidant system in plants were reversed by HT, suggesting that $H_2S$ plays an important role in mitigating Mn damage to plants. However, how $H_2S$ specifically affects plant antioxidant enzyme activities still needs further investigation.

## 5. Conclusions

$H_2S$ treatment mitigates the toxic effects of manganese stress on *M. hupehensis* seedlings. The mitigating effect of $H_2S$ on manganese stress likely involves several mechanisms: (1) $H_2S$ restores the absorption of essential minerals, such as Ca, Mg, Fe, and Zn, under manganese stress, thereby helping to maintain nutrient balance in plants. $H_2S$ also modulates the expression of MTP family genes, which regulate manganese transport, reduce toxic accumulation, and promote safe sequestration in leaves. (2) $H_2S$ enhances antioxidant enzyme activity and strengthens the ASA-GSH cycle, mitigating oxidative damage induced by manganese stress. (3) $H_2S$ restores key gas exchange parameters and chlorophyll content, improving photosynthesis and alleviating the adverse effects of manganese stress on plant growth. These findings are crucial for enhancing crop resilience under heavy metal stress, particularly in soils susceptible to manganese toxicity. Future research should focus on identifying specific signaling pathways and molecular targets involved in $H_2S$-mediated stress responses, which may provide new strategies for enhancing plant tolerance to other abiotic stresses.

**Author Contributions:** Conceptualization, software, validation, formal analysis, resources, data curation, writing—original draft preparation and writing—review and editing were performed by B.L. and B.W.; Methodology and investigation were performed by T.C.; Supervision, project administration and funding acquisition were performed by M.Z. All authors have read and agreed to the published version of the manuscript.

**Funding:** This research was funded by the Apple Research System (CARS-27): K3010723020.

**Data Availability Statement:** Data are contained within the article.

**Conflicts of Interest:** The authors declare no conflicts of interest.

## References

1. Li, J.; Jia, Y.; Dong, R.; Huang, R.; Liu, P.; Li, X.; Wang, Z.; Liu, G.; Chen, Z. Advances in the Mechanisms of Plant Tolerance to Manganese Toxicity. *Int. J. Mol. Sci.* **2019**, *20*, 5096. [CrossRef] [PubMed]
2. You, X.; Yang, L.-T.; Lu, Y.-B.; Li, H.; Zhang, S.-Q.; Chen, L.-S. Proteomic changes of Citrus roots in response to long-term manganese toxicity. *Trees* **2014**, *28*, 1383–1399. [CrossRef]
3. Millaleo, R.; Alvear, M.; Aguilera, P.; González-Villagra, J.; de la Luz Mora, M.; Alberdi, M.; Reyes-Díaz, M. Mn Toxicity Differentially Affects Physiological and Biochemical Features in Highbush Blueberry (*Vaccinium corymbosum* L.) Cultivars. *J. Soil. Sci. Plant Nutr.* **2019**, *20*, 795–805. [CrossRef]
4. Chu, H.H.; Car, S.; Socha, A.L.; Hindt, M.N.; Punshon, T.; Guerinot, M.L. The Arabidopsis MTP8 transporter determines the localization of manganese and iron in seeds. *Sci. Rep.* **2017**, *7*, 11024. [CrossRef] [PubMed]
5. Sparrow, L.A.; Uren, N.C. Manganese oxidation and reduction in soils: Effects of temperature, water potential, pH and their interactions. *Soil. Res.* **2014**, *52*, 483–494. [CrossRef]
6. Wang, F.; Ge, S.; Lyu, M.; Liu, J.; Li, M.; Jiang, Y.; Xu, X.; Xing, Y.; Cao, H.; Zhu, Z.; et al. DMPP reduces nitrogen fertilizer application rate, improves fruit quality, and reduces environmental cost of intensive apple production in China. *Sci. Total Environ.* **2022**, *802*, 149813. [CrossRef] [PubMed]
7. Wei, Y.; Han, R.; Xie, Y.; Jiang, C.; Yu, Y. Recent Advances in Understanding Mechanisms of Plant Tolerance and Response to Aluminum Toxicity. *Sustainability* **2021**, *13*, 1782. [CrossRef]
8. Wan, H.; Yang, F.; Zhuang, X.; Cao, Y.; He, J.; Li, H.; Qin, S.; Lyu, D. Malus rootstocks affect copper accumulation and tolerance in trees by regulating copper mobility, physiological responses, and gene expression patterns. *Environ. Pollut.* **2021**, *287*, 117610. [CrossRef]
9. Min, S.F.; Chang, Z.X.; Chen, L.; Feng, L.Z.; Tang, B.; Ping, Z.Z. Effects of manganese pollution on soil mincrobial community fuxtion and plant growth. *J. Environ. Prot. Ecol.* **2019**, *20*, 63–74.
10. Noor, I.; Sohail, H.; Zhang, D.; Zhu, K.; Shen, W.; Pan, J.; Hasanuzzaman, M.; Li, G.; Liu, J. Silencing of PpNRAMP5 improves manganese toxicity tolerance in peach (*Prunus persica*) seedlings. *J. Hazard. Mater.* **2023**, *454*, 131442. [CrossRef] [PubMed]
11. Ali, M.A.; Fahad, S.; Haider, I.; Ahmed, N.; Ahmad, S.; Hussain, S.; Arshad, M. Oxidative Stress and Antioxidant Defense in Plants Exposed to Metal/Metalloid Toxicity. In *Reactive Oxygen, Nitrogen and Sulfur Species in Plants*; John Wiley & Sons Ltd.: Hoboken, NJ, USA, 2019; pp. 353–370.

12. Liu, Y.; Chen, J.; Li, X.; Yang, S.; Wu, Z.; Xue, Y.; Chen, J. Physiological Mechanisms in Which Manganese Toxicity Inhibits Root Growth in Soybean. *J. Soil Sci. Plant Nutr.* **2023**, *23*, 4141–4156. [CrossRef]
13. Rosas, A.; Rengel, Z.; de la Luz Mora, M. Manganese supply and pH influence growth, carboxylate exudation and peroxidase activity of ryegrass and white clover. *J. Plant Nutr.* **2007**, *30*, 253–270. [CrossRef]
14. Chen, Z.; Sun, L.; Liu, P.; Liu, G.; Tian, J.; Liao, H. Malate Synthesis and Secretion Mediated by a Manganese-Enhanced Malate Dehydrogenase Confers Superior Manganese Tolerance in Stylosanthes guianensis. *Plant Physiol.* **2015**, *167*, 176–188. [CrossRef] [PubMed]
15. Song, R.; Li, Z.; Su, X.; Liang, M.; Li, W.; Tang, X.; Li, J.; Qiao, X. The Malus domestica metal tolerance protein MdMTP11.1 was involved in the detoxification of excess manganese in Arabidopsis thaliana. *J. Plant Physiol.* **2023**, *288*, 154056. [CrossRef] [PubMed]
16. Li, J.; Zheng, L.; Fan, Y.; Wang, Y.; Ma, Y.; Gu, D.; Lu, Y.; Zhang, S.; Chen, X.; Zhang, W. Pear metal transport protein PbMTP8.1 confers manganese tolerance when expressed in yeast and Arabidopsis thaliana. *Ecotoxicol. Environ. Saf.* **2021**, *208*, 111687. [CrossRef]
17. Ueno, D.; Sasaki, A.; Yamaji, N.; Miyaji, T.; Fujii, Y.; Takemoto, Y.; Moriyama, S.; Che, J.; Moriyama, Y.; Iwasaki, K.; et al. A polarly localized transporter for efficient manganese uptake in rice. *Nat. Plants* **2015**, *1*, 15170. [CrossRef] [PubMed]
18. Seneviratne, M.; Rajakaruna, N.; Rizwan, M.; Madawala, H.M.S.P.; Ok, Y.S.; Vithanage, M. Correction to: Heavy metal-induced oxidative stress on seed germination and seedling development: A critical review. *Environ. Geochem. Health* **2019**, *41*, 1813–1831. [CrossRef]
19. Arif, Y.; Hayat, S.; Yusuf, M.; Bajguz, A. Hydrogen sulfide: A versatile gaseous molecule in plants. *Plant Physiol. Biochem.* **2021**, *158*, 372–384. [CrossRef] [PubMed]
20. da-Silva, C.J.; Modolo, L.V. Hydrogen sulfide: A new endogenous player in an old mechanism of plant tolerance to high salinity. *Acta Bot. Bras.* **2017**, *32*, 150–160. [CrossRef]
21. Li, Z.G.; Fang, J.R.; Bai, S.J. Hydrogen sulfide signaling in plant response to temperature stress. *Front. Plant Sci.* **2024**, *15*, 1337250. [CrossRef]
22. Li, H.; Shi, J.; Wang, Z.; Zhang, W.; Yang, H. $H_2S$ pretreatment mitigates the alkaline salt stress on Malus hupehensis roots by regulating $Na^+/K^+$ homeostasis and oxidative stress. *Plant Physiol. Biochem.* **2020**, *156*, 233–241. [CrossRef] [PubMed]
23. He, H.; Li, Y.; He, L.F. The central role of hydrogen sulfide in plant responses to toxic metal stress. *Ecotoxicol. Environ. Saf.* **2018**, *157*, 403–408. [CrossRef] [PubMed]
24. Yu, Y.; Dong, J.; Li, R.; Zhao, X.; Zhu, Z.; Zhang, F.; Zhou, K.; Lin, X. Sodium hydrosulfide alleviates aluminum toxicity in Brassica napus through maintaining $H_2S$, ROS homeostasis and enhancing aluminum exclusion. *Sci. Total Environ.* **2023**, *858*, 160073. [CrossRef]
25. Kaya, C.; Ashraf, M.; Alyemeni, M.N.; Ahmad, P. Responses of nitric oxide and hydrogen sulfide in regulating oxidative defence system in wheat plants grown under cadmium stress. *Physiol. Plant* **2020**, *168*, 345–360. [CrossRef] [PubMed]
26. Kharbech, O.; Sakouhi, L.; Mahjoubi, Y.; Ben Massoud, M.; Debez, A.; Zribi, O.T.; Djebali, W.; Chaoui, A.; Mur, L.A.J. Nitric oxide donor, sodium nitroprusside modulates hydrogen sulfide metabolism and cysteine homeostasis to aid the alleviation of chromium toxicity in maize seedlings (*Zea mays* L.). *J. Hazard. Mater.* **2022**, *424*, 127302. [CrossRef]
27. Hongqiang, Y.; Kaixuan, D.; Wei, Z. Biology and Physiology of Malus Hupehensis for the Apogamic Plant Resource. *Acta Hortic.* **2008**, *769*, 441–447. [CrossRef]
28. DI, A. Copper enzymes in isolated chloroplasts Polyphenoloxidase in *Beta vulgaris*. *Plant Physiol.* **1949**, *24*, 1–15.
29. Lazcano-Ferrat, I.; Lovatt, C.J. Relationship between relative water content, nitrogen pools, and growth of *Phaseolus vulgaris* L. and *P. acutifolius* A. Gray during water deficit. *Crop Sci.* **1999**, *39*, 467–475. [CrossRef]
30. Maribel, L.; Dionisio-Sese, S.T. Antioxidant responses of rice seedlings to salinity stress. *Antioxid. Responses Rice Seedl. Salin. Stress* **1998**, *135*, 1–9.
31. Fang, T.; Cao, Z.; Li, J.; Shen, W.; Huang, L. Auxin-induced hydrogen sulfide generation is involved in lateral root formation in tomato. *Plant Physiol. Biochem.* **2014**, *76*, 44–51. [CrossRef]
32. Sheng, H.; Zeng, J.; Yan, F.; Wang, X.; Wang, Y.; Kang, H.; Fan, X.; Sha, L.; Zhang, H.; Zhou, Y. Effect of exogenous salicylic acid on manganese toxicity, mineral nutrients translocation and antioxidative system in polish wheat (*Triticum polonicum* L.). *Acta Physiol. Plant.* **2015**, *37*, 32. [CrossRef]
33. Rajpoot, R.; Srivastava, R.K.; Rani, A.; Pandey, P.; Dubey, R.S. Manganese-induced oxidative stress, ultrastructural changes, and proteomics studies in rice plants. *Protoplasma* **2021**, *258*, 319–335. [CrossRef] [PubMed]
34. Doncheva, S.; Poschenrieder, C.; Stoyanova, Z.; Georgieva, K.; Velichkova, M.; Barceló, J. Silicon amelioration of manganese toxicity in Mn-sensitive and Mn-tolerant maize varieties. *Environ. Exp. Bot.* **2009**, *65*, 189–197. [CrossRef]

35. Alamri, S.; Ali, H.M.; Khan, M.I.R.; Singh, V.P.; Siddiqui, M.H. Exogenous nitric oxide requires endogenous hydrogen sulfide to induce the resilience through sulfur assimilation in tomato seedlings under hexavalent chromium toxicity. *Plant Physiol. Biochem.* **2020**, *155*, 20–34. [CrossRef] [PubMed]
36. Mostofa, M.G.; Rahman, A.; Ansary, M.M.; Watanabe, A.; Fujita, M.; Tran, L.S. Hydrogen sulfide modulates cadmium-induced physiological and biochemical responses to alleviate cadmium toxicity in rice. *Sci. Rep.* **2015**, *5*, 14078. [CrossRef] [PubMed]
37. Li, G.; Shah, A.A.; Khan, W.U.; Yasin, N.A.; Ahmad, A.; Abbas, M.; Ali, A.; Safdar, N. Hydrogen sulfide mitigates cadmium induced toxicity in Brassica rapa by modulating physiochemical attributes, osmolyte metabolism and antioxidative machinery. *Chemosphere* **2021**, *263*, 127999. [CrossRef] [PubMed]
38. Duan, B.; Ma, Y.; Jiang, M.; Yang, F.; Ni, L.; Lu, W. Improvement of photosynthesis in rice (*Oryza sativa* L.) as a result of an increase in stomatal aperture and density by exogenous hydrogen sulfide treatment. *Plant Growth Regul.* **2014**, *75*, 33–44. [CrossRef]
39. Luo, S.; Liu, Z.; Wan, Z.; He, X.; Lv, J.; Yu, J.; Zhang, G. Foliar Spraying of NaHS Alleviates Cucumber Salt Stress by Maintaining $N^+/K^+$ Balance and Activating Salt Tolerance Signaling Pathways. *Plants* **2023**, *12*, 2450. [CrossRef] [PubMed]
40. Ekinci, M.; Turan, M.; Ors, S.; Dursun, A.; Yildirim, E. Improving salt tolerance of bean (*Phaseolus vulgaris* L.) with hydrogen sulfide. *Photosynthetica* **2023**, *61*, 25–36. [CrossRef]
41. Alejandro, S.; Holler, S.; Meier, B.; Peiter, E. Manganese in Plants: From Acquisition to Subcellular Allocation. *Front. Plant Sci.* **2020**, *11*, 300. [CrossRef]
42. Hou, L.; Wang, Z.; Gong, G.; Zhu, Y.; Ye, Q.; Lu, S.; Liu, X. Hydrogen Sulfide Alleviates Manganese Stress in Arabidopsis. *Int. J. Mol. Sci.* **2022**, *23*, 5046. [CrossRef]
43. Messant, M.; Hani, U.; Hennebelle, T.; Guerard, F.; Gakiere, B.; Gall, A.; Thomine, S.; Krieger-Liszkay, A. Manganese concentration affects chloroplast structure and the photosynthetic apparatus in Marchantia polymorpha. *Plant Physiol.* **2023**, *192*, 356–369. [CrossRef] [PubMed]
44. Zambrosi, F.C.B.; Mesquita, G.L.; Marchiori, P.E.R.; Tanaka, F.A.O.; Machado, E.C.; Ribeiro, R.V. Anatomical and physiological bases of sugarcane tolerance to manganese toxicity. *Environ. Exp. Bot.* **2016**, *132*, 100–112. [CrossRef]
45. Costa, G.B.; Simioni, C.; Ramlov, F.; Maraschin, M.; Chow, F.; Bouzon, Z.L.; Schmidt, É.C. Effects of manganese on the physiology and ultrastructure of *Sargassum cymosum*. *Environ. Exp. Bot.* **2017**, *133*, 24–34. [CrossRef]
46. Alsahli, A.A.; Bhat, J.A.; Alyemeni, M.N.; Ashraf, M.; Ahmad, P. Hydrogen Sulfide ($H_2S$) Mitigates Arsenic (As)-Induced Toxicity in Pea (*Pisum sativum* L.) Plants by Regulating Osmoregulation, Antioxidant Defense System, Ascorbate Glutathione Cycle and Glyoxalase System. *J. Plant Growth Regul.* **2020**, *40*, 2515–2531. [CrossRef]
47. González, A.; Lynch, J.P. Effects of manganese toxicity on leaf $CO_2$ assimilation of contrasting common bean genotypes. *Physiol. Plant.* **2006**, *101*, 872–880. [CrossRef]
48. Rizwan, M.; Mostofa, M.G.; Ahmad, M.Z.; Zhou, Y.; Adeel, M.; Mehmood, S.; Ahmad, M.A.; Javed, R.; Imtiaz, M.; Aziz, O.; et al. Hydrogen sulfide enhances rice tolerance to nickel through the prevention of chloroplast damage and the improvement of nitrogen metabolism under excessive nickel. *Plant Physiol. Biochem.* **2019**, *138*, 100–111. [CrossRef] [PubMed]
49. Chen, J.; Wu, F.H.; Wang, W.H.; Zheng, C.J.; Lin, G.H.; Dong, X.J.; He, J.X.; Pei, Z.M.; Zheng, H.L. Hydrogen sulphide enhances photosynthesis through promoting chloroplast biogenesis, photosynthetic enzyme expression, and thiol redox modification in Spinacia oleracea seedlings. *J. Exp. Bot.* **2011**, *62*, 4481–4493. [CrossRef] [PubMed]
50. Chen, X.; Ding, Y.; Yang, Y.; Song, C.; Wang, B.; Yang, S.; Guo, Y.; Gong, Z. Protein kinases in plant responses to drought, salt, and cold stress. *J. Integr. Plant Biol.* **2021**, *63*, 53–78. [CrossRef] [PubMed]
51. Jin, J.; Cui, H.; Lv, X.; Yang, Y.; Wang, Y.; Lu, X. Exogenous $CaCl_2$ reduces salt stress in sour jujube by reducing $Na^+$ and increasing $K^+$, $Ca^{2+}$, and $Mg^{2+}$ in different plant organs. *J. Hortic. Sci. Biotechnol.* **2016**, *92*, 98–106. [CrossRef]
52. Haider, M.U.; Hussain, M.; Farooq, M.; Nawaz, A. Zinc Nutrition for Improving the Productivity and Grain Biofortification of Mungbean. *J. Soil Sci. Plant Nutr.* **2020**, *20*, 1321–1335. [CrossRef]
53. Zhang, Z.; Fu, D.; Sun, Z.; Ju, C.; Miao, C.; Wang, Z.; Xie, D.; Ma, L.; Gong, Z.; Wang, C. Tonoplast-associated calcium signaling regulates manganese homeostasis in Arabidopsis. *Mol. Plant* **2021**, *14*, 805–819. [CrossRef]
54. Emamverdian, A.; Ding, Y.; Mokhberdoran, F.; Xie, Y. Heavy metal stress and some mechanisms of plant defense response. *Sci. World J.* **2015**, *2015*, 756120. [CrossRef] [PubMed]
55. Shahid, M.; Natasha; Khalid, S.; Abbas, G.; Niazi, N.K.; Murtaza, B.; Rashid, M.I.; Bibi, I. Redox Mechanisms and Plant Tolerance Under Heavy Metal Stress: Genes and Regulatory Networks. In *Plant Metallomics and Functional Omics: A System-Wide Perspective*; Sablok, G., Ed.; Springer International Publishing: Cham, Switzerland, 2019; pp. 71–105.
56. Mahmud, J.A.; Hasanuzzaman, M.; Nahar, K.; Bhuyan, M.; Fujita, M. Insights into citric acid-induced cadmium tolerance and phytoremediation in *Brassica juncea* L.: Coordinated functions of metal chelation, antioxidant defense and glyoxalase systems. *Ecotoxicol. Environ. Saf.* **2018**, *147*, 990–1001. [CrossRef]

57. Hasanuzzaman, M.; Nahar, K.; Hossain, M.S.; Mahmud, J.A.; Rahman, A.; Inafuku, M.; Oku, H.; Fujita, M. Coordinated Actions of Glyoxalase and Antioxidant Defense Systems in Conferring Abiotic Stress Tolerance in Plants. *Int. J. Mol. Sci.* **2017**, *18*, 200. [CrossRef] [PubMed]
58. Singh, V.P.; Singh, S.; Kumar, J.; Prasad, S.M. Hydrogen sulfide alleviates toxic effects of arsenate in pea seedlings through up-regulation of the ascorbate-glutathione cycle: Possible involvement of nitric oxide. *J. Plant Physiol.* **2015**, *181*, 20–29. [CrossRef] [PubMed]

horticulturae

MDPI

Article

# Functional Identification of the Isopentenyl Diphosphate Isomerase Gene from *Fritillaria unibracteata*

**Xinyi Yu [1], Jiao Chen [2], Han Yan [1], Xue Huang [1], Jieru Chen [1], Zichun Ma [1], Jiayu Zhou [1,*] and Hai Liao [1,*]**

[1] School of Life Science and Engineering, Southwest Jiaotong University, Chengdu 610031, China; yuxinyi205@163.com (X.Y.); yanhan15955737717@163.com (H.Y.); hxyellowsnow@163.com (X.H.); chenjieru0617@163.com (J.C.); mazichun020408@163.com (Z.M.)

[2] Deyang Food and Drug Safety Inspection Center, Deyang 618029, China; chenjiao1632024@163.com

* Correspondence: spinezhou@home.swjtu.edu.cn (J.Z.); ddliaohai@home.swjtu.edu.cn (H.L.); Tel./Fax: +86-28-87603202 (H.L.)

**Abstract:** Isopentenyl diphosphate isomerase (IPI) is a key enzyme in the synthesis of isoprenoids. In this paper, the in vivo biological activity of the *IPI* gene from *Fritillaria unibracteata* (*FuIPI*) was investigated. Combining a color complementation experiment with High-Performance Liquid Chromatography analysis showed that the *FuIPI* gene could accumulate *β*-carotene in *Escherichia coli*, and Glu190 was identified as a key residue for its catalytic activity. Bioinformatics analysis together with subcellular localization indicated that the FuIPI protein was localized in chloroplasts. Compared with wild-type *Arabidopsis thaliana*, *FuIPI* transgenic plants had higher abscisic acid content and strengthening tolerance to drought and salt stress. Overall, these results indicated that the *FuIPI* gene had substantial biological activity in vivo, hopefully laying a foundation for its further research and application in liliaceous ornamental and medicinal plants.

**Keywords:** isopentenyl diphosphate isomerase; color complementation experiment; subcellular localization; stress response; transgenetic plants

## 1. Introduction

The family *Liliaceae*, encompassing a variety of ornamental and medicinal plants, has attracted great attention from botanical researchers. Specifically, *Fritillaria unibracteata* Hsiao et K.C. Hsia. (*F. unibracteata*) has been used as a primary medicinal resource for *Fritillaria cirrhosa* D. Don [1], a traditional Chinese medicine, for thousands of years [2]. *F. unibracteata* contains various active ingredients with antitussive effects, such as isosteroidal alkaloids, steroidal alkaloids, terpenoids, and steroidal saponins [3–5]. In addition, it exhibits superior pharmaceutic quality among *Fritillaria cirrhosa* D. Don due to its antitussive and asthma-relieving effects with minimal toxicity and side effects [6]. However, the wild resources of *F. unibracteata* have become rare due to not only aggressive harvesting but also the low accumulation of active ingredients [7]. Meanwhile, it is challenging to cultivate *F. unibracteata* in low-altitude regions, possibly due to its specific adaptability [8]. Therefore, the imbalance between high demand and insufficient supply has triggered research on the biosynthetic pathway of the active ingredients in *F. unibracteata*, laying a foundation for the metabolic engineering of its active ingredients.

The biosynthetic pathways of steroidal alkaloids and terpenoids are primarily involved in the mevalonate-independent (MVA) pathway and the methyl-erythritol 4-phosphate (MEP) pathway [9]. Several biosynthetic enzymes, such as 1-deoxy-D-xylulose-5-phosphate synthase (DXS), 1-deoxy-D-xylulose-5-phosphate reductoisomerase (DXR), and lycopene *β*-cyclase [10,11], have been identified as key catalysts in the MVA and MEP pathways. The checkpoint of both pathways is the isomerization reaction of isopentenyl pyrophosphate (IPP) and dimethylallyl pyrophosphate (DMAPP), catalyzed by the enzyme isopentenyl diphosphate delta isomerase (IPI). Heterologous overexpression of exogenous IPIs in

*Escherichia coli* (*E. coli*) increases the production of carotene [12], lycopene [13], and astaxanthin [14], highlighting their positive roles in the accumulation of downstream products. Intriguingly, IPI activity demonstrated a linear relationship with isoprene emission factors in the leaves of oak species [15], indicating its possible involvement in isoprene biosynthesis by oak leaves. Moreover, overexpression of the *IPI* genes in *Eucommia ulmoides* improves trans-polyisoprene production [16]. Overall, it is reasonably predicted that IPI may be a key enzyme linking the MVA and MEP pathways.

In our previous work, we obtained the *IPI* gene from *F. unibracteata* (*FuIPI*) encoding a peptide of 274 residues [17]. An in vitro enzymatic assay revealed that the recombinant FuIPI protein could convert IPP to DMAPP [17]. Therefore, further confirmation of its activity in vivo is crucial to understand its biological function. In this study, *E. coli* was firstly used as a prokaryotic host to verify the in vivo activity of *FuIPI* through a color complementation experiment combined with High-Performance Liquid Chromatography (HPLC) analysis. Subsequently, the subcellular localization of *FuIPI* protein was determined to predict the metabolite trafficking pathway in plants. Then, the *FuIPI* transgenic *Arabidopsis thaliana* (*A. thaliana*) was constructed. The phenotypes and metabolite change in wild-type and transgenic plants were evaluated. As a result, its activity in a eukaryotic host lays a theoretical foundation for regulating the biosynthesis of active compounds in *F. unibracteata* at the molecular level. To our knowledge, this is the first report of the in vivo functional identification of genes encoding enzymes related to the MVA and MEP pathways from the family *Liliaceae* and it might provide information for biological research on *IPI* genes from other liliaceous plants.

## 2. Materials and Methods

### *2.1. Materials*

#### 2.1.1. Strain and Plasmid

*E. coli* containing recombinant plasmid pET28a-*FuIPI*, plasmid pTrc with pAC-BETA, and plasmid pBI121 with green fluorescent protein (GFP) label were stored in the laboratory. *Agrobacterium tumefaciens* (*A. tumefaciens*) GV3101 was purchased from Sangon Biotech Co., Ltd. (Shanghai, China) and stored at −80 °C in the laboratory. The primers used in the study were designed with Primer Premier 5.0 software and biosynthesized by TSINGKE Co., Ltd. (Beijing, China).

#### 2.1.2. Reagents and Plant Materials

Plasmid Mini Kit I D6943 (OMEGA Bio-tek, Guangzhou, China), 1-5$^{TM}$ 2× High Fidelity Master Mix (TSINGKE Co., Ltd., Beijing, China), Mut Express II Fast Mutagenesis Kit V2 (Vazyme Biotech Company, Ltd., Nanjing, China), Gel Extraction Kit D2500 (OMEGA Bio-tek, Guangzhou, China), BCA Protein Assay Kit (Sangon Biotech Co., Ltd., Shanghai, China), Prime STAR HS DNA Polymerase (Takara Biotechnology Co., Ltd., Dalian, China), and Plant hormone abscisic acid ELISA kit (Enzyme-linked Biotech Co., Ltd., Shanghai, China) were used. *E. coli* TOP10 strain, wild-type *A. thaliana* (ecotype Columbia-0), and *Nicotiana benthamiana* (*N. benthamiana*, ecotype LAB) seeds were kept in the laboratory.

#### 2.1.3. Experimental Apparatus

Veriti$^{TM}$96 Well Fast Thermal Cycler (Thermo Fisher, Waltham, MA, USA), HPLC (Agilent, Santa Clara, CA, USA), Laser confocal microscopy A1R+ (Nikon 1774059S, Tokyo, Japan).

### *2.2. Experimental Methods*

#### 2.2.1. Construction of pTrc-*FuIPI* Expression Plasmid and Expression Host

Using plasmid pET28a-*FuIPI* as a template, the *FuIPI* gene fragment was amplified by polymerase chain reaction (PCR) using a pair of primers, 5′-GC*GGCCGC*GAAAACCTGTAT TTTCAG-3′ (*Not*I cleavage site in italics, this forward primer is designed based on the 3–21 bp upstream sequence from the initiation codon ATG) and 5′-CG*GGATCC*CGTTAAAT

CAGTTTATGAATGG-3′ (*Bam*HI cleavage site in italics). Subsequently, the fragment was digested with *Not*I and *Bam*HI, gel-purified, and then cloned into plasmid pTrc to construct the recombinant plasmid pTrc-*FuIPI*. In addition, the recombinant plasmid pTrc-*FuIPI*, pTrc, and pAC-BETA were used to transform *E. coli* TOP10 strain to generate various recombinant strains, respectively.

2.2.2. Color Complementation Experiment

After verification by sequencing, four types of *E. coli* TOP10 strains, containing plasmid pTrc-*FuIPI*, plasmids pTrc and pAC-BETA, plasmid pAC-BETA, and plasmids pTrc-*FuIPI* and pAC-BETA, were cultured on Luria–Bertani (LB) solid medium at 37 °C for 48 h, respectively. The single colonies were then picked up and cultured on LB solid medium containing 50 μg/mL ampicillin (Ap) and 50 μg/mL chloramphenicol (Chl) at 37 °C for 48 h. Finally, the color depth of the colonies was observed and compared.

2.2.3. Determination of $\beta$-Carotene Content

Four types of bacterial solutions were transferred from the forementioned section to 50 mL of liquid LB medium and incubated at 37 °C overnight. Subsequently, 2–3 mL were taken and incubated at 28 °C and 200 rpm for 48 h to allow the $\beta$-carotene to fully accumulate. The $\beta$-carotene concentrations in *E. coli* TOP10 colonies were measured by HPLC analysis, as described previously [11]. The *E. coli* TOP10 cells were homogenized in methyl alcohol with 10% tetrahydrofuran and 0.01% dibutylhydroxytoluene, centrifuged, and filtered using a 0.45 μm microporous membrane [11]. For HPLC detection, the reversed-phase column Hypersil ODS2 (4.6 mm × 200 mm, 5 μm) was used with a mobile phase of methanol containing 10% tetrahydrofuran (volume fraction) and 0.0025% butylated hydroxytoluene. The injection volume was 10 μL with a flow rate of 1 mL/min. The column temperature was set at 30 °C and the detection wavelength was set at 448 nm.

2.2.4. Site Mutation at Glu190 Residue

As described by Song et al. [18], the codon GAA (Glu190) residue was replaced by GCG (Ala) residue using a pair of primers (5′-GAACATGCGCTGGATTATCTGCTGTTTATTGTT CG-3′ and 5′-GATAATCCAGCGCATGTTCACCCCATTTACC-3′). The site mutation was performed according to the site mutation kit using the plasmid pTrc-*FuIPI* as a template. The color complementation experiment of mutated pTrc-*FuIPI* E190A was then conducted using the same methods as for pTrc-*FuIPI*.

2.2.5. Subcellular Localization of FuIPI Protein

Using plasmid pET28a-*FuIPI* as a template, the *FuIPI* fragment was amplified by PCR with primers 5′-*GAATTC*ATGGCAGCCGGGAGCGTT-3′ (*Eco*RI cleavage site in italics) and 5′-CGC*GGATCC*GCGAATCAGTTTATGAATGG-3′ (*Bam*HI cleavage site in italics). In order to construct the fusion expression plasmid containing *FuIPI* and *GFP* fluorescent label, the stop codon (TAA) of the *FuIPI* gene was not included in the PCR product. Following the method described in Section 2.2.1, *FuIPI* was cloned into plasmid pBI121 (with GFP fluorescent label) and transformed into *A. tumefaciens*, which was then uniformly cultured on LB solid medium (containing 50 μg/mL gentamicin and 50 μg/mL kanamycin) at 30 °C for 2–3 days. Then, ten single colonies were randomly picked up and cultured for 24 h by shaking. After colony PCR verification, recombinant *A. tumefaciens* with plasmid pBI121-*FuIPI* was obtained.

A total of 2 mL of recombinant bacteria GV3101 containing pBI121-*FuIPI* was transferred and injected into the backs of *N. benthamiana* leaves and cultured for 2 days in low light. Fluorescence was observed using confocal laser microscopy. The excitation and emission wavelengths for chloroplast autofluorescence were set to 640 nm and 675 nm, respectively. The excitation and emission wavelengths for GFP were 488 nm and 510 nm, respectively.

2.2.6. Construction and Identification of *FuIPI* Transgenic *A. thaliana*

To analyze the functional roles of the *FuIPI* gene in transgenic plants, the *FuIPI* gene was overexpressed in wild-type *A. thaliana*. The *FuIPI* gene was first amplified by PCR using primers 5′-*GCGGGTCGACGGTACC*ATGGCAGCCGGGAGCGTT-3′ and 5′-*TAGACATATGGGTACC*TTAAATCAGTTTATGAATGGTTTTC-3′ (*Kpn*I cleavage site in italics), and subsequently ligated to obtain recombinant plasmid pCambia2301-*FuIPI* under the control of 35S promoter. The recombinant plasmid was transformed into *A. tumefaciens* and introduced into *A. thaliana* using the flower dip method [19]. The transgenic seeds were harvested and screened on 1/2 MS medium containing 50 μg /mL kanamycin. The lines showing positive growth were selected and cultivated in nutrient soil for two weeks. Leaves were used for the extraction of total DNA and RNA. DNA-based molecular verification of *FuIPI* transgenic plants was performed using primers 5′-CCTATTCTGCGTCTGCGTTCTT-3′ and 5′-TATCTGCAACTTCCTGCAGGG-3′. Meanwhile, the expression of the *FuIPI* gene in transgenic plants was analyzed by real-time PCR using primers 5′-AGCCATCCGCTGTATCGT-3′ and 5′-CATCGCTCGGTGCTTTAT-3′. The reference gene for real-time PCR was the *Arabidopsis 18S rRNA* gene (NC_003074.8) and the primers were 5′-CAGTCGGGGGCATTCGTATTT-3′ and 5′-CAGCCTTGCGACCATACTCC-3′.

2.2.7. Phenotypic Observation of *FuIPI* Transgenic Plants under Stress

Fourteen-day-old wild-type and *FuIPI* transgenic *A. thaliana* seedlings were subjected to natural drought and salt stress (120 mM NaCl solution), while the control group was watered with the same amount of water at the same frequency. After 21 days, the phenotypes of the plants were observed and analyzed, including plant height, fresh weight, leaf length, leaf width, etc. The materials were collected, frozen with liquid nitrogen, and stored at −80 °C.

2.2.8. Determination of Abscisic Acid (ABA) Content

Under no stress, drought stress, and salinity stress, 0.5 g of wild-type and *FuIPI* transgenic plants were weighed, ground with liquid nitrogen, extracted with 80% methanol at 4 °C for 16 h, and centrifuged at 10,000 rpm and 4 °C for 20 min, respectively. The extraction process was repeated twice, with the second extraction lasting 2 h. The collected supernatants were used to remove methanol by spin evaporation at 42 °C and subsequently extracted three times with ethyl acetate. The ABA fraction was dissolved in 0.8 mL methanol and assayed by an ELISA kit, according to the standard procedure.

## 3. Result

### 3.1. *Functional Identification of FuIPI Gene in E. coli*

Prior to the functional identification, sequence alignment between FuIPI protein and other IPIs with identified function was performed. As a result, based on the result of amino acid comparison and the classification of conserved structural motifs, the FuIPI protein was suggested to be included in the type I family. For instance, FuIPI displayed specific hits with two major motifs that are crucial for catalytic activities of the type I family. The first motif consisted of nine amino acids, ranging from Thr124 to Leu132. The other consensus motif with -WGEHE- and -DY- at the start and end points in the FuIPI protein was also detected from Trp186 to Tyr193. In addition, similar to other plant IPIs, FuIPI has a N-terminal extensions of 56 amino acids comparable with *E. coli* IDI (Figure 1).

The color complementation experiment was used for the functional identification of MEP pathway genes in various plants, including wheat [20] and *Amomum villosum* [21]. In this experiment, *E. coli* TOP10 cells along with cells containing a single plasmid of pTrc-*FuIPI* or pAC-BETA were used as negative controls. Two plasmids pTrc and pAC-BETA were co-transferred into *E. coli* TOP10 as positive controls. As a result, the regions 1, 4, and 5 (Figure 2), representing *E. coli* TOP10 cells, cells harboring plasmid pTrc-*FuIPI*, and cells harboring pAC-BETA, respectively, were unable to grow on LB medium with Ap and Chl due to the lack of the Ap resistance. Notably, the *E. coli* TOP10 cells containing plasmids pTrc-*FuIPI* and pAC-BETA in region 6 accumulated more yellow *β*-carotene than the positive controls (region 2) (Figure 2A). Overall, this visual method demonstrated that the *FuIPI* gene substantially accelerated *β*-carotene biosynthesis, confirming that *FuIPI* gene encoded the expected functional protein. Furthermore, the content of *β*-carotene in the *E. coli* TOP10 cells was determined by using HPLC. Neither *E. coli* TOP10 cells nor cells containing pTrc-*FuIPI* accumulated *β*-carotene, as they did not carry the plasmid pAC-BETA. A relatively small amount of *β*-carotene was observed in the *E. coli* TOP10 cells containing pAC-BETA (5.70 ± 0.094 μg/mL) and pTrc + pAC-BETA (16.56 ± 0.019 μg/mL), respectively. In contrast, the *E. coli* TOP10 cells co-transforming pTrc-*FuIPI* and pAC-BETA revealed the highest *β*-carotene content (29.30 ± 0.4 μg/mL). Together with the visualization and HPLC analyses, it was further confirmed that the *FuIPI* gene might be limiting for carotenoid biosynthesis, as also reported by Cunningham and Gantt [22].

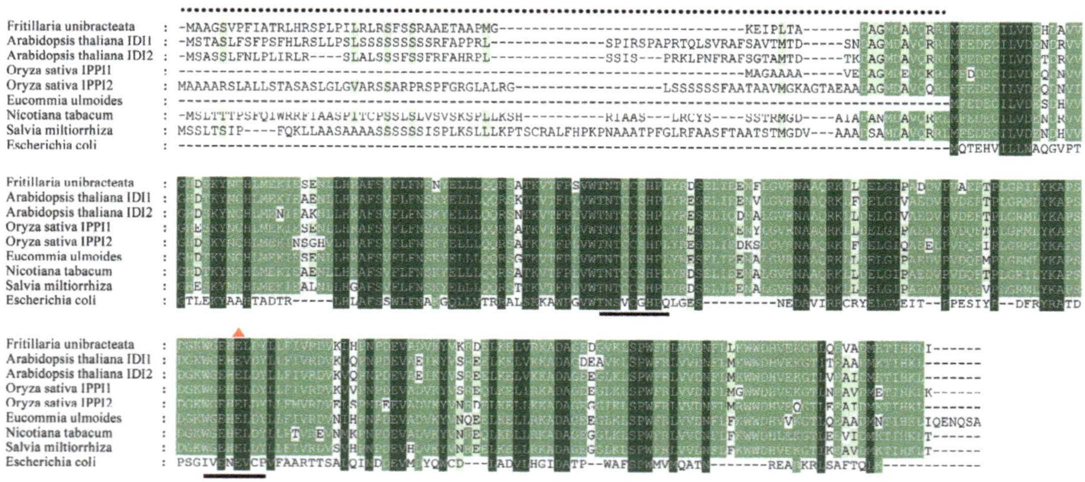


**Figure 1.** The amino acid sequence alignment of FuIPI with IPIs in other species. The amino acid accession numbers are as follows: *Arabidopsis thaliana* IDI1 At5g16440, *Arabidopsis thaliana* IDI2 At3g02780, *Salvia miltiorrhiza* EF635967, *Eucommia ulmoides* AB041629, *Nicotiana tabacum* BAB40973.1, *Oryza sativa* IPI1 AK065871, *Oryza sativa* IPI2 NM_001062082, and *E. coli* NP_417365. The conserved TNTCCSHPL and WGEHEXDY motifs were indicated by black lines, while the key Glu190 was indicated by regular triangles. The N-terminal extension was indicated by dashed line.

According to the sequence alignment and structural analysis, the Glu residue at position 190 was suggested to be a key residue in the active site of the IPI octahedron (Figure 1 in [17]). To test this, this reside was mutated to Ala residue. In the color complementation experiment (Figure 2A), *E. coli* TOP10 cells containing plasmids pTrc-*FuIPI* and pAC-BETA revealed the brightest color, followed by those containing plasmids pTrc-*FuIPI* E190A and pAC-BETA and, lastly, those with plasmids pTrc and pAC-BETA. This indicates that the E190A mutation caused a decrease in the accumulation of *β*-carotene. In addition, the *β*-carotene contents were detected using HPLC analysis (Figure 2B), the results of which indicate that the mutation caused a 15.30% reduction compared to cells containing plasmids pTrc-*FuIPI* and pAC-BETA. Overall, Glu190 residue was a key residue involved in the isomerization of FuIPI.

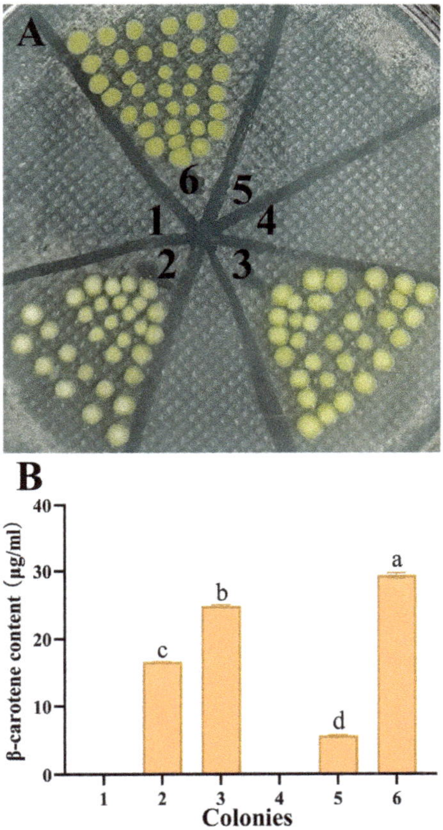


**Figure 2.** Color complementation experiment and HPLC detection of six *E. coli* TOP10 colonies. (**A**) Color complementation experiment; (**B**) $\beta$-carotene contents detected by HPLC. One-way ANOVA was used for analysis. All the experiments had at least three biological replicates. Regions 1, 2, 3, 4, 5, and 6 represent *E. coli* TOP10, *E. coli* TOP10 containing plasmids pTrc and pAC-BETA, *E. coli* TOP10 containing plasmids pTrc-*FuIPI* E190A and pAC-BETA, *E. coli* TOP10 containing plasmid pTrc-*FuIPI*, *E. coli* TOP10 containing plasmid pAC-BETA, and *E. coli* TOP 10 containing plasmids pTrc-*FuIPI* and pAC-BETA, respectively. Different lowercase letters represent significant differences between pairs ($p < 0.05$).

### 3.2. Subcellular Localization of FuIPI

iPSORT showed a putative mitochondrion-targeting peptide in the N-terminus of FuIPI, whereas ChloroP and Plant-mPLoc predicted that FuIPI is localized in chloroplasts with probabilities of 50% and 100%, respectively (Table 1). IPIs have been found to be localized in distinct compartments and vary among different plant species [23]. Considering the reaction catalyzed by FuIPI links MEP and MVA pathways and the subcellular localization of the MEP pathway in plants, it is a better prediction that FuIPI is localized in chloroplasts [24].

**Table 1.** Prediction of FuIPI subcellular localization.

| Software | Prediction Result |
|---|---|
| iPSORT | mitochondrial targeting or chloroplast transport peptide |
| Plant-mPLoc | chloroplast |

To verify this prediction, we constructed the recombinant plasmid PBI121-*FuIPI* labeled with GFP. Under laser confocal scanning, chloroplasts and GFP showed red and green fluorescence, respectively. As shown in Figure 3A, in the GFP control group, green fluorescence was uniformly distributed in the leaf cells of *N. benthamiana* without specificity. The subcellular localization of FuIPI protein was able to be revealed by the combined yellow fluorescence, indicating that FuIPI protein was localized in chloroplasts (Figure 3B) where the MEP pathway is localized.

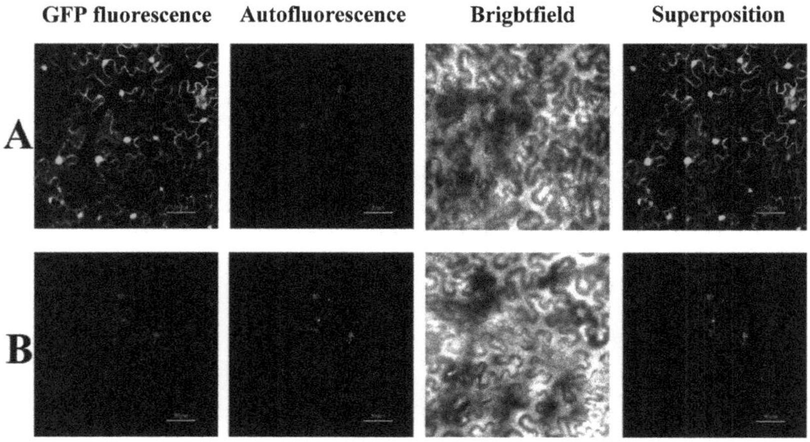


**Figure 3.** The subcellular localization of FuIPI in *N. benthamiana* leaves: (**A**) represents 35S: GFP, while (**B**) represents 35S: *FuIPI*-GFP.

### 3.3. Production of FuIPI Transgenic Plants

To verify the role of the *FuIPI* gene in influencing stress tolerance, recombinant plasmids containing *FuIPI* ORF, controlled by the CaMV 35S promoter, were constructed and transformed into *A. thaliana*. To validate the *FuIPI* transgenic plants, reverse PCR was performed using DNA as the template at first. As a result, the *FuIPI* fragment was detected in the *FuIPI* transgenic plants, while the wild-type plants exhibited negative PCR results (Figure 4A). In addition, the expression levels of the *FuIPI* gene were determined by real-time PCR using cDNA as the template. The *FuIPI* gene was substantially expressed in *FuIPI* transgenic plants, while it was not detected in wild-type plants (Figure 4B).

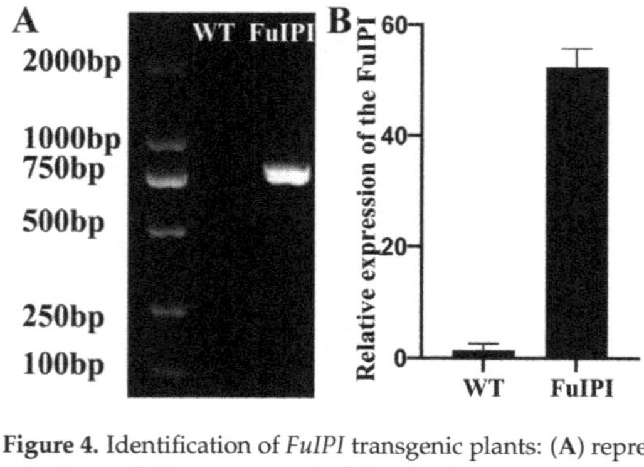


**Figure 4.** Identification of *FuIPI* transgenic plants: (**A**) represents the PCR result using DNA as the template and (**B**) represents the PCR result using cDNA as the template.

### 3.4. FuIPI Transgenic A. thaliana Strengthened Tolerance to Drought and Salinity Stress

The phenotypes of wild-type and *FuIPI* transgenic plants under various conditions were recorded to evaluate the function of the *FuIPI* gene in stress tolerance (Figure 5A).

Although no obvious morphological difference was observed between the wild-type and transgenic plants, the *FuIPI* transgenic plants exhibited a higher fresh weight (163.18%), number of base leaves (126.41%), and leaf length (151.95%) than those in wild-type plants under normal condition, respectively (Figure 5B–E). These results indicate that the *FuIPI* gene positively affects plant growth. Moreover, it was observed that drought caused damage to the wild-type plants, including leaf yellowing and reduced growth, while the *FuIPI* transgenic plants exhibited slight enhancement in drought tolerance due to their higher fresh weight, number of base leaves ($p < 0.05$), leaf length, and leaf width compared to wild-type plants (Figure 5B–E). Intriguingly, *FuIPI* transgenic plants revealed significant tolerance to salinity stress, as their leaves remained green and growth remained vigorous. Compared with those of wild-type plants, the leaf length, leaf width, and fresh weight of *FuIPI* transgenic plants were significantly increased by 1.88- ($p < 0.0001$), 2.06- ($p < 0.01$), and 1.65-fold ($p < 0.001$), respectively. This higher number of leaves and increased leaf area contributed to more efficient photosynthesis and enhanced plant metabolism. In summary, the *FuIPI* gene can enhance tolerance to drought and salinity stress.

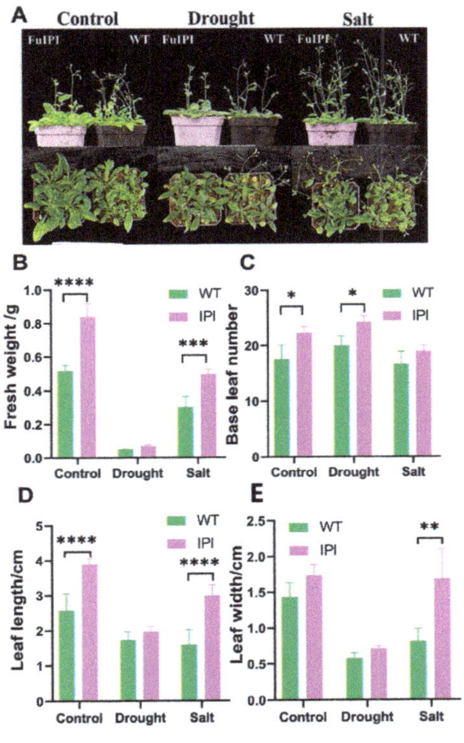


**Figure 5.** Phenotypic analysis of *A. thaliana* under various conditions. (**A**) Phenotypes of plants under various conditions. WT and FuIPI represent wild-type and *FuIPI* transgenic plants, respectively. The plants from left to right in Figure 5A represent those under no stress, drought stress, and salinity stress, respectively. (**B–E**) Represent various characteristics of plants, including fresh weight, base leaf number, leaf length, and leaf width, respectively. One-way ANOVA was used to perform statistical analysis. All the experiments had at least three biological replicates. *, **, *** and **** represented statistical significance of $p < 0.05$, $p < 0.01$, $p < 0.001$, and $p < 0.0001$, respectively.

### *3.5. Overexpression of the FuIPI Gene Enhanced the Accumulation of ABA Content*

The standard curve for ABA was determined with the equation $y = 0.0191x + 0.1018$ ($R^2 = 0.995$). Using this standard curve, the ABA content was measured. As shown in Figure 6, the ABA content in *FuIPI* transgenic plants was slightly higher than that in wild-type plants. Under drought conditions, the ABA content in *FuIPI* transgenic plants was 2.50 times that in wild-type plants ($p < 0.0001$), while the ABA concentration in *FuIPI* transgenic plants was also 1.50 times ($p < 0.05$) that in wild-type plants under salinity stress. Previous studies have confirmed that ABA is a stress-responsive plant hormone involved not only in tolerance to environmental stress [11] but also in plant development [25]. Therefore, these results indicated that the overexpression of the *FuIPI* gene could promote the accumulation of ABA, thereby improving the drought and salt tolerance of plants.

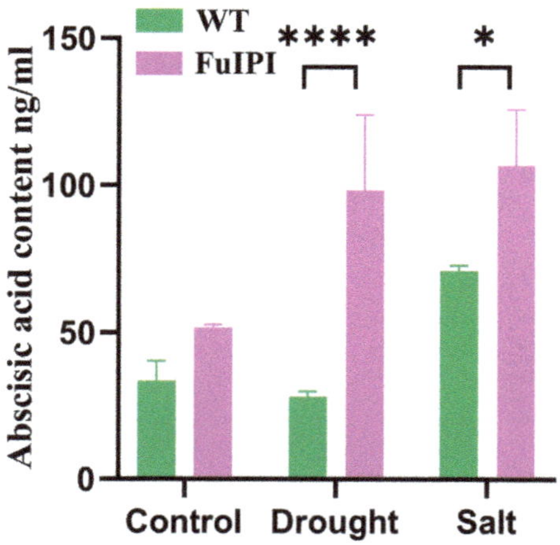


**Figure 6.** ABA content in plants under no stress, drought stress, and salinity stress. WT and FuIPI represent wild-type and *FuIPI* transgenic plants, respectively. One-way ANOVA was used for statistical analysis. All the experiments had at least three biological replicates. * and **** represented statistical significance of $p < 0.05$ and $p < 0.0001$, respectively.

## 4. Discussion

In plants, the biosynthesis of isoprenoid-type metabolites occurs in three main stages [9]. In the upstream stage, the basic five-carbon unit IPP is derived from the MVA pathway in cytoplasm or MEP pathway in plastids. IPP is then isomerized to DMAPP by the catalysis of IPI. In the middle stage, IPP and DMAPP are sequentially condensed to yield intermediates such as geranyl diphosphate (GPP), farnesyl diphosphate (FPP), and geranylgeranyl diphosphate (GGPP). Finally, in the downstream stage, these intermediates are converted into various isoprenoid-type metabolites, including steroidal alkaloid, artemisinin, chlorophyll, $\beta$-carotene, and ABA. Therefore, IPP and DMAPP serve as essential materials that connect the upstream MVA/MEP pathway with the downstream steroid biosynthesis pathway. IPI is proposed to catalyze a checkpoint reaction, providing the basic materials necessary for isoprenoid biosynthesis. We identified the in vivo function of the *FuIPI* gene through heterologous expression in prokaryotic and eukaryotic systems.

In line with previous studies, IPIs consist of two types (type I and II). To date, only type I IPIs have been found in plants, implying that the divergent functions of two types might occur before the differentiation of plants. We used two methods to confirm that *FuIPI* belonged to type I. At first, we compared the sequence identities of *FuIPI* with IPIs from other species. The top three species showing the highest similarity with *FuIPI* were *Zingiber officinale*, *Vigna unguiculata*, and *Glycine max*. The high similarity between *FuIPI* and other type-I IPIs from higher plants suggests that it is attributed to type I. Second, type-I IPIs have a conserved C residue in the TNTCCSHPL motif and a conserved E residue in the WGEHEXDY motif. These highly conserved residues are critical for the catalytic activity of the enzyme. FuIPI also contained the highly conserved C (124–132 aa) and E (186–193 aa) residues in the two motifs, strongly suggesting that it belongs to the type-I IPI family (Figure 1). As shown in Figure 2, in the color complementation experiment, the strains containing both pTrc-*FuIPI* and pAC-BETA appeared dark yellow on the medium, while those containing pTrc and pAC-BETA were light yellow, indicating that the expression of the *FuIPI* gene led to greater accumulation of *β*-carotene. HPLC analysis indicated that the *FuIPI* gene in *E. coli* enhanced *β*-carotent production by 5.14-fold higher than that in the control group.

Previous studies on IPI from *E. coli* have indicated that the active center of IPI bound to metal ions through a distorted octahedral coordination geometry composed of His25, His32, His69, Glu114, and Glu116 residues [26]. Among these, Glu116 (numbering based on *E. coli* IPI) was directly involved in the protonation of IPP to DMAPP. Therefore, the homologous Glu190 residue in *FuIPI* was hypothesized to be a key residue and was, thus, mutated to Ala. According to the color complementation test, the strain containing pTrc-*FuIPI* E190A was darker than those containing plasmids pTrc and pAC-BETA but lighter than those containing plasmids pTrc-*FuIPI* and pAC-BETA. The mutant resulted in a 15.30% decrease in *β*-carotene content compared with the *E. coli* strain containing plasmids pTrc-*FuIPI* and pAC-BETA. These results were consistent with our hypothesis that Glu190 in *FuIPI* was one of the key amino acids involved in the catalytic activity.

Many plants contained two type-I IPI isozymes with diversified subcellular localizations. Specifically, both IPIs from rice are localized in endoplasmic reticulum, peroxisomes, and mitochondria, whereas OsIPPI2 was also found in plastids [27]. *A. thaliana* has two IPIs that may be present in mitochondria and plastids, respectively [28], while IPIs from *Nicotiana tabacum* might function in cytosol and plastids [29], respectively. Intriguingly, most known plant type-I IPPs appeared to have an N-terminal extension that was associated with organellar localization [27–29]. Sequence analysis in Figure 1 indicated that the N-terminal extension of FuIPI shared an identity of more than 40% with that of OsIPPI2 (a plastid-localized IPI) while sharing only an identity of less than 20% with that of OsIPPI1 (a nonplastid-localized IPI) [27]. The subsequent bioinformatic prediction and subcellular localization of *FuIPI* confirmed its localization within the plastids, resembling that of IPIs from grape, *Nicotiana tabacum*, and rice [27,29,30]. The flow-controlled function of plastidial IPI for different downstream isoprenoid products, especially for those in plastids, has been highlighted by Pankratov et al. [31]. Hopefully, *FuIPI* could, thus, be a candidate gene for the metabolic engineering and synthetic biology of isoprenoids in plastids.

ABA has been implicated in plant stress tolerance, growth and development, and hormone responses [11,25]. By specifically binding to the *cis*-regulatory elements, such as DREBs and AREBs, in the promoters of genes involved in the ABA-related signaling pathway, ABA triggers various physiological processes and metabolic changes that enable whole tissues to acquire drought and salinity tolerance [32,33]. In plants, ABA is biosynthesized in plastids mainly from the *β*-carotene precursor, so an increase in *E. coli* overexpressing the *FuIPI* gene might enhance ABA biosynthesis. It was observed that the *FuIPI* transgenic plants contained an ABA content 2.50- and 1.50-fold higher than that in wild-type plants under drought and salinity stress, respectively. Considering the importance of ABA in the stress response of plants, it was particularly interesting to explore the phenotype of transgenic plants under these conditions. As a result, *FuIPI* transgenic plants exhibited slight enhancement in drought tolerance and significant enhancement in salinity tolerance compared with wild-type plants. Previous studies by our lab [11], along with findings from Marusig and Tombesi [34], have revealed that ABA correlates with improved drought and salinity tolerance in plants. Combined with current results and previous reports, the enhanced drought and salinity tolerance of transgenic plants could be attributed to its increased ABA content. Intriguingly, the *FuIPI* transgenic plants revealed higher biomass, including a greater fresh weight, higher number of base leaves, and longer leaf length than that of wild-type plants under no stress. This might imply that FuIPI would play a role in balancing plant growth and stress tolerance. A similar balance between plant growth and drought tolerance was also observed in plants overexpressing the *Cassia tora DXR1* gene, possibly due to the regulation of ABA and chlorophyll biosynthesis [11]. Furthermore, the *IPI* genes from *Aconitum carmichaelii*, *Dipsacus asperoides*, and *Artemisia annua* have been proposed to play roles in the biosynthesis of diterpenoid alkaloids [35], triterpenoid saponin [36], and artemisinin [37], respectively. Therefore, in vivo functional identification of the *FuIPI* gene supported its role in the biosynthesis of isoprenoids, including $\beta$-carotene and ABA. It is hypothesized that FuIPI might be applied to accumulate isoprenoid-derived compounds in medicinal plants, to improve the nutritional qualities of crops by enhancing the accumulation of compounds such as $\beta$-carotene, and to improve the yield of crops by increasing adaptability to environmental stress as well.

Although the phenotypes of *FuIPI* transgenic plants can be largely linked to altered ABA levels, future research remains necessary. First, in addition to the aerial tissues, the underground tissues should also be evaluated for the drought and salinity tolerance of *FuIPI* transgenic plants, given that ABA exerts a wide influence on various plant tissues. Second, a detailed investigation into the expression of downstream genes that are involved in the ABA-dependent signaling pathway is required to comprehensively understand the role of the *FuIPI* gene in drought and salinity tolerance. Moreover, in order to explore the potential application of the *FuIPI* gene in various fields, different chassis, such as crops and medicinal plants, will be used in upcoming research.

## 5. Conclusions

This study demonstrates that *FuIPI* is involved in the biosynthesis of isoprenoids by heterologous expression in both prokaryotic and eukaryotic cells. The site mutation confirmed that Glu190 was one key residue related to the catalytic reaction. The subcellular localization of FuIPI in chloroplasts provided a challenge to monitor the crosstalk between MVA and MEP pathways, which were associated with diverse biological processes. The tolerance to drought and salinity stress could be augmented by the overexpression of the *FuIPI* gene, as indicated by the improving accumulation of ABA content. Overall, our studies extend the IPI members of the family *Liliaceae*, which is rich with ornamental and medicinal plants. In addition, our current work will contribute to understanding the molecular mechanisms of promoting isoprenoid-related accumulation and enhancing stress tolerance via the overexpression of the *FuIPI* gene.

**Supplementary Materials:** The following supporting information can be downloaded at: https://www.mdpi.com/article/10.3390/horticulturae10080887/s1, Supplementary Material S1: codon-optimized FuIPI; Supplementary Material S2: The original gel of Figure 4; Supplementary Material S3: The original data of Figure 2B; Supplementary Material S4: The original data of Figure 4B; Supplementary Material S5: The original data of Figure 5B; Supplementary Material S6: The original data of Figure 6.

**Author Contributions:** X.Y., J.C. (Jiao Chen), H.Y., X.H., J.C. (Jieru Chen), Z.M., J.Z. and H.L. performed the experiments. X.Y., J.C. (Jiao Chen), H.Y., X.H., J.C. (Jieru Chen) and Z.M. analyzed the data. H.L. and J.Z. contributed to the materials and analysis tool. H.L. wrote the manuscript. All authors have read and agreed to the published version of the manuscript.

**Funding:** This work was partially funded by the National Natural Science Foundation of China (No. 32270410), Sichuan Science and Technology Program (No. 2018SZ0061), Innovation Program for Technology of Chengdu City (2022-YF05-01357-SN), Sichuan Administration of TCM program (No. 2021MS116).

**Data Availability Statement:** The full-length coding sequence (CDS) of FuIPI was submitted to GenBank databases with the accession number PQ058665. In this study, the codon-optimized FuIPI, the original gel of Figure 4A, and other original data are included in the Supplementary Materials.

**Conflicts of Interest:** The authors declare no conflicts of interest.

## References

1. Zhang, T.; Huang, S.; Song, S.; Zou, M.; Yang, T.; Wang, W.; Zhou, J.; Liao, H. Identification of evolutionary relationships and DNA markers in the medicinally important genus *Fritillaria* based on chloroplast genomics. *PeerJ* **2021**, *9*, e12612. [CrossRef]
2. Yang, L.; Zhang, M.; Yang, T.; Wai Ming, T.; Wai Gaun, T.K.; Ye, B. LC-MS/MS coupled with chemometric analysis as an approach for the differentiation of bulbus *Fritillaria unibracteata* and *Fritillaria ussuriensis*. *Phytochem. Anal.* **2021**, *32*, 957–969. [CrossRef] [PubMed]

3. Lin, G.; Li, P.; Li, S.L.; Chan, S.W. Chromatographic analysis of *Fritillaria* isosteroidal alkaloids, the active ingredients of Beimu, the antitussive traditional Chinese medicinal herb. *J. Chromatogr. A* **2001**, *935*, 321–338. [CrossRef] [PubMed]
4. Hu, Z.; Zong, J.F.; Yili, A.; Yu, M.H.; Aisa, H.A.; Hou, A.J. Isosteroidal alkaloids from the bulbs of *Fritillaria tortifolia*. *Fitoterapia* **2018**, *131*, 112–118. [CrossRef] [PubMed]
5. Wang, Y.; Hou, H.; Ren, Q.; Hu, H.; Yang, T.; Li, X. Natural drug sources for respiratory diseases from *Fritillaria*: Chemical and biological analyses. *Chin. Med.* **2021**, *16*, 40. [CrossRef]
6. Zhang, Y.; Han, H.; Li, D.; Fan, Y.; Liu, M.; Ren, H.; Liu, L. Botanical characterization, phytochemistry, biosynthesis, pharmacology clinical application, and breeding techniques of the Chinese herbal medicine *Fritillaria unibracteata*. *Front. Pharmacol.* **2024**, *15*, 1428037. [CrossRef]
7. Zhao, D.; Wang, J.; Dai, W.; Ye, K.; Chen, J.; Lai, Q.; Li, H.; Zhong, B.; Yu, X. Effects of climate warming and human activities on the distribution patterns of *Fritillaria unibracteata* in eastern Qinghai-Tibetan Plateau. *Sci. Rep.* **2023**, *13*, 15770. [CrossRef]
8. Jiang, R.; Zou, M.; Qin, Y.; Tan, G.; Huang, S.; Quan, H.; Zhou, J.; Liao, H. Modeling of the Potential Geographical Distribution of Three *Fritillaria* Species Under Climate Change. *Front. Plant Sci.* **2022**, *12*, 749838. [CrossRef]
9. Liao, H.; Quan, H.; Huang, B.; Ji, H.; Zhang, T.; Chen, J.; Zhou, J. Integrated transcriptomic and metabolomic analysis reveals the molecular basis of tissue-specific accumulation of bioactive steroidal alkaloids in *Fritillaria unibracteata*. *Phytochemistry* **2023**, *214*, 113831. [CrossRef]
10. Moreno, J.C.; Mi, J.; Agrawal, S.; Kössler, S.; Turečková, V.; Tarkowská, D.; Thiele, W.; Al-Babili, S.; Bock, R.; Schöttler, M.A. Expression of a carotenogenic gene allows faster biomass production by redesigning plant architecture and improving photosynthetic efficiency in tobacco. *Plant J.* **2020**, *103*, 1967–1984. [CrossRef]
11. Tian, C.; Quan, H.; Jiang, R.; Zheng, Q.; Huang, S.; Tan, G.; Yan, C.; Zhou, J.; Liao, H. Differential roles of Cassia tora 1-deoxy-D-xylulose-5-phosphate synthase and 1-deoxy-D-xylulose-5-phosphate reductoisomerase in trade-off between plant growth and drought tolerance. *Front. Plant Sci.* **2023**, *14*, 1270396. [CrossRef]
12. Liao, Z.; Chen, M.; Yang, Y.; Yang, C.; Fu, Y.; Zhang, Q.; Wang, Q. A new isopentenyl diphosphate isomerase gene from sweet potato: Cloning, characterization and color complementation. *Biologia* **2008**, *63*, 221–226. [CrossRef]
13. Zhang, X.; Guan, H.; Dai, Z.; Guo, J.; Shen, Y.; Cui, G.; Gao, W.; Huang, L. Functional Analysis of the Isopentenyl Diphosphate Isomerase of *Salvia miltiorrhiza* via Color Complementation and RNA Interference. *Molecules* **2015**, *20*, 20206–20218. [CrossRef] [PubMed]
14. Wang, C.W.; Oh, M.K.; Liao, J.C. Engineered isoprenoid pathway enhances astaxanthin production in *Escherichia coli*. *Biotechnol. Bioeng.* **1999**, *62*, 235–241. [CrossRef]
15. Brüggemann, N.; Schnitzler, J.P. Relationship of isopentenyl diphosphate (IDP) isomerase activity to isoprene emission of oak leaves. *Tree Physiol.* **2002**, *22*, 1011–1018. [CrossRef] [PubMed]
16. Chen, R.; Harada, Y.; Bamba, T.; Nakazawa, Y.; Gyokusen, K. Overexpression of an isopentenyl diphosphate isomerase gene to enhance trans-polyisoprene production in Eucommia ulmoides Oliver. *BMC. Biotechnol.* **2012**, *12*, 78. [CrossRef]
17. Chen, J.; Song, S.; Tang, J.; Xin, J.; Zhang, Q.; Zhang, H.; Chen, X.; Zhou, J.Y.; Liao, H. Cloning and functional analysis of IPI gene from *Fritillaria unibracteata* Hsiao et K.C.Hsia. *Acta Pharm. Sin.* **2023**, *58*, 447–453. [CrossRef]
18. Song, S.; Chen, A.; Zhu, J.; Yan, Z.; An, Q.; Zhou, J.; Liao, H.; Yu, Y. Structure basis of the caffeic acid O-methyltransferase from *Ligusiticum chuanxiong* to understand its selective mechanism. *Int. J. Biol. Macromol.* **2022**, *194*, 317–330. [CrossRef]
19. Qin, Y.; Li, Q.; An, Q.; Li, D.; Huang, S.; Zhao, Y.; Chen, W.; Zhou, J.; Liao, H. A phenylalanine ammonia lyase from *Fritillaria unibracteata* promotes drought tolerance by regulating lignin biosynthesis and SA signaling pathway. *Int. J. Biol. Macromol.* **2022**, *213*, 574–588. [CrossRef]
20. Flowerika; Alok, A.; Kumar, J.; Thakur, N.; Pandey, A.; Pandey, A.K.; Upadhyay, S.K.; Tiwari, S. Characterization and Expression Analysis of Phytoene Synthase from Bread Wheat (*Triticum aestivum* L.). *PLoS ONE* **2016**, *11*, e0162443. [CrossRef]
21. Yang, J.; Adhikari, M.N.; Liu, H.; Xu, H.; He, G.; Zhan, R.; Wei, J.; Chen, W. Characterization and functional analysis of the genes encoding 1-deoxy-D-xylulose-5-phosphate reductoisomerase and 1-deoxy-D-xylulose-5-phosphate synthase, the two enzymes in the MEP pathway, from *Amomum villosum* Lour. *Mol. Biol. Rep.* **2012**, *39*, 8287–8296. [CrossRef]
22. Cunningham, F.X., Jr.; Gantt, E. A portfolio of plasmids for identification and analysis of carotenoid pathway enzymes: Adonis aestivalis as a case study. *Photosynth. Res.* **2007**, *92*, 245–259. [CrossRef]
23. Okada, K.; Kasahara, H.; Yamaguchi, S.; Kawaide, H.; Kamiya, Y.; Nojiri, H.; Yamane, H. Genetic evidence for the role of isopentenyl diphosphate isomerases in the mevalonate pathway and plant development in *Arabidopsis*. *Plant Cell. Physiol.* **2008**, *49*, 604–616. [CrossRef]
24. Llamas, E.; Pulido, P.; Rodriguez-Concepcion, M. Interference with plastome gene expression and Clp protease activity in *Arabidopsis* triggers a chloroplast unfolded protein response to restore protein homeostasis. *PLoS. Genet.* **2017**, *13*, e1007022. [CrossRef]
25. Yu, T.; Yang, Y.; Wang, H.; Qian, W.; Hu, Y.; Gao, S.; Liao, H. The Variations of C/N/P Stoichiometry, Endogenous Hormones, and Non-Structural Carbohydrate Contents in *Micheliamaudiae* 'Rubicunda' Flower at Five Development Stages. *Horticulturae* **2023**, *9*, 1198. [CrossRef]
26. Zhang, C.; Liu, L.; Xu, H.; Wei, Z.; Wang, Y.; Lin, Y.; Gong, W. Crystal structures of human IPP isomerase: New insights into the catalytic mechanism. *J. Mol. Biol.* **2007**, *366*, 1437–1446. [CrossRef] [PubMed]

27. Jin, X.; Baysal, C.; Gao, L.; Medina, V.; Drapal, M.; Ni, X.; Sheng, Y.; Shi, L.; Capell, T.; Fraser, P.D.; et al. The subcellular localization of two isopentenyl diphosphate isomerases in rice suggests a role for the endoplasmic reticulum in isoprenoid biosynthesis. *Plant Cell Rep.* **2020**, *39*, 119–133. [CrossRef]
28. Phillips, M.A.; D'Auria, J.C.; Gershenzon, J.; Pichersky, E. The *Arabidopsis thaliana* type I Isopentenyl Diphosphate Isomerases are targeted to multiple subcellular compartments and have overlapping functions in isoprenoid biosynthesis. *Plant Cell* **2008**, *20*, 677–696. [CrossRef]
29. Nakamura, A.; Shimada, H.; Masuda, T.; Ohta, H.; Takamiya, K. Two distinct isopentenyl diphosphate isomerases in cytosol and plastid are differentially induced by environmental stresses in tobacco. *FEBS Lett.* **2001**, *506*, 61–64. [CrossRef] [PubMed]
30. Chen, T.; Xu, T.; Wang, J.; Zhang, T.; Yang, J.; Feng, L.; Song, T.; Yang, J.; Wu, Y. Transcriptomic and free monoterpene analyses of aroma reveal that isopentenyl diphosphate isomerase inhibits monoterpene biosynthesis in grape (*Vitis vinifera* L.). *BMC Plant Biol.* **2024**, *24*, 595. [CrossRef]
31. Pankratov, I.; McQuinn, R.; Schwartz, J.; Bar, E.; Fei, Z.; Lewinsohn, E.; Zamir, D.; Giovannoni, J.J.; Hirschberg, J. Fruit carotenoid-deficient mutants in tomato reveal a function of the plastidial isopentenyl diphosphate isomerase (IDI1) in carotenoid biosynthesis. *Plant J.* **2016**, *88*, 82–94. [CrossRef] [PubMed]
32. Fujita, Y.; Yoshida, T.; Yamaguchi-Shinozaki, K. Pivotal role of the AREB/ABF-SnRK2 pathway in ABRE-mediated transcription in response to osmotic stress in plants. *Physiol. Plant* **2013**, *147*, 15–27. [CrossRef] [PubMed]
33. Muhammad Aslam, M.; Waseem, M.; Jakada, B.H.; Okal, E.J.; Lei, Z.; Saqib, H.S.A.; Yuan, W.; Xu, W.; Zhang, Q. Mechanisms of Abscisic Acid-Mediated Drought Stress Responses in Plants. *Int. J. Mol. Sci.* **2022**, *23*, 1084. [CrossRef]
34. Marusig, D.; Tombesi, S. Abscisic Acid Mediates Drought and Salt Stress Responses in Vitis vinifera-A Review. *Int. J. Mol. Sci.* **2020**, *21*, 8648. [CrossRef]
35. Hu, Y.; Chen, L.; Huang, L.; Wang, G. The expression of AcIDI1 reveals diterpenoid alkaloids' allocation strategies in the roots of *Aconitum carmichaelii* Debx. *Gene* **2024**, *920*, 148529. [CrossRef]
36. Pan, J.; Huang, C.; Yao, W.; Niu, T.; Yang, X.; Wang, R. Full-length transcriptome, proteomics and metabolite analysis reveal candidate genes involved triterpenoid saponin biosynthesis in *Dipsacus asperoides*. *Front. Plant. Sci.* **2023**, *14*, 1134352. [CrossRef]
37. Deng, Y.A.; Li, L.; Peng, Q.; Feng, L.F.; Yang, J.F.; Zhan, R.T.; Ma, D.M. Isolation and characterization of AaZFP1, a C2H2 zinc finger protein that regulates the AaIPPI1 gene involved in artemisinin biosynthesis in *Artemisia annua*. *Planta* **2022**, *255*, 122. [CrossRef]

horticulturae

Article

# Evaluating the Cold Tolerance of *Stenotaphrum* Trin Plants by Integrating Their Performance at Both Fall Dormancy and Spring Green-Up

**Jia Qu [1,†], Dong-Li Hao [2,†], Jin-Yan Zhou [3], Jing-Bo Chen [2], Dao-Jin Sun [2], Jian-Xiu Liu [2], Jun-Qin Zong [2,*] and Zhi-Yong Wang [1,*]**

1 Sanya Nanfan Research Institute, College of Tropical Agriculture and Forestry, Hainan University, Haikou 570228, China; 22220953000028@hainanu.edu.cn
2 The National Forestry and Grassland Administration Engineering Research Center for Germplasm Innovation and Utilization of Warm-Season Turfgrasses, Jiangsu Key Laboratory for the Research and Utilization of Plant Resources, Institute of Botany, Jiangsu Province and Chinese Academy of Sciences (Nanjing Botanical Garden Mem. Sun Yat-Sen), Nanjing 210014, China; haodongli@jib.ac.cn (D.-L.H.); chenjingbo@jib.ac.cn (J.-B.C.); sundaojin@jib.ac.cn (D.-J.S.); turfunit@cnbg.net (J.-X.L.)
3 Department of Agronomy and Horticulture, Jiangsu Vocational College of Agriculture and Forest, Jurong 212400, China; zhoujinyan@jsafc.edu.cn
* Correspondence: zongjunqin@jib.ac.cn (J.-Q.Z.); wangzhiyong@hainanu.edu.cn (Z.-Y.W.)
† These authors contributed equally to this work.

**Abstract:** Owing to the poor cold tolerance of *Stenotaphrum* Trin and the urgent need for shade-tolerant grass species in temperate regions of East China, this study evaluated the cold tolerance of 55 *Stenotaphrum* accessions, aiming to provide shade-tolerant materials for temperate regions. A fine cold-tolerant turfgrass should have both the advantages of delayed fall dormancy and early spring green-up. However, previous research on the cold resistance of turfgrass has mainly focused on the performance of the spring green-up, with less attention paid to the fall dormancy, which has affected the ornamental and application value of turfgrass. This study first dynamically investigated the leaf colour of each accession during the fall dormancy and the coverage during the spring green-up and evaluated the cold resistance of the accession through membership functions and cluster analysis. Significant differences in the cold resistance were found with the assignment of breeding lines to four categories. The weak correlation ($R^2$ = 0.1682) between leaf colour during the fall dormancy and coverage during the spring green-up indicates that using the performance of a single period to represent the cold resistance of accessions is not appropriate. To test whether using the laboratory-based LT50 and stolon regrowth rating analysis can replace the above-improved method, we conducted a related analysis and found that the fit between these two methods is very poor. This phenomenon is attributed to the poor correlation between the laboratory-based parameters and the pot-investigated data. Therefore, this study presents a cold resistance evaluation method for *Stenotaphrum* that integrates performance in both the fall dormancy and spring green-up periods. This improved evaluation method cannot be simplified by the growth performance of a single period or replaced by using laboratory-based LT50 and stolon regrowth tests. With the help of this improved method, several excellent cold tolerance accessions (ST003, S13, and S12) were identified for temperate regions of East China.

**Keywords:** perennial plants; *Stenotaphrum* Trin; fall dormancy; spring green-up; cold tolerance evaluation

## 1. Introduction

The tropical turfgrass *Stenotaphrum* Trin has both feed and ornamental value [1]. Due to its much greater shade tolerance than that of other warm-season turfgrasses, *Stenotaphrum* species are preferred for use in shaded landscapes in tropical and subtropical regions [2,3].

However, its poor cold tolerance severely restricts its use in temperate regions [4–6]. Given the need for shade-tolerant turfgrasses in the temperate regions of East China, the use of Stenotraphrum requires the identification and use of cold-resistant germplasm for acceptance and application by the public.

Unlike annual plants such as wheat that do not undergo the fall dormancy process, perennial turfgrass with strong cold resistance should have both the advantages of delayed fall dormancy and early spring green-up [7–9]. Field and laboratory evaluations are commonly used for evaluating cold resistance. Field evaluation is considered as a relatively accurate method of evaluating cold resistance. However, field evaluations mainly focus on the performance of green-up in the following year, with less attention paid to the performance of fall dormancy [10,11]. Electrolyte leakage and tissue regrowth measurements are common laboratory methods used to assess cold tolerance in turfgrass species [5,12,13]. Electrolyte leakage has been shown to poorly represent the actual cold tolerance [14–16]. Measurements of tissue regrowth have also been shown to be poorly correlated with cold resistance [4,17,18]. These methods are likely not reflective of actual cold tolerance because they do not account for a plant's ability to retain colour in the fall or an ability to green-up earlier in the spring. Therefore, the objective of this research was to measure both fall dormancy and spring green-up to assess the cold tolerance of Stenotaphrum breeding lines and to compare this assessment to electrolyte leakage and stolon regrowth following exposure to cold.

*Stenotaphrum* plants can be divided into seven categories (https://powo.science.kew.org/taxon/urn:lsid:ipni.org:names:19069-1#source-KB, accessed on 20 June 2024). (1) *Stenotaphrum helferi*. the native range of this species is South China to Peninsula Malaysia, Philippines. It is a perennial or rhizomatous geophyte and grows primarily in the wet tropical biome. (2) *Stenotaphrum clavigerum*. The native range of this species is Aldabra to Assumption. It is an annual and grows primarily in the seasonally dry tropical biome. (3) *Stenotaphrum dimidiatum*. The native range of this species is coastal Kenya to South Africa, West Indian Ocean, and the Indian Subcontinent to Peninsula Malaysia. It is a perennial or rhizomatous geophyte and grows primarily in the seasonally dry tropical biome. (4) *Stenotaphrum micranthum*. The native range of this species is Southeast Tanzania to the West Indian Ocean, Spratly Islands, and East Malaysia to the Pacific Ocean. It is an annual and grows primarily in the seasonally dry tropical biome. (5) *Stenotaphrum oostachyum*. The native range of this species is Northwest and Central Madagascar. It is a perennial or rhizomatous geophyte and grows primarily in the seasonally dry tropical biome. (6) *Stenotaphrum secundatum*. The native range of this species is Southeast U.S.A. to eastern and southern South America and West Tropical Africa to Chad. It is a perennial or rhizomatous geophyte and grows primarily in the seasonally dry tropical biome. It is used as animal food and a medicine and has uses environmentally and for food. (7) *Stenotaphrum unilaterale*. The native range of this species is Central Madagascar. It is a perennial or rhizomatous geophyte and grows primarily in the seasonally dry tropical biome. The distribution of *Stenotaphrum* plants in China is mainly in South China. The use of which is also mainly in the tropical and subtropical regions of China. No dominant variety was found. Significant variations in cold tolerance among the species of *Stenotaphrum* plants [5,6] have been detected. Although the cold tolerance has been investigated on many accessions from Europe and North America [19–22], the large-scale cold tolerance evaluation of *Stenotaphrum* plants from China is still lacking. This notion is further supported by the finding that *Stenotaphrum* research undertaken by Chinese scholars mainly focuses on molecular maker analysis, physiology, and pathology, with little attention paid to the cold tolerance evaluation of *Stenotaphrum* plants from China [23–30]. This study used 55 accessions of *Stenotaphrum* plants, the composition of which was mainly wild accessions from China, aiming to establish a suitable cold tolerance evaluation method for *Stenotaphrum* plants and identify several accessions that can be applied to the temperate regions of East China.

Specifically, this study first dynamically investigated the leaf colour during the fall dormancy and the coverage during the spring green-up of 55 *Stenotaphrum* Trin resources and evaluated their cold resistance by integrating the two-index using a membership function. By analysing the correlation between autumn/winter leaf colour and next year's green-up coverage, as well as comparing the growth performance of a single growth period with the ranking of a comprehensive evaluation, whether this method can be simplified by the growth performance of the fall dormancy or spring green-up is clarified. Furthermore, the cold resistance of these resources was evaluated through the laboratory-based cold resistance evaluation methods (leaf LT50 and stolon regrowth analysis) and was ranked and clustered through membership functions and cluster analysis. Whether field evaluation methods can be replaced by laboratory evaluation methods was determined by comparing the differences between laboratory cold resistance evaluation data and field evaluation data. By conducting a correlation analysis between field evaluation data and laboratory evaluation data, the reasons for the differences between field evaluation and laboratory evaluation were clarified. Through the above series of experiments, this study ultimately presented a method to compensate for the shortcomings of the current evaluation methods for the cold resistance of turfgrass and clarified whether this improved method can be simplified or replaced by other methods.

## 2. Materials and Methods

### 2.1. Accession Information

Eighty domestic and foreign accessions of *Stenotaphrum* were collected for cold tolerance evaluation. These accessions came from China (63), the United States (10), South Africa (3), Zimbabwe (1), Australia (1), Argentina (1), and Vanuatu (1). At one year of planting in Nanjing, China, only 55 accessions had survived, i.e., 25 accessions did not survive the winter. Among the 55 surviving accessions, 41 were sourced from China, 8 from the United States, 2 from South Africa, 1 from Zimbabwe, 1 from Australia, 1 from Argentina, and 1 from Vanuatu (Supplementary Table S1). Specifically in China, the strains were obtained from Fujian Province (14), Hainan Province (14), Yunnan Province (5), Guangxi Province (5), and Guangdong Province (3) and conformed to the distribution characteristics of *Stenotaphrum* (https://www.iplant.cn/z/frps/33150, accessed on 15 June 2024). Subsequently, the cold resistance of these 55 accessions was evaluated.

### 2.2. Plant Growth Conditions

The 50 accessions for the potted experiment were planted in flower pots with a bottom diameter of 18 cm. The flower pot was filled with half soil and half sand. Each accession contained three replicates. All the accessions were cultivated in the turfgrass nursery of Nanjing Botanical Garden Mem. Sun Yat-Sen, China. Compound fertilizer was applied once a month [31]. The dosage used was 0.5 g/pot for the pot experiments. The lowest temperature in winter was −10 °C. The detailed climate information is listed in Supplementary Table S2.

### 2.3. Investigation of Leaf Colour during the Fall Dormancy and Coverage during the Spring Green-Up

From the end of October 2022 to the middle of December 2022, photos of potted accessions were taken during the fall dormancy. The interval at which the images were taken was 7 days. The leaf colour was scored according to the GBT30395-2013 standard as previous reported [32–35].

From the end of April 2023 to the beginning of June 2023, photos of potted accessions were taken during the green-up. The rate of green-up was obtained through visual inspection [32–34,36].

### 2.4. LT50

The experiment was conducted in September 2023. According to previous methods [37], the lethal temperature killing 50% of the plants (LT50) was determined by the electrolyte leakage method, and each treatment included 4 replicates. Briefly, the leaves of healthy plants were removed and rinsed 3 times with deionized water, after which the water that clung to the leaf surface was removed with filter paper. The leaves were cut to a length of approximately 0.5 cm and divided into 5 parts (2 g each). The samples were treated in a cryogenic circulator (Polyscience Company, Warrington, PA, USA), followed by an overnight pre-culture at 4 °C. The low-temperature gradients were set to 4, −1, −6, −11, and −16 °C. After thawing at 4 °C, the electrolyte was extracted by adding 20 mL of deionized water. The conductivity was measured with a conductivity meter (Shanghai Leici Instrument, Shanghai, China) before and after the samples were boiled in a water bath for 15 min.

$$\text{electrolyte leakage} = \frac{\text{conductivity before boiling water bath}}{\text{conductivity after boiling water bath}} * 100\%$$

The LT50 was obtained by fitting the logistic growth equation:

$$\mathrm{Y} = \frac{\mathrm{YM} * \mathrm{Y0}}{(\mathrm{YM} - \mathrm{Y0}) * \exp(-\mathrm{k} * \mathrm{x}) + \mathrm{Y0}}$$

where Y is the electrolyte leakage, x is the temperature, and YM and Y0 are the maximal and initial electrolyte leakage. K is the rate constant. The LT50 is the x value where Y equals 50%.

### *2.5. Stolon Regrowth Experiments*

The experiment was conducted in September 2023. Using the methods in [38], stolon regrowth was evaluated via the number of stolons regenerated after low-temperature treatment. The stolons with 5 nodes were treated with 5 temperature gradients (8, 3, −2, −7, and −12 °C). Each treatment included 3 replicates (10 stolons per replicate). After thawing at 4 °C for 24 h, the stolons were planted in plugs filled with half soil and half sand. After 7 days of cultivation in the laboratory, the number of regrowing stolons was determined.

### *2.6. Membership Function Analysis*

The comprehensive cold tolerance evaluation was carried out using the membership function method [39–41]. Briefly, the subordination function value of the average greenness, average coverage, and total relative regrowth rates were calculated by the following equation:

$$\mathrm{Uij} = \frac{\mathrm{Xij} - \mathrm{Ximin}}{\mathrm{Ximax} - \mathrm{Ximin}}$$

whereas the subordination function value of the LT50 was calculated by the following equation:

$$\mathrm{Uij} = 1 - \frac{\mathrm{Xij} - \mathrm{Ximin}}{\mathrm{Ximax} - \mathrm{Ximin}}$$

i: a certain material, j: the index, U: the membership grade, Xij: the measured value of the index in a certain material, Ximin: the minimum value of the index in material i, and Ximax: the maximum value of the index in material i.

### *2.7. Data Statistics and Graphing*

The cluster analysis was performed by using squared Euclidean distance coefficient and linkage between groups cluster method in SPSS 19.0 software. Correlation analysis was conducted using Graphpad Prism 9.5 software. For the tables, the data presented are the means ± standard error of at least three replicates. The differences among accessions in a same column was assessed using SPSS 19.0 software and a significant difference is indicated by different letters.

## 3. Results

### 3.1. Investigation of Leaf Greenness during the Fall Dormancy and the Coverage during the Spring Green-Up Using the Pot Experiment

Since only 55 accessions survived a winter among the 80 accessions (Supplementary Table S1), detailed cold tolerance evaluations were conducted on the surviving 55 accessions. The investigation of leaf greenness began at the end of October and lasted until late December. As autumn and winter continued, the leaf greenness of *Stenotaphrum* gradually decreased and all the accessions progressed to a withered yellow state at the end of the monitoring period (Table 1). To objectively reflect the trend of changes in greenness, we averaged the data from 8 time points and obtained the average greenness. This indicator was subsequently used to rank the greenness of the 55 accessions in autumn and winter. The greater the average greenness was, the longer the green period was during the fall dormancy. The average greenness of these accessions varied greatly, ranging from 4.625 to 1.375. The three accessions with the highest average greenness were ST003, S12, and 674925-1. The three accessions with the lowest average greenness were S01, S27, and S28.

The investigation of coverage began at the end of April and lasted until early June. As the green-up period progressed, the grass gradually turned green, and by the end of monitoring, most accessions had approached a complete green-up state (Table 2). To objectively reflect the trend of this green-up change, we averaged the greening coverage at four-time points and obtained the average coverage. By using this indicator, the average coverage of the 55 accessions during the green-up was ranked. A higher average coverage indicates the faster green-up ability of an accession. The average coverage varied greatly among the accessions, ranging from 85% to 9%. The three accessions with the highest average coverage were 291594, S62, and S13. The three accessions with the lowest average coverage were S02, S10, and S28.

**Table 1.** Leaf greenness during the fall dormancy.

| Name | Leaf Greenness | | | | | | | | | | | | | | | | Av. Greenness | |
|---|---|---|---|---|---|---|---|---|---|---|---|---|---|---|---|---|---|---|
| | 29 October | | 5 November | | 12 November | | 19 November | | 26 November | | 5 December | | 10 December | | 17 December | | | |
| ST003 | 7.000 ± 0.577 | b | 7.000 ± 0.577 | a | 6.000 ± 0.289 | a | 5.000 ± 0.000 | a | 4.000 ± 0.289 | a | 3.000 ± 0.000 | a | 3.000 ± 0.000 | a | 2.000 ± 0.577 | a | 4.625 ± 0.289 | a |
| S12 | 7.000 ± 0.000 | b | 7.000 ± 0.000 | a | 6.000 ± 0.289 | a | 5.000 ± 0.289 | a | 4.000 ± 0.000 | a | 3.000 ± 0.577 | a | 2.000 ± 0.289 | b | 1.000 ± 0.000 | b | 4.375 ± 0.180 | ab |
| 674925-1 | 7.000 ± 0.000 | b | 7.000 ± 0.000 | a | 6.000 ± 0.289 | a | 4.000 ± 0.577 | b | 3.000 ± 0.000 | b | 3.000 ± 0.000 | a | 2.000 ± 0.000 | b | 1.000 ± 0.000 | b | 4.125 ± 0.108 | bc |
| 674925-3 | 7.000 ± 0.000 | b | 7.000 ± 0.000 | a | 6.000 ± 0.289 | a | 4.000 ± 0.000 | b | 3.000 ± 0.000 | b | 3.000 ± 0.000 | a | 2.000 ± 0.289 | b | 1.000 ± 0.000 | b | 4.125 ± 0.072 | bc |
| 410361 | 8.000 ± 0.577 | a | 7.000 ± 0.577 | a | 6.000 ± 0.289 | a | 4.000 ± 0.000 | b | 3.000 ± 0.000 | b | 2.000 ± 0.000 | b | 1.000 ± 0.000 | c | 1.000 ± 0.000 | b | 4.000 ± 0.180 | cd |
| S004 | 7.000 ± 0.000 | b | 6.000 ± 0.000 | b | 5.000 ± 0.000 | b | 4.000 ± 0.289 | b | 3.000 ± 0.000 | b | 3.000 ± 0.000 | a | 2.000 ± 0.289 | b | 1.000 ± 0.000 | b | 3.875 ± 0.072 | cde |
| S04 | 7.000 ± 0.577 | b | 6.000 ± 0.000 | b | 5.000 ± 0.000 | b | 4.000 ± 0.000 | b | 3.000 ± 0.000 | b | 3.000 ± 0.000 | a | 2.000 ± 0.000 | b | 1.000 ± 0.000 | b | 3.875 ± 0.072 | cde |
| S08 | 6.000 ± 0.000 | b | 6.000 ± 0.000 | b | 6.000 ± 0.000 | a | 4.000 ± 0.577 | b | 3.000 ± 0.289 | b | 3.000 ± 0.289 | a | 2.000 ± 0.289 | b | 1.000 ± 0.000 | b | 3.875 ± 0.180 | cde |
| S25 | 6.000 ± 0.000 | b | 6.000 ± 0.000 | b | 6.000 ± 0.000 | a | 4.000 ± 0.577 | b | 3.000 ± 0.289 | b | 3.000 ± 0.289 | a | 2.000 ± 0.289 | b | 1.000 ± 0.000 | b | 3.875 ± 0.180 | cde |
| S39 | 7.000 ± 0.000 | b | 6.000 ± 0.000 | b | 5.000 ± 0.000 | b | 4.000 ± 0.000 | b | 3.000 ± 0.289 | b | 3.000 ± 0.289 | a | 2.000 ± 0.000 | b | 1.000 ± 0.000 | b | 3.875 ± 0.072 | cde |
| S48 | 7.000 ± 0.000 | b | 6.000 ± 0.000 | b | 6.000 ± 0.000 | a | 4.000 ± 0.577 | b | 3.000 ± 0.289 | b | 2.000 ± 0.000 | b | 2.000 ± 0.000 | b | 1.000 ± 0.000 | b | 3.875 ± 0.108 | cde |
| S13 | 7.000 ± 0.000 | b | 6.000 ± 0.000 | b | 5.000 ± 0.000 | b | 4.000 ± 0.000 | b | 3.000 ± 0.000 | b | 2.000 ± 0.000 | b | 2.000 ± 0.000 | b | 1.000 ± 0.000 | b | 3.75 ± 0.000 | de |
| S26 | 7.000 ± 0.000 | b | 6.000 ± 0.000 | b | 5.000 ± 0.000 | b | 4.000 ± 0.000 | b | 3.000 ± 0.000 | b | 2.000 ± 0.000 | b | 2.000 ± 0.000 | b | 1.000 ± 0.000 | b | 3.75 ± 0.000 | de |
| S31 | 6.000 ± 0.000 | b | 6.000 ± 0.000 | b | 5.000 ± 0.000 | b | 4.000 ± 0.000 | b | 3.000 ± 0.000 | b | 3.000 ± 0.000 | a | 2.000 ± 0.000 | b | 1.000 ± 0.000 | b | 3.75 ± 0.000 | de |
| S58 | 7.000 ± 0.577 | b | 6.000 ± 0.289 | b | 5.000 ± 0.000 | b | 4.000 ± 0.000 | b | 2.000 ± 0.000 | c | 2.000 ± 0.000 | b | 2.000 ± 0.000 | b | 1.000 ± 0.000 | b | 3.625 ± 0.108 | ef |
| 647924-2 | 7.000 ± 0.289 | b | 6.000 ± 0.000 | b | 5.000 ± 0.289 | b | 3.000 ± 0.000 | c | 2.000 ± 0.000 | c | 2.000 ± 0.000 | b | 1.000 ± 0.000 | c | 1.000 ± 0.000 | b | 3.375 ± 0.072 | fg |
| S14 | 6.000 ± 0.000 | b | 5.000 ± 0.000 | c | 4.000 ± 0.000 | c | 4.000 ± 0.000 | b | 3.000 ± 0.289 | b | 2.000 ± 0.289 | b | 2.000 ± 0.289 | b | 1.000 ± 0.000 | b | 3.375 ± 0.108 | fg |
| S52 | 6.000 ± 0.577 | b | 5.000 ± 0.000 | c | 5.000 ± 0.000 | b | 3.000 ± 0.000 | c | 3.000 ± 0.000 | b | 2.000 ± 0.000 | b | 2.000 ± 0.000 | b | 1.000 ± 0.000 | b | 3.375 ± 0.072 | fg |
| S007 | 6.000 ± 0.000 | b | 5.000 ± 0.000 | c | 4.000 ± 0.000 | c | 3.000 ± 0.000 | c | 3.000 ± 0.000 | b | 2.000 ± 0.000 | b | 2.000 ± 0.000 | b | 1.000 ± 0.000 | b | 3.25 ± 0.000 | gh |
| S10 | 6.000 ± 0.289 | b | 5.000 ± 0.000 | c | 4.000 ± 0.577 | c | 3.000 ± 0.577 | c | 3.000 ± 0.000 | b | 2.000 ± 0.000 | b | 2.000 ± 0.000 | b | 1.000 ± 0.000 | b | 3.25 ± 0.180 | gh |
| 291594 | 7.000 ± 0.000 | b | 5.000 ± 0.577 | c | 4.000 ± 0.000 | c | 3.000 ± 0.000 | c | 2.000 ± 0.000 | c | 2.000 ± 0.000 | b | 1.000 ± 0.000 | c | 1.000 ± 0.000 | b | 3.125 ± 0.072 | ghi |
| 300129 | 7.000 ± 0.000 | b | 5.000 ± 0.000 | c | 4.000 ± 0.000 | c | 3.000 ± 0.000 | c | 2.000 ± 0.000 | c | 2.000 ± 0.000 | b | 1.000 ± 0.000 | c | 1.000 ± 0.000 | b | 3.125 ± 0.000 | ghi |
| 300130 | 7.000 ± 0.000 | b | 5.000 ± 0.577 | c | 4.000 ± 0.289 | c | 3.000 ± 0.289 | c | 2.000 ± 0.289 | c | 2.000 ± 0.289 | b | 1.000 ± 0.000 | c | 1.000 ± 0.000 | b | 3.125 ± 0.253 | ghi |
| S22 | 6.000 ± 0.000 | b | 6.000 ± 0.000 | b | 4.000 ± 0.000 | c | 3.000 ± 0.000 | c | 2.000 ± 0.000 | c | 2.000 ± 0.000 | b | 1.000 ± 0.000 | c | 1.000 ± 0.000 | b | 3.125 ± 0.000 | ghi |
| S37 | 5.000 ± 0.577 | c | 4.000 ± 0.000 | d | 4.000 ± 0.000 | c | 3.000 ± 0.000 | c | 3.000 ± 0.000 | b | 2.000 ± 0.000 | b | 2.000 ± 0.000 | b | 1.000 ± 0.000 | b | 3.125 ± 0.072 | ghi |
| 410364 | 6.000 ± 0.000 | b | 5.000 ± 0.000 | c | 4.000 ± 0.000 | c | 3.000 ± 0.000 | c | 2.000 ± 0.000 | c | 2.000 ± 0.000 | b | 1.000 ± 0.000 | c | 1.000 ± 0.000 | b | 3.000 ± 0.000 | hij |
| 674924-2 | 6.000 ± 0.289 | b | 5.000 ± 0.289 | c | 4.000 ± 0.000 | c | 3.000 ± 0.289 | c | 2.000 ± 0.289 | c | 2.000 ± 0.000 | b | 1.000 ± 0.000 | c | 1.000 ± 0.000 | b | 3.000 ± 0.182 | hij |
| S11 | 6.000 ± 0.000 | b | 4.000 ± 0.289 | d | 4.000 ± 0.289 | c | 3.000 ± 0.289 | c | 2.000 ± 0.000 | c | 2.000 ± 0.000 | b | 2.000 ± 0.000 | b | 1.000 ± 0.000 | b | 3.000 ± 0.108 | hij |
| S23 | 6.000 ± 0.000 | b | 5.000 ± 0.000 | c | 4.000 ± 0.000 | c | 3.000 ± 0.000 | c | 2.000 ± 0.000 | c | 2.000 ± 0.000 | b | 1.000 ± 0.000 | c | 1.000 ± 0.000 | b | 3.000 ± 0.000 | hij |
| S53 | 6.000 ± 0.000 | b | 4.000 ± 0.000 | d | 4.000 ± 0.000 | c | 3.000 ± 0.000 | c | 2.000 ± 0.000 | c | 2.000 ± 0.000 | b | 2.000 ± 0.000 | b | 1.000 ± 0.000 | b | 3.000 ± 0.000 | hij |
| 509038 | 6.000 ± 0.000 | b | 5.000 ± 0.000 | c | 3.000 ± 0.000 | d | 3.000 ± 0.000 | c | 2.000 ± 0.000 | c | 2.000 ± 0.000 | b | 1.000 ± 0.000 | c | 1.000 ± 0.000 | b | 2.875 ± 0.000 | ijk |
| S006 | 5.000 ± 0.577 | c | 4.000 ± 0.000 | d | 4.000 ± 0.000 | c | 3.000 ± 0.289 | c | 2.000 ± 0.289 | c | 2.000 ± 0.289 | b | 2.000 ± 0.289 | b | 1.000 ± 0.000 | b | 2.875 ± 0.218 | ijk |
| S46 | 5.000 ± 0.000 | c | 4.000 ± 0.000 | d | 4.000 ± 0.000 | c | 3.000 ± 0.000 | c | 2.000 ± 0.000 | c | 2.000 ± 0.000 | b | 2.000 ± 0.000 | b | 1.000 ± 0.000 | b | 2.875 ± 0.000 | ijk |
| S54 | 6.000 ± 0.577 | b | 4.000 ± 0.000 | d | 4.000 ± 0.000 | c | 3.000 ± 0.000 | c | 2.000 ± 0.000 | c | 2.000 ± 0.000 | b | 1.000 ± 0.000 | c | 1.000 ± 0.000 | b | 2.875 ± 0.072 | ijk |
| 410363 | 5.000 ± 0.000 | c | 5.000 ± 0.000 | c | 3.000 ± 0.000 | d | 3.000 ± 0.000 | c | 2.000 ± 0.000 | c | 2.000 ± 0.000 | b | 1.000 ± 0.000 | c | 1.000 ± 0.000 | b | 2.750 ± 0.000 | jkl |
| S21 | 5.000 ± 0.000 | c | 4.000 ± 0.000 | d | 4.000 ± 0.000 | c | 3.000 ± 0.000 | c | 2.000 ± 0.000 | c | 2.000 ± 0.000 | b | 1.000 ± 0.000 | c | 1.000 ± 0.000 | b | 2.750 ± 0.000 | jkl |
| S29 | 5.000 ± 0.000 | c | 5.000 ± 0.000 | c | 3.000 ± 0.289 | d | 3.000 ± 0.289 | c | 2.000 ± 0.000 | c | 2.000 ± 0.000 | b | 1.000 ± 0.000 | c | 1.000 ± 0.000 | b | 2.750 ± 0.072 | jkl |
| S38 | 5.000 ± 0.000 | c | 5.000 ± 0.000 | c | 3.000 ± 0.000 | d | 2.000 ± 0.000 | d | 2.000 ± 0.000 | c | 2.000 ± 0.000 | b | 2.000 ± 0.000 | b | 1.000 ± 0.000 | b | 2.750 ± 0.000 | jkl |
| S44 | 4.000 ± 0.000 | d | 4.000 ± 0.000 | d | 3.000 ± 0.000 | d | 3.000 ± 0.000 | c | 2.000 ± 0.000 | b | 2.000 ± 0.000 | b | 2.000 ± 0.000 | b | 1.000 ± 0.000 | b | 2.750 ± 0.000 | jkl |
| S62 | 5.000 ± 0.577 | c | 4.000 ± 0.289 | d | 4.000 ± 0.289 | c | 3.000 ± 0.289 | c | 2.000 ± 0.000 | c | 2.000 ± 0.000 | b | 1.000 ± 0.000 | c | 1.000 ± 0.000 | b | 2.750 ± 0.180 | jkl |
| S30 | 5.000 ± 0.000 | c | 4.000 ± 0.000 | d | 3.000 ± 0.000 | d | 3.000 ± 0.000 | c | 2.000 ± 0.000 | c | 2.000 ± 0.000 | b | 1.000 ± 0.000 | c | 1.000 ± 0.000 | b | 2.625 ± 0.000 | klm |
| S47 | 6.000 ± 0.000 | b | 5.000 ± 0.577 | c | 4.000 ± 0.000 | c | 2.000 ± 0.000 | d | 1.000 ± 0.000 | d | 1.000 ± 0.000 | c | 1.000 ± 0.000 | c | 1.000 ± 0.000 | b | 2.625 ± 0.072 | klm |
| S49 | 5.000 ± 0.577 | c | 4.000 ± 0.000 | d | 4.000 ± 0.000 | c | 3.000 ± 0.000 | c | 2.000 ± 0.000 | c | 1.000 ± 0.000 | c | 1.000 ± 0.000 | c | 1.000 ± 0.000 | b | 2.625 ± 0.072 | klm |
| S55 | 4.000 ± 0.000 | d | 4.000 ± 0.000 | d | 3.000 ± 0.000 | d | 3.000 ± 0.000 | c | 2.000 ± 0.000 | c | 2.000 ± 0.000 | b | 1.000 ± 0.000 | c | 1.000 ± 0.000 | b | 2.5 ± 0.000 | lmn |
| S005 | 6.000 ± 0.000 | b | 4.000 ± 0.000 | d | 3.000 ± 0.000 | d | 2.000 ± 0.000 | d | 1.000 ± 0.000 | d | 1.000 ± 0.000 | c | 1.000 ± 0.000 | c | 1.000 ± 0.000 | b | 2.375 ± 0.000 | mn |
| S50 | 4.000 ± 0.000 | d | 4.000 ± 0.000 | d | 3.000 ± 0.577 | d | 3.000 ± 0.000 | c | 2.000 ± 0.000 | c | 1.000 ± 0.000 | c | 1.000 ± 0.000 | c | 1.000 ± 0.000 | b | 2.375 ± 0.072 | mn |
| G001 | 4.000 ± 0.000 | d | 4.000 ± 0.000 | d | 3.000 ± 0.000 | d | 2.000 ± 0.000 | d | 2.000 ± 0.000 | c | 2.000 ± 0.000 | b | 1.000 ± 0.000 | c | 1.000 ± 0.000 | b | 2.375 ± 0.000 | mn |
| S45 | 4.000 ± 0.000 | d | 3.000 ± 0.577 | e | 2.000 ± 0.000 | e | 2.000 ± 0.000 | d | 2.000 ± 0.000 | c | 2.000 ± 0.000 | b | 2.000 ± 0.000 | b | 1.000 ± 0.000 | b | 2.250 ± 0.072 | n |
| S51 | 5.000 ± 0.577 | c | 4.000 ± 0.000 | d | 3.000 ± 0.000 | d | 2.000 ± 0.000 | d | 1.000 ± 0.000 | d | 1.000 ± 0.000 | c | 1.000 ± 0.000 | c | 1.000 ± 0.000 | b | 2.250 ± 0.072 | n |

**Table 1.** *Cont.*

| Name | Leaf Greenness | | | | | | | | | | | | | | | | Av. Greenness | |
|---|---|---|---|---|---|---|---|---|---|---|---|---|---|---|---|---|---|---|
| | 29 October | | 5 November | | 12 November | | 19 November | | 26 November | | 5 December | | 10 December | | 17 December | | | |
| S57 | 4.000 ± 0.577 | d | 3.000 ± 0.289 | e | 3.000 ± 0.289 | d | 2.000 ± 0.000 | d | 2.000 ± 0.000 | c | 2.000 ± 0.000 | b | 1.000 ± 0.000 | c | 1.000 ± 0.000 | b | 2.250 ± 0.144 | n |
| S56 | 4.000 ± 0.289 | d | 3.000 ± 0.000 | e | 3.000 ± 0.000 | d | 2.000 ± 0.000 | d | 1.000 ± 0.000 | d | 1.000 ± 0.000 | c | 1.000 ± 0.000 | c | 1.000 ± 0.000 | b | 2.000 ± 0.036 | n |
| S02 | 3.000 ± 0.577 | e | 2.000 ± 0.000 | f | 2.000 ± 0.000 | e | 1.000 ± 0.000 | e | 1.000 ± 0.000 | d | 1.000 ± 0.000 | c | 1.000 ± 0.000 | c | 1.000 ± 0.000 | b | 1.500 ± 0.072 | o |
| S01 | 2.000 ± 0.000 | f | 2.000 ± 0.000 | f | 2.000 ± 0.000 | e | 1.000 ± 0.000 | e | 1.000 ± 0.000 | d | 1.000 ± 0.000 | c | 1.000 ± 0.000 | c | 1.000 ± 0.000 | b | 1.375 ± 0.000 | o |
| S27 | 3.000 ± 0.000 | e | 2.000 ± 0.000 | f | 1.000 ± 0.000 | f | 1.000 ± 0.000 | e | 1.000 ± 0.000 | d | 1.000 ± 0.000 | c | 1.000 ± 0.000 | c | 1.000 ± 0.000 | b | 1.375 ± 0.000 | o |
| S28 | 3.000 ± 0.000 | e | 2.000 ± 0.000 | f | 1.000 ± 0.000 | f | 1.000 ± 0.000 | e | 1.000 ± 0.000 | d | 1.000 ± 0.000 | c | 1.000 ± 0.000 | c | 1.000 ± 0.000 | b | 1.375 ± 0.000 | o |

Note: the word "Average" was abbreviated as Av. The significant differences among accessions in a same column is indicated by different letters.

**Table 2.** Coverage survey at the spring green-up.

| Name | Cover (%) | | | | | | | | Av. Cover (%) | |
|---|---|---|---|---|---|---|---|---|---|---|
| | 28 April | | 7 May | | 27 May | | 6 June | | | |
| 291594 | 60.000 ± 0.000 | a | 80.000 ± 0.012 | a | 98.000 ± 0.015 | abc | 100.000 ± 0.000 | a | 84.500 ± 0.007 | a |
| S62 | 60.000 ± 0.010 | a | 70.000 ± 0.000 | b | 100.000 ± 0.000 | a | 100.000 ± 0.000 | a | 82.500 ± 0.003 | ab |
| S13 | 60.000 ± 0.006 | a | 65.000 ± 0.020 | c | 98.000 ± 0.010 | abc | 100.000 ± 0.000 | a | 80.750 ± 0.009 | bc |
| S58 | 60.000 ± 0.000 | a | 58.000 ± 0.020 | e | 100.000 ± 0.000 | a | 100.000 ± 0.000 | a | 79.500 ± 0.002 | cd |
| S57 | 55.000 ± 0.020 | b | 63.000 ± 0.010 | cd | 93.000 ± 0.017 | def | 100.000 ± 0.000 | a | 77.750 ± 0.011 | de |
| S08 | 50.000 ± 0.000 | c | 55.000 ± 0.029 | e | 99.000 ± 0.000 | ab | 100.000 ± 0.000 | a | 76.000 ± 0.007 | ef |
| S006 | 50.000 ± 0.029 | c | 53.000 ± 0.015 | ef | 92.000 ± 0.017 | defg | 100.000 ± 0.000 | a | 73.750 ± 0.015 | fg |
| S39 | 35.000 ± 0.000 | f | 60.000 ± 0.015 | d | 100.000 ± 0.000 | a | 100.000 ± 0.000 | a | 73.750 ± 0.004 | fg |
| S47 | 40.000 ± 0.020 | e | 55.000 ± 0.012 | e | 100.000 ± 0.000 | a | 100.000 ± 0.000 | a | 73.750 ± 0.008 | fg |
| S007 | 45.000 ± 0.000 | d | 54.000 ± 0.01 | ef | 90.000 ± 0.000 | fg | 100.000 ± 0.000 | a | 72.250 ± 0.003 | gh |
| S25 | 35.000 ± 0.000 | f | 55.000 ± 0.000 | e | 93.000 ± 0.015 | def | 100.000 ± 0.000 | a | 70.750 ± 0.004 | hi |
| S30 | 35.000 ± 0.029 | f | 55.000 ± 0.029 | e | 90.000 ± 0.000 | fg | 100.000 ± 0.000 | a | 70.000 ± 0.014 | hi |
| S005 | 35.000 ± 0.000 | f | 50.000 ± 0.000 | fg | 94.000 ± 0.029 | cdef | 100.000 ± 0.000 | a | 69.750 ± 0.007 | i |
| 647924-2 | 30.000 ± 0.000 | g | 55.000 ± 0.000 | e | 90.000 ± 0.000 | fg | 100.000 ± 0.000 | a | 68.750 ± 0.000 | i |
| S22 | 35.000 ± 0.000 | f | 50.000 ± 0.000 | fg | 90.000 ± 0.000 | fg | 100.000 ± 0.000 | a | 68.750 ± 0.000 | i |
| S48 | 35.000 ± 0.000 | f | 50.000 ± 0.000 | fg | 90.000 ± 0.000 | fg | 100.000 ± 0.000 | a | 68.750 ± 0.000 | i |
| 674925-1 | 25.000 ± 0.029 | h | 60.000 ± 0.000 | d | 88.000 ± 0.029 | gh | 100.000 ± 0.000 | a | 68.250 ± 0.014 | ij |
| S51 | 25.000 ± 0.000 | h | 52.000 ± 0.015 | ef | 96.000 ± 0.015 | abcd | 100.000 ± 0.000 | a | 68.250 ± 0.008 | ij |
| S29 | 32.000 ± 0.015 | fg | 50.000 ± 0.000 | fg | 90.000 ± 0.000 | fg | 100.000 ± 0.000 | a | 68.000 ± 0.004 | ijk |
| S52 | 30.000 ± 0.000 | g | 43.000 ± 0.015 | hi | 98.000 ± 0.012 | abc | 100.000 ± 0.000 | a | 67.750 ± 0.007 | ijk |
| S54 | 30.000 ± 0.000 | g | 50.000 ± 0.000 | fg | 90.000 ± 0.000 | fg | 100.000 ± 0.000 | a | 67.500 ± 0.000 | ijk |
| S37 | 30.000 ± 0.000 | g | 55.000 ± 0.029 | e | 88.000 ± 0.015 | efg | 96.000 ± 0.020 | b | 67.250 ± 0.022 | ijk |
| S26 | 30.000 ± 0.000 | g | 45.000 ± 0.029 | h | 90.000 ± 0.000 | fg | 100.000 ± 0.000 | a | 66.250 ± 0.007 | jkl |
| S23 | 20.000 ± 0.000 | j | 50.000 ± 0.000 | fg | 94.000 ± 0.015 | cdef | 100.000 ± 0.000 | a | 66.000 ± 0.004 | jkl |
| S56 | 40.000 ± 0.000 | e | 47.000 ± 0.018 | gh | 85.000 ± 0.000 | h | 90.000 ± 0.000 | c | 65.500 ± 0.004 | kl |
| S31 | 20.000 ± 0.000 | j | 45.000 ± 0.000 | h | 96.000 ± 0.015 | abcd | 100.000 ± 0.000 | a | 65.250 ± 0.004 | kl |
| S50 | 25.000 ± 0.000 | h | 40.000 ± 0.000 | ij | 96.000 ± 0.020 | abcd | 100.000 ± 0.000 | a | 65.250 ± 0.005 | lm |
| 674924-2 | 15.000 ± 0.000 | k | 45.000 ± 0.029 | h | 100.000 ± 0.000 | a | 100.000 ± 0.000 | a | 65.000 ± 0.007 | lm |
| S14 | 20.000 ± 0.000 | j | 60.000 ± 0.000 | d | 80.000 ± 0.000 | i | 100.000 ± 0.000 | a | 65.000 ± 0.000 | lmn |
| S38 | 20.000 ± 0.000 | j | 40.000 ± 0.000 | ij | 100.000 ± 0.000 | a | 100.000 ± 0.000 | a | 65.000 ± 0.000 | lmn |
| ST003 | 25.000 ± 0.029 | h | 35.000 ± 0.029 | k | 100.000 ± 0.000 | a | 100.000 ± 0.000 | a | 65.000 ± 0.014 | lmn |
| S12 | 40.000 ± 0.000 | e | 38.000 ± 0.015 | jk | 80.000 ± 0.000 | i | 100.000 ± 0.000 | a | 64.500 ± 0.004 | lmn |
| S04 | 10.000 ± 0.000 | lm | 45.000 ± 0.029 | h | 96.000 ± 0.010 | abcd | 100.000 ± 0.000 | a | 62.750 ± 0.009 | mno |
| S55 | 20.000 ± 0.000 | j | 37.000 ± 0.015 | jk | 94.000 ± 0.006 | cdef | 100.000 ± 0.000 | a | 62.750 ± 0.005 | no |
| S49 | 15.000 ± 0.000 | k | 40.000 ± 0.000 | ij | 90.000 ± 0.000 | fg | 100.000 ± 0.000 | a | 61.250 ± 0.000 | op |
| G001 | 15.000 ± 0.000 | k | 35.000 ± 0.000 | k | 95.000 ± 0.000 | bcde | 100.000 ± 0.000 | a | 61.250 ± 0.000 | op |
| S53 | 2.000 ± 0.010 | op | 38.000 ± 0.012 | jk | 99.000 ± 0.006 | ab | 100.000 ± 0.000 | a | 59.750 ± 0.007 | p |
| S46 | 10.000 ± 0 | lm | 20.000 ± 0.000 | n | 96.000 ± 0.010 | abcd | 100.000 ± 0.000 | a | 56.500 ± 0.003 | q |
| 410363 | 8.000 ± 0.015 | mn | 25.000 ± 0.000 | m | 80.000 ± 0.000 | i | 100.000 ± 0.000 | a | 53.250 ± 0.004 | r |
| 300130 | 0.000 ± 0.000 | p | 35.000 ± 0.000 | k | 75.000 ± 0.000 | j | 100.000 ± 0.000 | a | 52.500 ± 0.000 | r |
| 509038 | 12 ± 0.015 | kl | 18.000 ± 0.015 | no | 80.000 ± 0.000 | i | 100.000 ± 0.000 | a | 52.500 ± 0.008 | r |
| 410364 | 1.000 ± 0.000 | p | 30.000 ± 0.000 | l | 70.000 ± 0.000 | k | 100.000 ± 0.000 | a | 50.250 ± 0.000 | s |
| 674925-3 | 5.000 ± 0.000 | no | 20.000 ± 0.000 | n | 70.000 ± 0.000 | k | 100.000 ± 0.000 | a | 48.750 ± 0.000 | s |
| S21 | 15.000 ± 0.029 | k | 30.000 ± 0.029 | l | 60.000 ± 0.000 | l | 85.000 ± 0.000 | d | 47.500 ± 0.014 | s |
| 410361 | 10.000 ± 0.000 | lm | 30.000 ± 0.000 | l | 55.000 ± 0.029 | m | 80.000 ± 0.000 | e | 43.750 ± 0.007 | t |
| S004 | 2.000 ± 0.010 | op | 10.000 ± 0.000 | p | 70.000 ± 0.000 | k | 90.000 ± 0.000 | c | 43.000 ± 0.003 | t |
| S45 | 10.000 ± 0.000 | lm | 25.000 ± 0.029 | m | 55.000 ± 0.029 | m | 65.000 ± 0.029 | f | 38.750 ± 0.022 | u |
| S11 | 5.000 ± 0.000 | no | 5.000 ± 0.000 | q | 45.000 ± 0.000 | n | 60.000 ± 0.000 | g | 28.750 ± 0.000 | v |
| S44 | 5.000 ± 0.000 | no | 7.000 ± 0.006 | pq | 30.000 ± 0.000 | o | 40.000 ± 0.000 | h | 20.500 ± 0.001 | w |
| 300129 | 7.000 ± 0.017 | mn | 15.000 ± 0.000 | o | 20.000 ± 0.000 | p | 35.000 ± 0.000 | i | 19.250 ± 0.004 | w |
| S01 | 2.000 ± 0.000 | op | 8.000 ± 0.015 | pq | 25.000 ± 0.000 | o | 40.000 ± 0.000 | h | 18.750 ± 0.004 | w |
| S27 | 2.000 ± 0.010 | op | 8.000 ± 0.010 | pq | 30.000 ± 0.000 | o | 35.000 ± 0.000 | i | 18.750 ± 0.005 | w |
| S02 | 2.000 ± 0.000 | op | 6.000 ± 0.006 | pq | 15.000 ± 0.000 | q | 30.000 ± 0.000 | j | 13.250 ± 0.001 | x |
| S10 | 0.000 ± 0.000 | p | 5.000 ± 0.000 | q | 15.000 ± 0.000 | q | 25.000 ± 0.000 | k | 11.250 ± 0.000 | y |
| S28 | 1.000 ± 0.000 | p | 5.000 ± 0.000 | q | 10.000 ± 0.000 | r | 20.000 ± 0.000 | l | 9.000 ± 0.000 | y |

Note: the word "Average" was abbreviated as Av. The significant differences among accessions in a same column is indicated by different letters.

### 3.2. Membership Function Analysis and Cluster Analysis Based on Pot Experiment Results

The data for the leaf colour during the fall dormancy and the turfgrass coverage during the spring green-up was analysed using membership function analysis. The score obtained was used to rank the cold tolerance. The cold tolerance varied greatly among the accessions, ranging from 0.87 to 0. The three accessions with the highest cold tolerance were ST003, S13, and S12. The three accessions with the lowest cold tolerance were S27, S02, and

S28 (Table 3). The picture of these six accessions at fall dormancy and spring green-up was presented in Supplementary Figure S1. Cluster analysis revealed that these accessions were divided into two categories: cold tolerant and cold sensitive (Figure 1). The cold tolerance category can be further subdivided into two subcategories: super cold-tolerance and middle cold-tolerance. The cold sensitive category can be subdivided into two subcategories: super cold-sensitive and middle cold-sensitive (Figure 1). The accessions ranking 1–13 are mainly concentrated in the super cold-tolerant subcategory. The accessions ranking 14–46 are mainly concentrated in middle cold-tolerant subcategory. Those that ranked 47–51 are gathered in the middle cold-sensitive subcategory, and those that ranked 52–55 are gathered in the super cold-sensitive subcategory (Figure 1).

**Table 3.** Membership function analysis based on pot experiment results. Note: the word "Average" was abbreviated as Av. The significant differences among accessions in a same column is indicated by different letters.

| Accession | Av. Greenness | | Av. Coverage | | Score | | Rank |
|---|---|---|---|---|---|---|---|
| ST003 | 1.000 ± 0.000 | a | 0.741 ± 0.012 | m | 0.871 ± 0.006 | a | 1 |
| S13 | 0.742 ± 0.066 | cd | 0.950 ± 0.004 | c | 0.846 ± 0.032 | ab | 2 |
| S12 | 0.927 ± 0.027 | ab | 0.735 ± 0.001 | m | 0.831 ± 0.014 | abc | 3 |
| S08 | 0.771 ± 0.013 | cd | 0.887 ± 0.002 | e | 0.829 ± 0.005 | abcd | 4 |
| 674925-1 | 0.853 ± 0.043 | bc | 0.784 ± 0.012 | ijk | 0.819 ± 0.015 | abcd | 5 |
| S39 | 0.777 ± 0.047 | cd | 0.857 ± 0.002 | f | 0.817 ± 0.025 | abcd | 6 |
| S58 | 0.697 ± 0.028 | de | 0.925 ± 0.006 | d | 0.811 ± 0.017 | bcd | 7 |
| S25 | 0.771 ± 0.013 | cd | 0.817 ± 0.002 | gh | 0.794 ± 0.007 | bcde | 8 |
| S48 | 0.775 ± 0.036 | cd | 0.791 ± 0.006 | ijk | 0.783 ± 0.021 | cde | 9 |
| 291594 | 0.543 ± 0.026 | fgh | 1.000 ± 0.000 | a | 0.771 ± 0.013 | de | 10 |
| S26 | 0.742 ± 0.066 | cd | 0.758 ± 0.003 | lm | 0.750 ± 0.031 | ef | 11 |
| S04 | 0.777 ± 0.047 | cd | 0.711 ± 0.006 | n | 0.744 ± 0.021 | ef | 12 |
| S31 | 0.742 ± 0.066 | cd | 0.745 ± 0.001 | m | 0.743 ± 0.034 | ef | 13 |
| S007 | 0.586 ± 0.052 | efg | 0.837 ± 0.004 | fg | 0.712 ± 0.028 | fg | 14 |
| 647924-2 | 0.621 ± 0.033 | ef | 0.791 ± 0.006 | ijk | 0.706 ± 0.021 | fgh | 15 |
| S52 | 0.621 ± 0.033 | ef | 0.778 ± 0.001 | kl | 0.699 ± 0.015 | fgh | 16 |
| S62 | 0.419 ± 0.018 | hijkl | 0.973 ± 0.005 | b | 0.696 ± 0.006 | fgh | 17 |
| 674925-3 | 0.855 ± 0.054 | bc | 0.526 ± 0.004 | rs | 0.691 ± 0.029 | fghi | 18 |
| S14 | 0.619 ± 0.021 | ef | 0.741 ± 0.006 | m | 0.681 ± 0.014 | ghij | 19 |
| S22 | 0.547 ± 0.049 | fgh | 0.791 ± 0.006 | ijk | 0.669 ± 0.028 | ghijk | 20 |
| S37 | 0.543 ± 0.026 | fgh | 0.782 ± 0.022 | jk | 0.662 ± 0.002 | ghijkl | 21 |
| S006 | 0.443 ± 0.029 | hijk | 0.857 ± 0.012 | f | 0.651 ± 0.021 | hijklm | 22 |
| 410361 | 0.810 ± 0.016 | cd | 0.460 ± 0.005 | t | 0.635 ± 0.005 | ijklmn | 23 |
| S23 | 0.508 ± 0.045 | fghij | 0.754 ± 0.001 | lm | 0.631 ± 0.023 | jklmn | 24 |
| S47 | 0.394 ± 0.057 | ijklm | 0.857 ± 0.002 | f | 0.626 ± 0.027 | jklmn | 25 |
| S54 | 0.464 ± 0.019 | ghijk | 0.774 ± 0.006 | kl | 0.619 ± 0.013 | klmn | 26 |
| S004 | 0.777 ± 0.047 | cd | 0.450 ± 0.001 | t | 0.613 ± 0.024 | klmno | 27 |
| 674924-2 | 0.484 ± 0.015 | ghij | 0.741 ± 0.003 | m | 0.613 ± 0.008 | klmno | 28 |
| S29 | 0.425 ± 0.015 | hijkl | 0.781 ± 0.001 | jk | 0.603 ± 0.008 | lmnop | 29 |
| S30 | 0.390 ± 0.035 | jklmn | 0.807 ± 0.012 | hi | 0.599 ± 0.011 | mnop | 30 |
| S53 | 0.508 ± 0.045 | fghij | 0.672 ± 0.002 | o | 0.590 ± 0.021 | mnopq | 31 |
| S57 | 0.265 ± 0.021 | op | 0.910 ± 0.007 | d | 0.587 ± 0.014 | nopq | 32 |
| S38 | 0.429 ± 0.038 | hijkl | 0.741 ± 0.006 | m | 0.585 ± 0.022 | nopqr | 33 |
| S005 | 0.312 ± 0.028 | lmno | 0.804 ± 0.002 | hij | 0.558 ± 0.012 | opqrs | 34 |
| S46 | 0.468 ± 0.042 | ghijk | 0.629 ± 0.002 | p | 0.549 ± 0.022 | pqrs | 35 |
| 300130 | 0.519 ± 0.033 | fghi | 0.576 ± 0.005 | q | 0.548 ± 0.014 | pqrs | 36 |
| S49 | 0.386 ± 0.012 | jklmn | 0.692 ± 0.006 | no | 0.539 ± 0.009 | qrs | 37 |
| S55 | 0.351 ± 0.031 | klmno | 0.711 ± 0.001 | n | 0.531 ± 0.015 | qrs | 38 |
| 410364 | 0.508 ± 0.045 | fghij | 0.546 ± 0.004 | r | 0.527 ± 0.025 | rst | 39 |
| S51 | 0.269 ± 0.001 | no | 0.784 ± 0.003 | ijk | 0.527 ± 0.001 | rst | 40 |
| S50 | 0.308 ± 0.005 | lmno | 0.745 ± 0.002 | m | 0.526 ± 0.003 | rst | 41 |
| 509038 | 0.468 ± 0.042 | ghijk | 0.576 ± 0.005 | q | 0.522 ± 0.018 | st | 42 |
| 410363 | 0.429 ± 0.038 | hijkl | 0.586 ± 0.001 | q | 0.508 ± 0.019 | st | 43 |
| G001 | 0.312 ± 0.028 | lmno | 0.692 ± 0.006 | no | 0.502 ± 0.017 | st | 44 |
| S56 | 0.193 ± 0.006 | p | 0.748 ± 0.001 | m | 0.471 ± 0.003 | t | 45 |
| S21 | 0.429 ± 0.038 | hijkl | 0.509 ± 0.014 | s | 0.469 ± 0.012 | t | 46 |
| S11 | 0.502 ± 0.011 | fghij | 0.261 ± 0.002 | v | 0.381 ± 0.006 | u | 47 |
| 300129 | 0.547 ± 0.049 | fgh | 0.135 ± 0.004 | w | 0.341 ± 0.022 | uv | 48 |
| S45 | 0.277 ± 0.047 | mno | 0.393 ± 0.025 | u | 0.335 ± 0.011 | uv | 49 |
| S10 | 0.576 ± 0.004 | gf | 0.029 ± 0.001 | y | 0.302 ± 0.002 | v | 50 |

**Table 3.** *Cont.*

| Accession | Av. Greenness | | Av. Coverage | | Score | | Rank |
|---|---|---|---|---|---|---|---|
| S44 | 0.429 ± 0.038 | hijkl | 0.152 ± 0.001 | w | 0.291 ± 0.019 | v | 51 |
| S01 | 0.000 ± 0.000 | q | 0.129 ± 0.003 | wx | 0.064 ± 0.001 | w | 52 |
| S27 | 0.000 ± 0.000 | q | 0.129 ± 0.005 | x | 0.064 ± 0.002 | w | 53 |
| S02 | 0.035 ± 0.019 | q | 0.056 ± 0.001 | x | 0.045 ± 0.011 | wx | 54 |
| S28 | 0.000 ± 0.000 | q | 0.000 ± 0.000 | z | 0.000 ± 0.000 | x | 55 |

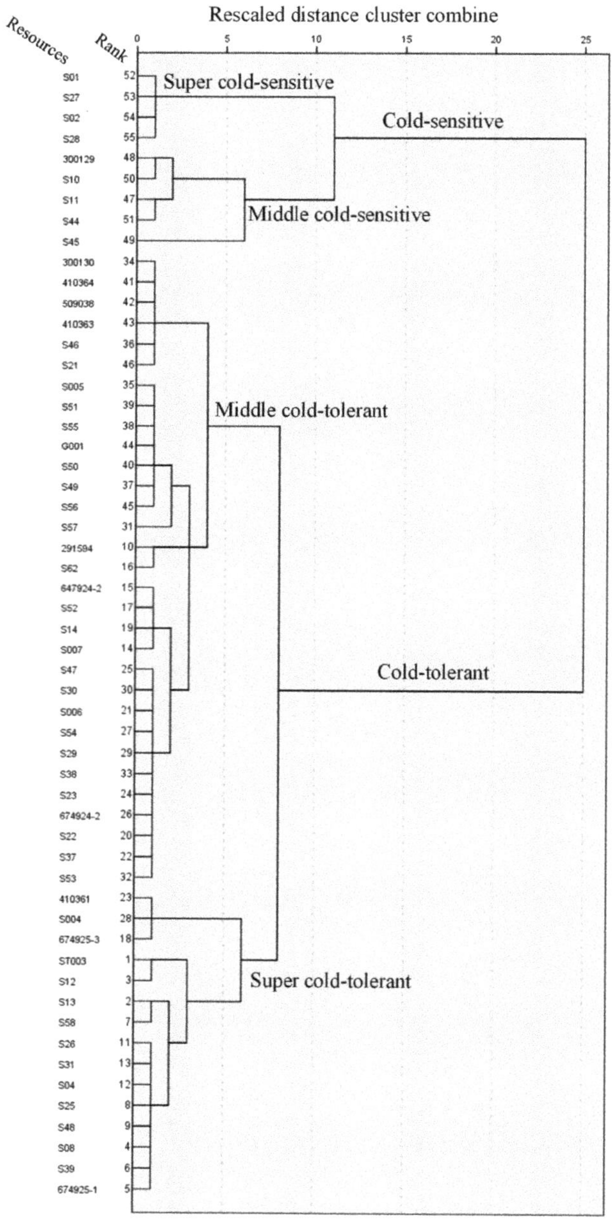


**Figure 1.** Cluster analysis based on pot experiment results.

The correlation analysis revealed that a positive correlation existed between autumn/winter leaf greenness and the turfgrass coverage during the spring green-up (Figure 2). Accessions with delayed fall dormancy will have an early spring green-up rate. However, the poor $R^2$ (0.1682) indicates a weak correlation, suggesting that it is inappropriate to simplify the cold resistance evaluation by investigating the growth performance of a single period. This result was further confirmed by the discrepancy between the ranking of autumn leaf colour or spring coverage rate and the ranking of comprehensive evaluation (Table 3).

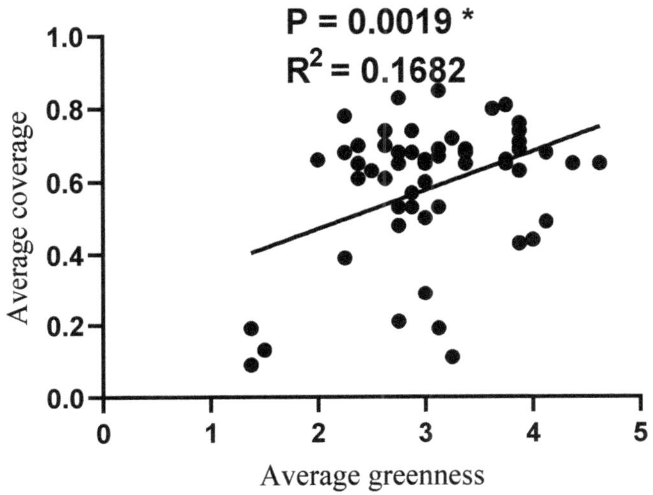


**Figure 2.** Correlation between the average greenness during the fall dormancy and average coverage during the spring green-up. The P and $R^2$ are the fitted parameters. The symbol * indicates a correlation.

### *3.3. Laboratory-Based Leaf LT50 and Stolon Regrowth Rate Analysis*

To investigate whether the pot evaluation can be replaced by the laboratory-based cold resistance evaluation methods, we conducted leaf LT50 and stolon regrowth rate analysis. The cold tolerance of the leaves was measured through an electrolyte leakage test (Table 4). Logistic equation fitting was performed on the relative conductivity of each accession, and the fitting degree ($R^2$) of each equation was greater than 0.86, indicating that the equation fit was good and that the obtained LT50 was highly reliable. A more negative LT50 value indicates a greater cold tolerance ability for leaves. A significant difference in leaf LT50 was detected among the accessions, ranging from −9.32 to 3.45. The three accessions with the most favourable LT50 values were S39, S005, and 410364. The three accessions with the lowest LT50 values were S11, S006, and S27.

Since the spring green-up of *Stenotaphrum* depends on the budding of the stolons, the regrowth ability of the stolons was evaluated. For all the accessions, as the temperature during the low-temperature pre-treatment decreased, the ability of the stolons to regrow significantly decreased. At −7 °C, most accessions lost their ability to regrow, while at −12 °C, all accessions lost their ability to regrow (Table 5). Given the significant differences in the ability of the accessions to regrow at relatively normal temperatures (8 °C), the regrowth ability of stolons at 8 °C was used as a control and the regrowth ability of stolons under other temperature treatments was standardized. The total relative regrowth rate of stolons at low temperatures was subsequently obtained by averaging the total relative regrowth rate of stolons at −3 °C, −2 °C, and −7 °C. A higher total relative regrowth rate indicates a greater ability for stolons to regrow in the next year. The total relative regrowth rate varied greatly among the accessions, ranging from 52% to 13%. The top three accessions in terms of total relative regrowth rate were S30, S39, and S25. The three accessions with the lowest total relative regrowth rates were S27, S10, and 300129.

**Table 4.** Leaf LT50.

| Name | Fitted Equation | $R^2$ | LT50 |
|---|---|---|---|
| S39 | Y = 0.0001214257/(0.726933e(0.9834X) + 0.000167) | 0.9962 | −9.322 |
| S005 | Y = 0.00009260336/(0.99510695e(1.062X) + 0.00009305) | 0.9936 | −8.744 |
| 410364 | Y = 0.000057860928/(0.99514186e(1.128X) + 0.00005814) | 0.9938 | −8.651 |
| 674925-1 | Y = 0.00012980552/(0.9599648e(1.041X) + 0.0001352) | 0.9953 | −8.598 |
| S22 | Y = 0.009776992/(0.87537e(0.5697X) + 0.01103) | 0.9388 | −8.130 |
| S25 | Y = 0.0060481639/(0.856899e(0.6375X) + 0.007001) | 0.9382 | −8.039 |
| S30 | Y = 0.000074015504/(0.98732504e(1.254X) + 0.00007496) | 0.9932 | −7.584 |
| S14 | Y = 0.00101304/(1.003992e(0.9263X) + 0.001008) | 0.9303 | −7.442 |
| S13 | Y = 0.001925836/(1.010097e(0.8403X) + 0.001903) | 0.9207 | −7.431 |
| S55 | Y = 0.00072324/(1.0072825e(0.9924X) + 0.0007175) | 0.9509 | −7.286 |
| 509038 | Y = 0.0319708/(0.9969e(0.518X) + 0.0311) | 0.9763 | −6.588 |
| S47 | Y = 0.244601/(0.9515e(0.2011X) + 0.2105) | 0.951 | −6.105 |
| s46 | Y = 0.034527/(0.98615e(0.5549X)+0.03385) | 0.9939 | −6.005 |
| 410363 | Y = 0.0050289084/(0.886966e(1.136X) + 0.005634) | 0.9972 | −4.666 |
| S57 | Y = 0.004862/(0.995138e(1.165X) + 0.004862) | 0.9906 | −4.567 |
| 300130 | Y = 0.002255/(0.997745e(1.455X) + 0.002255) | 0.996 | −4.187 |
| G001 | Y = 0.05019384/(0.93299e(0.7484X) + 0.05101) | 0.9973 | −3.926 |
| S51 | Y = 0.11036856/(0.841e(0.5361X) + 0.1154) | 0.9734 | −3.875 |
| S58 | Y = 0.2392359/(0.8063e(0.3166X) + 0.2307) | 0.9029 | −3.726 |
| ST003 | Y = 0.14605676/(0.8111e(−0.4706X) + 0.1517) | 0.9755 | −3.726 |
| 674925-3 | Y = 0.251808/(0.788e(0.306X) + 0.244) | 0.9965 | −3.628 |
| 647924-2 | Y = 0.2639386/(0.7887e(−0.2983X) + 0.2533) | 0.9962 | −3.537 |
| S04 | Y = 0.01177386/(0.9775e(1.343X) + 0.0119) | 0.9986 | −3.298 |
| S02 | Y = 0.02462/(0.97538e(1.119X) + 0.02462) | 0.999 | −3.287 |
| S26 | Y = 0.29094953/(0.6728e(0.2717X) + 0.2993) | 0.945 | −3.192 |
| S29 | Y = 0.081644706/(0.87528e(0.778X) + 0.08502) | 0.9446 | −3.103 |
| S31 | Y = 0.00738243/(0.982543e(1.627X) + 0.007457) | 0.9968 | −3.012 |
| 410361 | Y = 0.019615484/(0.97498e(1.302X) + 0.01972) | 0.9963 | −3.004 |
| S44 | Y = 0.0661362/(0.8303e(0.8921X) + 0.0732) | 0.9909 | −2.962 |
| S48 | Y = 0.038806316/(0.94997e(1.092X) + 0.03923) | 0.9937 | −2.938 |
| S49 | Y = 0.016183542/(0.96979e(−1.436X)+0.01641) | 0.975 | −2.860 |
| S01 | Y = 0.06156948/(0.92834e(0.9526X) + 0.06216) | 0.9834 | −2.858 |
| 674924-2 | Y = 0.12187968/(0.8518e(0.6895X) + 0.1248) | 0.9736 | −2.855 |
| S23 | Y = 0.049398912/(0.93632e(1.055X) + 0.05008) | 0.9893 | −2.801 |
| S52 | Y = 0.4486097/(0.8117e(0.1628X) + 0.3773) | 0.9669 | −2.736 |
| S08 | Y = 0.16656668/(0.8332e(0.6524X) + 0.1666) | 0.9569 | −2.467 |
| S004 | Y = 0.1205039/(0.8749e(0.8163X) + 0.121) | 0.9124 | −2.433 |
| S53 | Y = 0.054600213/(0.94213e(1.266X) + 0.05477) | 0.9529 | −2.252 |
| 291594 | Y = 0.37764/(0.689e(0.2548X) + 0.36) | 0.977 | −2.181 |
| S007 | Y = 0.033322732/(0.96213e(1.551X) + 0.03347) | 0.9852 | −2.171 |
| S38 | Y = 0.23126787/(0.7486e(0.5685X) + 0.2351) | 0.9767 | −2.095 |
| S12 | Y = 0.18992891/(0.809e(0.736X) + 0.1901) | 0.9566 | −1.971 |
| S54 | Y = 0.27675508/(0.5583e(0.4981X) + 0.3164) | 0.9954 | −1.719 |
| S45 | Y = 0.3547992/(0.6842e(0.3752X) + 0.3448) | 0.9328 | −1.676 |
| 300129 | Y = 0.335682/(0.6909e(0.4776X) + 0.3291) | 0.9623 | −1.471 |
| S62 | Y = 0.393151/(0.6483e(0.3226X) + 0.3817) | 0.9171 | −1.461 |
| S37 | Y = 0.4515098/(0.6016e(0.2602X) + 0.4354) | 0.9082 | −0.968 |
| S50 | Y = 0.4506476/(0.5882e(0.2879X) + 0.4388) | 0.9715 | −0.835 |
| S56 | Y = 0.4829058/(0.5944e(0.275X) + 0.4586) | 0.8743 | −0.576 |
| S28 | Y = 0.50284/(0.5565e(−0.2231X) + 0.4835) | 0.9411 | −0.285 |
| S10 | Y = 0.5437773/(0.5721e(0.2231X) + 0.5049) | 0.8281 | 0.081 |
| S21 | Y = 0.5490298/(0.5151e(0.2924X) + 0.5269) | 0.9247 | 0.353 |
| S11 | Y = 0.5569336/(0.5256e(0.2513X) + 0.5284) | 0.9165 | 0.429 |
| S006 | Y = 0.6057768/(0.4544e(0.207X) + 0.5836) | 0.9358 | 1.562 |
| S27 | Y = 0.7050042/(0.4224e(0.1686X) + 0.6546) | 0.8635 | 3.447 |

**Table 5.** Stolon regrowth ability post-low-temperature treatment.

| Name | Regrowth Ratio (%) | | | | | Relative Regrowth Rate (%) | | | Total Relative Regrowth Rate (%) |
|---|---|---|---|---|---|---|---|---|---|
| | 8 °C | 3 °C | −2 °C | −7 °C | −12 °C | 3 °C | 2 °C | −7 °C | |
| S30 | 46.000 ± 0.033 abc | 36.000 ± 0.033 bc | 26.000 ± 0.033 ab | 10.000 ± 0.000 a | 0.000 ± 0.000 | 78.000 ± 0.032 cd | 57.000 ± 0.054 abc | 22.000 ± 0.021 a | 52.000 ± 0.058 a |
| S39 | 50.000 ± 0.057 ab | 40.000 ± 0.100 ab | 26.000 ± 0.066 ab | 10.000 ± 0.100 a | 0.000 ± 0.000 | 80.000 ± 0.036 bc | 52.000 ± 0.045 bcdef | 20.000 ± 0.036 ab | 51.000 ± 0.014 ab |
| S25 | 40.000 ± 0.100 cd | 36.000 ± 0.033 bc | 20.000 ± 0.000 abcd | 3.000 ± 0.033 ab | 0.000 ± 0.000 | 90.000 ± 0.052 a | 50.000 ± 0.078 cdefg | 8.000 ± 0.069 abc | 49.000 ± 0.064 bc |
| S55 | 40.000 ± 0.000 cd | 36.000 ± 0.066 bc | 23.000 ± 0.033 abc | 0.000 ± 0.000 b | 0.000 ± 0.000 | 90.000 ± 0.032 a | 58.000 ± 0.021 ab | 0.000 ± 0.031 c | 49.000 ± 0.036 bc |
| S13 | 53.000 ± 0.033 a | 46.000 ± 0.033 a | 26.000 ± 0.088 ab | 3.000 ± 0.033 ab | 0.000 ± 0.000 | 87.000 ± 0.058 ab | 49.000 ± 0.065 defg | 6.000 ± 0.066 bc | 47.000 ± 0.067 c |
| 410364 | 46.000 ± 0.088 abc | 36.000 ± 0.033 bc | 26.000 ± 0.066 ab | 3.000 ± 0.033 ab | 0.000 ± 0.000 | 78.000 ± 0.088 cd | 57.000 ± 0.021 abc | 7.000 ± 0.025 abc | 47.000 ± 0.065 c |
| S14 | 50.000 ± 0.057 ab | 40.000 ± 0.056 ab | 30.000 ± 0.057 a | 0.000 ± 0.000 b | 0.000 ± 0.000 | 80.000 ± 0.065 bc | 60.000 ± 0.054 a | 0.000 ± 0.000 c | 47.000 ± 0.069 c |
| S31 | 30.000 ± 0.033 ef | 26.000 ± 0.033 def | 16.000 ± 0.033 abcde | 0.000 ± 0.000 b | 0.000 ± 0.000 | 87.000 ± 0.037 ab | 53.000 ± 0.088 abcde | 0.000 ± 0.000 c | 47.000 ± 0.056 c |
| S22 | 50.000 ± 0.057 ab | 40.000 ± 0.056 ab | 20.000 ± 0.032 abcd | 6.000 ± 0.033 ab | 0.000 ± 0.000 | 80.000 ± 0.041 bc | 40.000 ± 0.056 hijk | 12.000 ± 0.032 abc | 44.000 ± 0.044 d |
| 674925-1 | 46.000 ± 0.066 abc | 30.000 ± 0.098 cde | 20.000 ± 0.057 abcd | 10.000 ± 0.100 a | 0.000 ± 0.000 | 65.000 ± 0.012 ghi | 43.000 ± 0.032 ghij | 22.000 ± 0.021 a | 43.000 ± 0.021 d |
| 291594 | 30.000 ± 0.000 ef | 26.000 ± 0.088 def | 13.000 ± 0.033 abcde | 0.000 ± 0.000 b | 0.000 ± 0.000 | 87.000 ± 0.037 ab | 43.000 ± 0.036 ghij | 0.000 ± 0.000 c | 43.000 ± 0.032 d |
| S58 | 30.000 ± 0.057 ef | 23.000 ± 0.033 efg | 16.000 ± 0.033 abcde | 0.000 ± 0.000 b | 0.000 ± 0.000 | 77.000 ± 0.077 cde | 53.000 ± 0.088 abcde | 0.000 ± 0.000 c | 43.000 ± 0.032 d |
| ST003 | 40.000 ± 0.100 cd | 36.000 ± 0.033 bc | 16.000 ± 0.033 abcde | 0.000 ± 0.000 b | 0.000 ± 0.000 | 90.000 ± 0.032 a | 40.000 ± 0.022 hijk | 0.000 ± 0.000 c | 43.000 ± 0.024 d |
| 674925-3 | 36.000 ± 0.066 de | 26.000 ± 0.033 def | 20.000 ± 0.057 abcd | 0.000 ± 0.000 b | 0.000 ± 0.000 | 72.000 ± 0.032 def | 56.000 ± 0.032 abcd | 0.000 ± 0.000 c | 43.000 ± 0.024 d |
| S006 | 30.000 ± 0.000 ef | 20.000 ± 0.023 fh | 16.000 ± 0.033 abcde | 0.000 ± 0.000 b | 0.000 ± 0.000 | 67.000 ± 0.054 fghi | 53.000 ± 0.088 abcde | 0.000 ± 0.000 c | 40.000 ± 0.054 e |
| 410363 | 43.000 ± 0.066 bcd | 30.000 ± 0.100 cde | 20.000 ± 0.057 abcd | 0.000 ± 0.000 b | 0.000 ± 0.000 | 70.000 ± 0.035 efgh | 47.000 ± 0.023 efgh | 0.000 ± 0.000 c | 39.000 ± 0.054 ef |
| S56 | 26.000 ± 0.033 fg | 20.000 ± 0.100 fh | 10.000 ± 0.057 bcde | 0.000 ± 0.000 b | 0.000 ± 0.000 | 77.000 ± 0.078 cd | 38.000 ± 0.041 ijkl | 0.000 ± 0.000 c | 38.000 ± 0.045 efg |
| S005 | 50.000 ± 0.100 ab | 33.000 ± 0.033 bcd | 23.000 ± 0.033 abc | 0.000 ± 0.000 b | 0.000 ± 0.000 | 66.000 ± 0.065 fghi | 46.000 ± 0.054 fgh | 0.000 ± 0.000 c | 37.000 ± 0.023 fg |
| S47 | 50.000 ± 0.100 ab | 36.000 ± 0.133 bc | 20.000 ± 0.100 abcd | 0.000 ± 0.000 b | 0.000 ± 0.000 | 72.000 ± 0.033 defg | 40.000 ± 0.022 hijk | 0.000 ± 0.000 c | 37.000 ± 0.065 fg |
| G001 | 30.000 ± 0.320 ef | 23.000 ± 0.088 efg | 10.000 ± 0.057 bcde | 0.000 ± 0.000 b | 0.000 ± 0.000 | 77.000 ± 0.057 cde | 33.000 ± 0.014 klm | 0.000 ± 0.000 c | 37.000 ± 0.047 fg |
| S50 | 36.000 ± 0.066 de | 26.000 ± 0.033 def | 13.000 ± 0.066 abcde | 0.000 ± 0.000 b | 0.000 ± 0.000 | 72.000 ± 0.023 def | 36.000 ± 0.036 kl | 0.000 ± 0.000 c | 36.000 ± 0.054 g |
| S51 | 36.000 ± 0.066 de | 26.000 ± 0.033 def | 13.000 ± 0.066 abcde | 0.000 ± 0.000 b | 0.000 ± 0.000 | 72.000 ± 0.054 def | 36.000 ± 0.054 kl | 0.000 ± 0.000 c | 36.000 ± 0.054 g |
| S54 | 30.000 ± 0.054 ef | 26.000 ± 0.033 def | 6.666 ± 0.033 cde | 0.000 ± 0.000 b | 0.000 ± 0.000 | 87.000 ± 0.036 ab | 20.000 ± 0.059 n | 0.000 ± 0.000 c | 36.000 ± 0.064 g |
| S02 | 36.000 ± 0.133 de | 26.000 ± 0.066 def | 10.000 ± 0.057 bcde | 0.000 ± 0.000 b | 0.000 ± 0.000 | 72.000 ± 0.023 def | 28.000 ± 0.023 m | 0.000 ± 0.000 c | 33.000 ± 0.033 h |
| S08 | 36.000 ± 0.033 de | 30.000 ± 0.098 cde | 6.000 ± 0.066 cde | 0.000 ± 0.000 b | 0.000 ± 0.000 | 83.000 ± 0.054 abc | 17.000 ± 0.032 n | 0.000 ± 0.000 c | 33.000 ± 0.057 h |
| S23 | 36.000 ± 0.066 de | 26.000 ± 0.066 def | 10.000 ± 0.057 bcde | 0.000 ± 0.000 b | 0.000 ± 0.000 | 72.000 ± 0.023 def | 28.000 ± 0.036 m | 0.000 ± 0.000 c | 33.000 ± 0.036 h |
| S37 | 20.000 ± 0.021 g | 10.000 ± 0.021 h | 10.000 ± 0.100 bcde | 0.000 ± 0.000 b | 0.000 ± 0.000 | 50.000 ± 0.054 m | 50.000 ± 0.078 cdefg | 0.000 ± 0.000 c | 33.000 ± 0.035 h |
| S44 | 36.000 ± 0.033 de | 20.000 ± 0.100 fh | 16.000 ± 0.033 abcde | 0.000 ± 0.000 b | 0.000 ± 0.000 | 56.000 ± 0.036 klm | 44.000 ± 0.064 ghi | 0.000 ± 0.000 c | 33.000 ± 0.025 h |
| S48 | 30.000 ± 0.057 ef | 20.000 ± 0.000 fh | 10.000 ± 0.054 bcde | 0.000 ± 0.000 b | 0.000 ± 0.000 | 67.000 ± 0.023 fghi | 33.000 ± 0.014 klm | 0.000 ± 0.000 c | 33.000 ± 0.014 h |
| 300130 | 43.000 ± 0.033 bcd | 26.000 ± 0.066 def | 16.000 ± 0.088 abcd | 0.000 ± 0.000 b | 0.000 ± 0.000 | 60.000 ± 0.036 ijk | 37.000 ± 0.056 jkl | 0.000 ± 0.000 c | 33.000 ± 0.025 h |
| 410361 | 36.000 ± 0.033 de | 20.000 ± 0.057 fh | 13.000 ± 0.033 abcde | 0.000 ± 0.000 b | 0.000 ± 0.000 | 56.000 ± 0.021 klm | 36.000 ± 0.032 kl | 0.000 ± 0.000 c | 31.000 ± 0.023 hi |
| S29 | 36.000 ± 0.033 de | 20.000 ± 0.100 fh | 13.000 ± 0.033 abcde | 0.000 ± 0.000 b | 0.000 ± 0.000 | 56.000 ± 0.025 klm | 36.000 ± 0.025 kl | 0.000 ± 0.000 c | 31.000 ± 0.065 hi |
| S49 | 36.000 ± 0.133 de | 20.000 ± 0.023 fh | 13.000 ± 0.033 abcde | 0.000 ± 0.000 b | 0.000 ± 0.000 | 56.000 ± 0.025 klm | 36.000 ± 0.021 kl | 0.000 ± 0.000 c | 31.000 ± 0.021 hi |
| 509038 | 46.000 ± 0.033 abc | 26.000 ± 0.033 def | 16.000 ± 0.033 abcde | 0.000 ± 0.000 b | 0.000 ± 0.000 | 57.000 ± 0.065 klm | 35.000 ± 0.057 kl | 0.000 ± 0.000 c | 30.000 ± 0.035 hij |
| S46 | 46.000 ± 0.066 abc | 26.000 ± 0.088 def | 16.000 ± 0.088 abcde | 0.000 ± 0.000 b | 0.000 ± 0.000 | 57.000 ± 0.041 klm | 35.000 ± 0.36 kl | 0.000 ± 0.000 c | 30.000 ± 0.025 hij |
| S57 | 40.000 ± 0.100 cd | 23.000 ± 0.033 efg | 13.000 ± 0.033 abcde | 0.000 ± 0.000 b | 0.000 ± 0.000 | 58.000 ± 0.023 jkl | 33.000 ± 0.014 lm | 0.000 ± 0.000 c | 30.000 ± 0.054 hij |
| 674924-2 | 36.000 ± 0.033 de | 26.000 ± 0.133 def | 6.000 ± 0.066 cde | 0.000 ± 0.000 b | 0.000 ± 0.000 | 72.000 ± 0.045 def | 17.000 ± 0.025 n | 0.000 ± 0.000 c | 30.000 ± 0.036 hij |
| S004 | 30.000 ± 0.000 ef | 16.000 ± 0.033 g | 10.000 ± 0.057 bcde | 0.000 ± 0.000 b | 0.000 ± 0.000 | 53.000 ± 0.069 lm | 33.000 ± 0.045 klm | 0.000 ± 0.000 c | 29.000 ± 0.031 ij |
| S01 | 30.000 ± 0.000 ef | 16.000 ± 0.033 g | 10.000 ± 0.057 bcde | 0.000 ± 0.000 b | 0.000 ± 0.000 | 53.000 ± 0.047 lm | 33.000 ± 0.025 klm | 0.000 ± 0.000 c | 29.000 ± 0.025 ij |
| S21 | 30.000 ± 0.000 ef | 20.000 ± 0.013 fh | 6.000 ± 0.033 cde | 0.000 ± 0.000 b | 0.000 ± 0.000 | 67.000 ± 0.014 fghi | 20.000 ± 0.056 n | 0.000 ± 0.000 c | 29.000 ± 0.021 ij |
| S26 | 30.000 ± 0.000 ef | 16.000 ± 0.033 g | 10.000 ± 0.000 bcde | 0.000 ± 0.000 b | 0.000 ± 0.000 | 53.000 ± 0.021 lm | 33.000 ± 0.014 klm | 0.000 ± 0.000 c | 29.000 ± 0.021 ij |
| S38 | 30.000 ± 0.100 ef | 20.000 ± 0.023 fh | 6.000± 0.033 cde | 0.000 ± 0.000 b | 0.000 ± 0.000 | 67.000 ± 0.023 fghi | 20.000 ± 0.056 n | 0.000 ± 0.000 c | 29.000 ± 0.023 ij |
| S52 | 30.000 ± 0.100 ef | 16.000 ± 0.033 g | 10.000 ± 0.023 bcde | 0.000 ± 0.000 b | 0.000 ± 0.000 | 53.000 ± 0.056 lm | 33.000 ± 0.014 klm | 0.000 ± 0.000 c | 29.000 ± 0.025 ij |
| 647924-2 | 36.000 ± 0.033 de | 16.000 ± 0.033 g | 13.000 ± 0.033 abcde | 0.000 ± 0.000 b | 0.000 ± 0.000 | 44.000 ± 0.025 m | 36.000 ± 0.054 kl | 0.000 ± 0.000 c | 27.000 ± 0.014 j |
| S53 | 40.000 ± 0.052 cd | 16.000 ± 0.033 g | 16.000 ± 0.066 abcde | 0.000 ± 0.000 b | 0.000 ± 0.000 | 40.000 ± 0.036 mn | 40.000 ± 0.036 hijk | 0.000 ± 0.000 c | 27.000 ± 0.036 j |
| S62 | 20.000 ± 0.021 g | 16.000 ± 0.033 g | 0.000 ± 0.000 e | 0.000 ± 0.000 b | 0.000 ± 0.000 | 80.000 ± 0.054 bc | 0.000 ± 0.000 p | 0.000 ± 0.000 c | 27.000 ± 0.054 j |
| S007 | 36.000 ± 0.033 de | 16.000 ± 0.033 g | 10.000 ± 0.100 bcde | 0.000 ± 0.000 b | 0.000 ± 0.000 | 44.000 ± 0.023 m | 28.000 ± 0.025 m | 0.000 ± 0.000 c | 24.000 ± 0.021 k |
| S12 | 36.000 ± 0.133 de | 23.000 ± 0.066 efg | 3.000 ± 0.033 de | 0.000 ± 0.000 b | 0.000 ± 0.000 | 64.000 ± 0.036 hij | 8.000 ± 0.036 o | 0.000 ± 0.000 c | 24.000 ± 0.036 k |
| S28 | 30.000 ± 0.054 ef | 20.000 ± 0.045 fh | 0.000 ± 0.000 e | 0.000 ± 0.000 b | 0.000 ± 0.000 | 67.000 ± 0.025 fghi | 0.000 ± 0.000 p | 0.000 ± 0.000 c | 22.000 ± 0.000 k |

**Table 5.** *Cont.*

| Name | Regrowth Ratio (%) | | | | | | | | | Relative Regrowth Rate (%) | | | | | | Total Relative Regrowth Rate (%) | |
|---|---|---|---|---|---|---|---|---|---|---|---|---|---|---|---|---|---|
| | 8 °C | | 3 °C | | −2 °C | | −7 °C | | −12 °C | 3 °C | | 2 °C | | −7 °C | | | |
| S45 | 30.000 ± 0.100 | ef | 20.000 ± 0.100 | fh | 0.000 ± 0.000 | e | 0.000 ± 0.000 | b | 0.000 ± 0.000 | 67.000 ± 0.056 | fghi | 0.000 ± 0.000 | p | 0.000 ± 0.000 | c | 22.000 ± 0.056 | k |
| S04 | 40.000 ± 0.052 | cd | 20.000 ± 0.000 | fh | 6.000 ± 0.033 | cde | 0.000 ± 0.000 | b | 0.000 ± 0.000 | 50.000 ± 0.054 | m | 15.000 ± 0.025 | n | 0.000 ± 0.000 | c | 22.000 ± 0.021 | k |
| S11 | 26.000 ± 0.066 | fg | 16.000 ± 0.033 | g | 0.000 ± 0.000 | e | 0.000 ± 0.000 | b | 0.000 ± 0.000 | 62.000 ± 0.023 | ijk | 0.000 ± 0.000 | p | 0.000 ± 0.000 | c | 21.000 ± 0.023 | k |
| S27 | 26.000 ± 0.066 | fg | 16.000 ± 0.033 | g | 0.000 ± 0.000 | e | 0.000 ± 0.000 | b | 0.000 ± 0.000 | 62.000 ± 0.056 | ijk | 0.000 ± 0.000 | p | 0.000 ± 0.000 | c | 21.000 ± 0.056 | k |
| S10 | 26.000 ± 0.066 | fg | 13.000 ± 0.033 | gh | 0.000 ± 0.000 | e | 0.000 ± 0.000 | b | 0.000 ± 0.000 | 50.000 ± 0.054 | m | 0.000 ± 0.000 | p | 0.000 ± 0.000 | c | 17.000 ± 0.054 | l |
| 300129 | 26.000 ± 0.120 | fg | 10.000 ± 0.000 | h | 0.000 ± 0.000 | e | 0.000 ± 0.000 | b | 0.000 ± 0.000 | 38.000 ± 0.000 | n | 0.000 ± 0.000 | p | 0.000 ± 0.000 | c | 13.000 ± 0.000 | m |

Note: The significant differences among accessions in a same column is indicated by different letters.

### 3.4. Membership Function Analysis and Cluster Analysis Based on Laboratory Results

The score and rank based on the laboratory results were obtained by membership function analysis (Table 6). A significant difference between the laboratory-based rank and the pot-evaluated rank was detected, and the coincidence degree between them was poor (Tables 3 and 6). Cluster analysis revealed that these accessions could be divided into two categories and three subcategories based on the laboratory results (Figure 3). Specifically, the accessions ranking 1–9 are clustered in one subcategory (cold-tolerant). The accessions ranking 10–50 are clustered in one subcategory (middle cold-sensitive), while those rankings 51–55 are clustered in one subcategory (super cold-sensitive) (Figure 3). A very poor consistency with the clustering results between the pot evaluation and the laboratory-based evaluation was found (Figures 1 and 3).

**Table 6.** Membership function analysis based on laboratory results. The significant differences among accessions in a same column is indicated by different letters.

| Accession | LT50 | Total Relative Regrowth Rate | | Score | | Rank |
|---|---|---|---|---|---|---|
| S39 | 1.000 ± 0.000 | 0.974 ± 0.014 | ab | 0.987 ± 0.007 | a | 1 |
| S30 | 0.863 ± 0.000 | 1.000 ± 0.000 | a | 0.931 ± 0.000 | b | 2 |
| S25 | 0.899 ± 0.000 | 0.923 ± 0.001 | bc | 0.911 ± 0.001 | bc | 3 |
| 410364 | 0.947 ± 0.000 | 0.871 ± 0.026 | c | 0.909 ± 0.013 | bc | 4 |
| S55 | 0.841 ± 0.000 | 0.922 ± 0.026 | bc | 0.881 ± 0.013 | cd | 5 |
| S14 | 0.852 ± 0.000 | 0.872 ± 0.023 | c | 0.862 ± 0.011 | d | 6 |
| S13 | 0.852 ± 0.000 | 0.872 ± 0.012 | c | 0.862 ± 0.006 | d | 7 |
| 674925-1 | 0.943 ± 0.000 | 0.768 ± 0.031 | d | 0.856 ± 0.015 | d | 8 |
| S22 | 0.906 ± 0.000 | 0.795 ± 0.022 | d | 0.851 ± 0.011 | d | 9 |
| S005 | 0.954 ± 0.000 | 0.616 ± 0.023 | fg | 0.785 ± 0.011 | e | 10 |
| S31 | 0.505 ± 0.000 | 0.871 ± 0.031 | c | 0.688 ± 0.015 | f | 11 |
| S47 | 0.748 ± 0.000 | 0.615 ± 0.005 | fg | 0.681 ± 0.002 | fg | 12 |
| S58 | 0.561 ± 0.000 | 0.771 ± 0.062 | d | 0.666 ± 0.031 | fg | 13 |
| ST003 | 0.561 ± 0.000 | 0.769 ± 0.003 | d | 0.665 ± 0.001 | fg | 14 |
| 674925-3 | 0.554 ± 0.000 | 0.769 ± 0.023 | d | 0.661 ± 0.011 | fg | 15 |
| 410363 | 0.635 ± 0.000 | 0.667 ± 0.024 | ef | 0.651 ± 0.012 | g | 16 |
| 509038 | 0.785 ± 0.000 | 0.436 ± 0.019 | ijk | 0.611 ± 0.009 | h | 17 |
| 291594 | 0.441 ± 0.000 | 0.768 ± 0.041 | d | 0.604 ± 0.021 | h | 18 |
| G001 | 0.577 ± 0.000 | 0.615 ± 0.021 | fg | 0.596 ± 0.011 | h | 19 |
| S46 | 0.740 ± 0.000 | 0.435 ± 0.008 | ijk | 0.587 ± 0.004 | h | 20 |
| S51 | 0.573 ± 0.000 | 0.591 ± 0.023 | g | 0.581 ± 0.011 | hi | 21 |
| 300130 | 0.597 ± 0.000 | 0.512 ± 0.012 | h | 0.555 ± 0.006 | ij | 22 |
| S57 | 0.627 ± 0.000 | 0.436 ± 0.018 | ijk | 0.531 ± 0.009 | jk | 23 |
| S02 | 0.527 ± 0.000 | 0.512 ± 0.012 | h | 0.521 ± 0.006 | jkl | 24 |
| S44 | 0.501 ± 0.000 | 0.513 ± 0.021 | hi | 0.507 ± 0.011 | klm | 25 |
| S48 | 0.501 ± 0.000 | 0.512 ± 0.007 | hi | 0.506 ± 0.003 | klm | 26 |
| S23 | 0.489 ± 0.000 | 0.513 ± 0.022 | hi | 0.501 ± 0.011 | klmn | 27 |
| S54 | 0.404 ± 0.000 | 0.590 ± 0.021 | g | 0.497 ± 0.011 | klmn | 28 |
| S49 | 0.493 ± 0.000 | 0.486 ± 0.007 | hij | 0.491 ± 0.003 | lmno | 29 |
| S08 | 0.463 ± 0.000 | 0.512 ± 0.007 | hi | 0.487 ± 0.003 | lmno | 30 |
| S29 | 0.512 ± 0.000 | 0.462 ± 0.021 | hijk | 0.487 ± 0.011 | lmno | 31 |
| 410361 | 0.505 ± 0.000 | 0.461 ± 0.007 | hijk | 0.483 ± 0.003 | mnopq | 32 |
| S26 | 0.519 ± 0.000 | 0.436 ± 0.018 | ijk | 0.478 ± 0.009 | mnopq | 33 |
| S56 | 0.315 ± 0.000 | 0.641 ± 0.005 | efg | 0.478 ± 0.002 | mnopq | 34 |
| 674924-2 | 0.493 ± 0.000 | 0.435 ± 0.012 | ijk | 0.464 ± 0.006 | nopqr | 35 |
| S50 | 0.335 ± 0.000 | 0.589 ± 0.006 | g | 0.462 ± 0.003 | nopqr | 36 |
| 647924-2 | 0.546 ± 0.000 | 0.359 ± 0.017 | l | 0.453 ± 0.008 | opqrs | 37 |
| S01 | 0.493 ± 0.000 | 0.410 ± 0.021 | jkl | 0.452 ± 0.011 | pqrs | 38 |
| S52 | 0.484 ± 0.000 | 0.411 ± 0.018 | kl | 0.447 ± 0.009 | qrs | 39 |
| S004 | 0.461 ± 0.000 | 0.410 ± 0.018 | kl | 0.435 ± 0.009 | rst | 40 |
| S38 | 0.434 ± 0.000 | 0.435 ± 0.012 | ijk | 0.434 ± 0.006 | rst | 41 |
| S37 | 0.345 ± 0.000 | 0.513 ± 0.021 | hi | 0.429 ± 0.011 | rst | 42 |
| S006 | 0.147 ± 0.000 | 0.693 ± 0.071 | e | 0.420 ± 0.035 | st | 43 |
| S53 | 0.446 ± 0.000 | 0.358 ± 0.013 | l | 0.402 ± 0.006 | tu | 44 |
| S04 | 0.528 ± 0.000 | 0.231 ± 0.003 | m | 0.379 ± 0.001 | uv | 45 |
| S62 | 0.384 ± 0.000 | 0.358 ± 0.009 | l | 0.371 ± 0.004 | uv | 46 |
| S007 | 0.439 ± 0.000 | 0.283 ± 0.033 | m | 0.361 ± 0.016 | v | 47 |
| S12 | 0.424 ± 0.000 | 0.281 ± 0.013 | m | 0.353 ± 0.006 | vw | 48 |
| S21 | 0.242 ± 0.000 | 0.410 ± 0.012 | kl | 0.326 ± 0.006 | w | 49 |
| S45 | 0.401 ± 0.000 | 0.231 ± 0.011 | m | 0.315 ± 0.005 | w | 50 |

**Table 6.** *Cont.*

| Accession | LT50 | Total Relative Regrowth Rate | | Score | | Rank |
|---|---|---|---|---|---|---|
| S28 | 0.292 ± 0.000 | 0.231 ± 0.017 | m | 0.261 ± 0.008 | x | 51 |
| S11 | 0.236 ± 0.000 | 0.205 ± 0.003 | m | 0.221 ± 0.001 | xy | 52 |
| 300129 | 0.385 ± 0.000 | 0.000 ± 0.000 | o | 0.192 ± 0.000 | y | 53 |
| S10 | 0.263 ± 0.000 | 0.102 ± 0.001 | n | 0.183 ± 0.001 | y | 54 |
| S27 | 0.000 ± 0.000 | 0.346 ± 0.182 | m | 0.173 ± 0.091 | y | 55 |

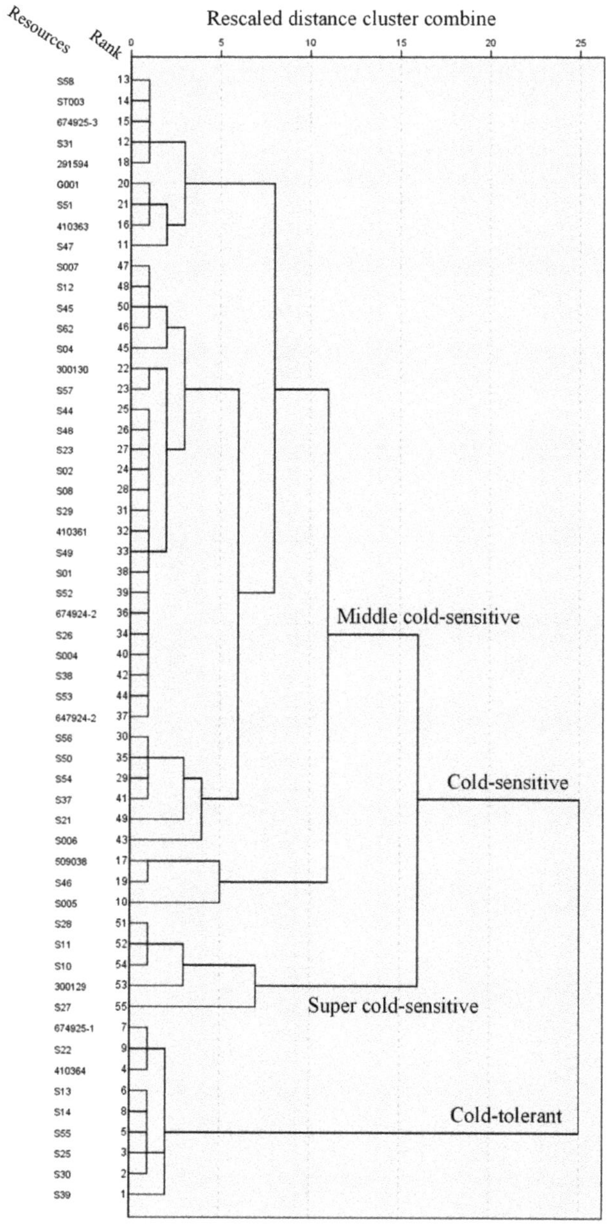


**Figure 3.** Cluster analysis based on laboratory results.

### *3.5. Correlation Analysis between the Laboratory-Based Data and the Pot Analysis Data*

A correlation analysis was conducted between LT50 and other data. It was found that LT50 was not correlated with the average greenness during the autumn/winter nor average coverage during the spring green-up but was correlated with the total relative regrowth rate (Figure 4).

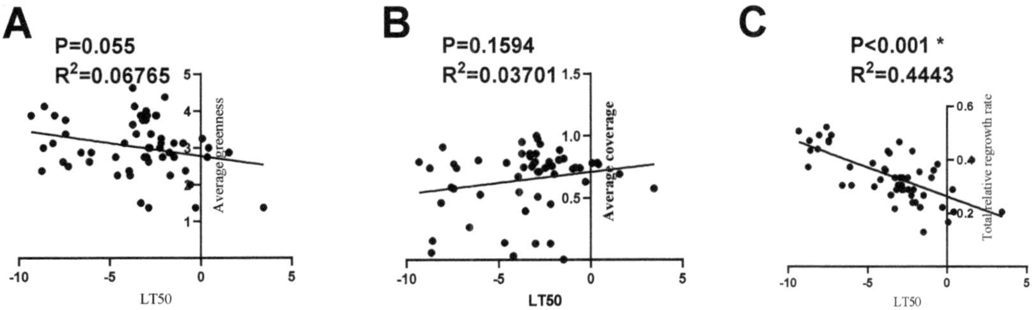


**Figure 4.** Correlation between LT50 and other parameters. (**A**) Correlation between LT50 and average greenness. (**B**) Correlation between LT50 and average coverage. (**C**) Correlation between LT50 and total relative regrowth rate. The P and $R^2$ are the fitted parameters. The symbol * indicates a correlation.

A correlation analysis was conducted between the total relative regrowth rate and other data. It was found that the total relative regrowth rate was not correlated with average greenness but was positively correlated with average coverage during the spring green-up (Figure 5).

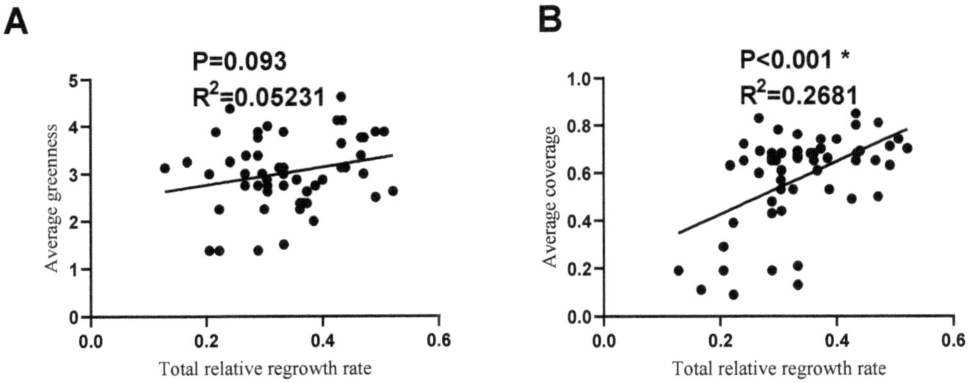


**Figure 5.** Correlation between total relative regrowth rate and other parameters. (**A**) Correlation between total relative regrowth rate and average greenness. (**B**) Correlation between total relative regrowth rate and average coverage. The P and $R^2$ are the fitted parameters. The symbol * indicates a correlation.

## 4. Discussion

### *4.1. Establishment of a Method for Evaluating the Cold Tolerance of Stenotaphrum by Integrating Its Performance at Both the Fall Dormancy and the Spring Green-Up*

It is assumed that the cold resistance of the leaves of perennial turfgrass is crucial when it first encounters autumn/winter cold, while the regrowth ability of stolons is crucial for rapid spring green-up. Briefly, when plants encounter chilling injury, their leaves strive to maintain photosynthesis by prolonging the green period and synthesizing energy to resist chilling injury. Secondly, the leaves transport photosynthate to the stolon to store energy for spring green-up. Leaves with a strong cold resistance are more conducive to the implementation of the above process. The bud from the stolon was the initial site for spring green-up. Therefore, the cold tolerance of perennial turfgrass relies on both the leaves and stolons. This mechanism is completely different from the mechanism of cold resistance in annual plants, which do not experience fall dormancy and spring green-up [42–44]. Moreover, the use of an evaluation system for the cold resistance of annual plants is inappropriate for evaluating the cold resistance of perennials [45–48]. A successful cold-tolerant perennial grass plant needs to be characterized by both delayed fall dormancy and early spring green-up. In this study, through the dynamic investigation of changes in leaf colour in autumn/winter and green coverage in the following year and the membership function analysis and cluster analysis, the growth performance of 55 *Stenotaphrum* accessions in these two stages was evaluated (Tables 1–3; Figure 1). This improved evaluation method compensates for the limitation of the previous cold tolerance evaluation method, which mainly focused on the spring green-up stage and paid less attention to fall dormancy [10,11]. The cold-tolerant accessions selected by this improved method have more ornamental value and are easier for the public to accept. The results showing that the autumn/winter leaf colour and the spring green-up coverage has a weak correlation and that the rank fit between a single period and these two periods is poor indicates that the growth performance of a single period cannot replace the overall evaluation (Figure 2; Table 3). The use of technical methods can only regulate the growth performance of one period (e.g., fall dormancy) and does not affect the growth performance of another period (e.g., spring green-up), supporting the above results [49,50].

### *4.2. This Improved Evaluation Method Cannot Be Replaced by Laboratory Evaluation*

The laboratory-based cold tolerance results and the pot cold tolerance results are inconsistent (Tables 3 and 6; Figures 1 and 3). There was no correlation between LT50 and the autumn leaf colour or the spring green-up coverage. The total relative regrowth was not correlated with the autumn/winter leaf colour and only correlated with the next year's green coverage (Figures 4 and 5). This explains why LT50 and stolon regrowth data cannot fully reflect the cold resistance of *Stenotaphrum* accessions. The inconsistency of the LT50 or the stolon regrowth data to the cold tolerance found in other perennial grass might be also partially attributed to this reason [16,17].

The stolon regrowth rate is positively correlated with the spring green-up coverage (Figure 5), which is in agreement with reports on other turfgrasses [5,51]. These data can to some extent reflect the situation of spring green-up of turfgrass. However, caution should be taken when using this method for cold tolerance evaluation in the future due to the weak correlation.

### *4.3. Screening of Several Excellent Cold-Tolerant Accessions That Can Be Directly Used in Temperate Regions of China*

Through the comprehensive evaluation method, we selected excellent-cold-resistance accessions, which were represented by ST003, S13, and S12 (Table 3). These excellent accessions were obtained from Sydney, Australia, Wenchang, Hainan (China), and Tengchong, Yunnan (China) (Supplementary Table S1). Although several studies have evaluated the cold resistance of *Stenotaphrum* plants, most have focused on accessions from Europe and North America [19–22]. This study evaluated the cold resistance of *Stenotaphrum* plants mainly using Chinese accessions. Due to cold tolerance being an important factor limiting plant geographical distribution [52] and cold tolerance being evaluated in temperate regions in China, the selected excellent-cold-resistance accessions can be directly applied locally. In the future, we plan to further evaluate the shade tolerance of these cold-tolerant accessions and screen for excellent shade-tolerant grass accessions that can be applied in temperate regions.

## 5. Conclusions

A turfgrass with good cold resistance should have both the advantage of delayed fall dormancy and early spring green-up. Based on the situation that the previous cold tolerance evaluation of turfgrass mainly focused on the spring green-up and paid less attention to its performance in the fall dormancy, this study integrates the performance of these two stages by dynamically investigating the autumn/winter leaf colour and next year's coverage and uses membership functions and cluster analysis to comprehensively evaluate the cold resistance of 55 *Stenotaphrum* accessions. This method cannot be simplified by the performance of one period, nor can it be replaced by indicator measurements conducted in the laboratory. The establishment of this method compensates for the shortcomings of previous methods for evaluating the cold tolerance of turfgrass. With the help of this improved method, we have screened several excellent-cold-tolerance accessions (ST003, S13, and S12) for the temperate regions of East China.

**Supplementary Materials:** The following supporting information can be downloaded at: https://www.mdpi.com/article/10.3390/horticulturae10070761/s1, Table S1: Aaccession information; Table S2: Extreme temperature and rainfall in Nanjing during the experimental process; Figure S1: The performance of six accessions with extreme cold tolerance and cold sensitivity.

**Author Contributions:** J.-Q.Z. and Z.-Y.W. designed the work. J.Q. and D.-L.H. carried out the experiments. D.-L.H., J.-Q.Z. and Z.-Y.W. wrote the paper. J.-Y.Z., J.-B.C., D.-J.S. and J.-X.L. revised the paper. All authors contributed to the article. All authors have read and agreed to the published version of the manuscript.

**Funding:** The authors are grateful for the financial support provided by the Forestry Science and Technology Innovation and Promotion Project of Jiangsu Province (LYKJ[2023]17), the Jiangsu Provincial Double-Innovation Doctor Program (Grant No. JSSCBS20221643), the Jiangsu Institute of Botany Talent Fund (Grant No. JIBTF202210), and the Program for the Young Innovative Talents of Jiangsu Vocational College of Agriculture and Forest (Grant No. 2021kj26).

**Data Availability Statement:** The original contributions presented in the study are included in the article/Supplementary Materials, further inquiries can be directed to the corresponding authors.

**Conflicts of Interest:** The authors declare no competing interests.

## References

1. Rhaetia, R.A.; Revathy, K.; Priya, M.M.; Shankar, S.R. Nutritive and proximate analysis on five weedy grasses for their potential use as fodder. *AgroLife Sci.* **2020**, *9*, 26–34. Available online: http://agrolifejournal.usamv.ro/index.php/agrolife/article/view/258 (accessed on 18 June 2024).
2. Cai, X.; Trenholm, L.E.; Kruse, J.; Sartain, J.B. Response of 'Captiva'St. Augustinegrass to Shade and Potassium. *HortScience* **2011**, *46*, 1400–1403. [CrossRef]
3. Wherley, B.; Chandra, A.; Genovesi, A.; Kearns, M.; Pepper, T.; Thomas, J. Developmental Response of St. Augustinegrass Cultivars and Experimental Lines in Moderate and Heavy Shade. *HortScience* **2013**, *48*, 1047–1051. [CrossRef]
4. Li, R.; Qu, R.; Bruneau, A.H.; Livingston, D.P. Selection for Freezing Tolerance in St. Augustinegrass through Somaclonal Variation and Germplasm Evaluation. *Plant Breed.* **2010**, *129*, 417–421. [CrossRef]
5. Kimball, J.A.; Tuong, T.D.; Arellano, C.; Iii, D.P.L.; Milla-Lewis, S.R. Assessing Freeze-Tolerance in St. Augustinegrass: Temperature Response and Evaluation Methods. *Euphytica* **2017**, *213*, 110. [CrossRef]
6. Moseley, D.O.; Trappe, J.M.; Milla-Lewis, S.R.; Chandra, A.; Kenworthy, K.E.; Liu, W.; Patton, A.J. Characterizing the Growth and Winter Survival of Commercially Available and Experimental Genotypes of St. Augustinegrass. *Crop Sci.* **2021**, *61*, 3097–3109. [CrossRef]
7. Li, Z.; Li, X.; He, F. Non-Structural Carbohydrates Contributed to Cold Tolerance and Regeneration of *Medicago sativa* L. *Planta* **2023**, *257*, 116. [CrossRef] [PubMed]
8. Rimi, F.; Macolino, S.; Richardson, M.D.; Karcher, D.E.; Leinauer, B. Influence of Three Nitrogen Fertilization Schedules on Bermudagrass and Seashore Paspalum: I. Spring Green-up and Fall Color Retention. *Crop Sci.* **2013**, *53*, 1161–1167. [CrossRef]
9. Schiavon, M.; Leinauer, B.; Sevastionova, E.; Serena, M.; Maier, B. Warm-season Turfgrass Quality, Spring Green-up, and Fall Color Retention under Drip Irrigation. *Appl. Turfgrass Sci.* **2011**, *8*, 1–9. [CrossRef]
10. Dunne, J.C.; Tuong, T.D.; Livingston, D.P.; Reynolds, W.C.; Milla-Lewis, S.R. Field and Laboratory Evaluation of Bermudagrass Germplasm for Cold Hardiness and Freezing Tolerance. *Crop Sci.* **2019**, *59*, 392–399. [CrossRef]

11. Hinton, J.D.; Livingston, D.P.; Miller, G.L.; Peacock, C.H.; Tuong, T. Freeze Tolerance of Nine Zoysiagrass Cultivars Using Natural Cold Acclimation and Freeze Chambers. *HortScience* **2012**, *47*, 112–115. [CrossRef]
12. Gopinath, L.; Moss, J.Q.; Wu, Y. Quantifying Freeze Tolerance of Hybrid Bermudagrasses Adapted for Golf Course Putting Greens. *HortScience* **2021**, *56*, 478–480. [CrossRef]
13. Ebdon, J.S.; Gagne, R.A.; Manley, R.C. Comparative Cold Tolerance in Diverse Turf Quality Genotypes of Perennial Ryegrass. *HortScience* **2002**, *37*, 826–830. [CrossRef]
14. Maier, F.P.; Lang, N.S.; Fry, J.D. Evaluation of an Electrolyte Leakage Technique to Predict St. Augustinegrass Freezing Tolerance. *HortScience* **1994**, *29*, 316–318. [CrossRef]
15. Fry, J.; Lang, N.S.; Clifton, R.G.P.; Maier, F.P. Freezing Tolerance and Carbohydrate Content of Low-temperature-acclimated and Nonacclimated Centipedegrass. *Crop Sci.* **1993**, *33*, 1051–1055. [CrossRef]
16. Cardona, C.A.; Duncan, R.R.; Lindstrom, O. Low Temperature Tolerance Assessment in Paspalum. *Crop Sci.* **1997**, *37*, 1283–1291. [CrossRef]
17. Milla-Lewis, S.R.; Kimball, J.A.; Claure, T.E.; Tuong, T.D.; Arellano, C.; Livingston, D.P., III. Freezing Tolerance and the Histology of Recovering Nodes in St. Augustinegrass. *Intl. Turfgrass Soc. J.* **2013**, *12*, 523–530.
18. Maier, F.P.; Lang, N.S.; Fry, J.D. Freezing Tolerance of Three St. Augustinegrass Cultivars as Affected by Stolon Carbohydrate and Water Content. *J. Am. Soc. Hortic. Sci.* **1994**, *119*, 473–476. [CrossRef]
19. Kimball, J.A.; Tuong, T.D.; Arellano, C.; Livingston, D.P.; Milla-Lewis, S.R. Linkage Analysis and Identification of Quantitative Trait Loci Associated with Freeze Tolerance and Turf Quality Traits in St. Augustinegrass. *Mol. Breed.* **2018**, *38*, 67. [CrossRef]
20. Kimball, J.A.; Isleib, T.G.; Reynolds, W.C.; Zuleta, M.C.; Milla-Lewis, S.R. Combining Ability for Winter Survival and Turf Quality Traits in St. Augustinegrass. *HortScience* **2016**, *51*, 810–815. [CrossRef]
21. Webster, D.E.; Ebdon, J.S. Effects of Nitrogen and Potassium Fertilization on Perennial Ryegrass Cold Tolerance during Deacclimation in Late Winter and Early Spring. *HortScience* **2005**, *40*, 842–849. [CrossRef]
22. Fry, J.D.; Lang, N.S.; Clifton, R.G.P. Freezing Resistance and Carbohydrate Composition of 'Floratam' St. Augustinegrass. *HortScience* **1991**, *26*, 1537–1539. [CrossRef]
23. Luo, Y.; Zhang, X.; Xu, J.; Zheng, Y.; Pu, S.; Duan, Z.; Li, Z.; Liu, G.; Chen, J.; Wang, Z. Phenotypic and Molecular Marker Analysis Uncovers the Genetic Diversity of the Grass Stenotaphrum Secundatum. *BMC Genet.* **2020**, *21*, 86. [CrossRef] [PubMed]
24. Song, Y.-B.; Yu, F.-H.; Li, J.-M.; Keser, L.H.; Fischer, M.; Dong, M.; Van Kleunen, M. Plant Invasiveness Is Not Linked to the Capacity of Regeneration from Small Fragments: An Experimental Test with 39 Stoloniferous Species. *Biol. Invasions* **2013**, *15*, 1367–1376. [CrossRef]
25. Li, Q.; Nothnagel, E.A. Absence of Variable Fluorescence from Guard Cell Chloroplasts of *Stenotaphrum Secundatum*. *Plant Physiol.* **1988**, *86*, 429–434. [CrossRef] [PubMed]
26. Liu, T.; Chen, D.; Liu, Z.; Wang, Z.Y.; Hu, J.S. First Report of Pyricularia Leaf Spot on St. Augustine Grass (*Stenotaphrum secundatum*) in China. *Plant Dis.* **2018**, *102*, 1666. [CrossRef]
27. Mei, S.S.; Wang, Z.Y.; Zhang, J.; Rong, W. First Report of Leaf Blight on *Stenotaphrum Secundatum* Caused by *Nigrospora Osmanthi* in China. *Plant Dis.* **2019**, *103*, 1783. [CrossRef]
28. Zong, J.; Li, L.; Yao, X.; Chen, J.; Wang, H.; Zhao, X.; Liu, J. Performance of Five Typical Warm-season Turfgrasses and Their Influence on Soil Bacterial Community under a Simulated Tropical Coral Island Environment. *Land Degrad. Dev.* **2021**, *32*, 3920–3929. [CrossRef]
29. Wu, Y.; Liao, L.; Wang, Z.; He, L. The Complete Plastid Genome of *Stenotaphrum Subulatum Trin.* (Panicoideae) and Phylogenetic Analysis. *Mitochondrial DNA Part B* **2020**, *5*, 1378–1380. [CrossRef]
30. Wang, Z.; Raymer, P.; Chen, Z. Isolation and Characterization of Microsatellite Markers for Stenotaphrum Trin. Using 454 Sequencing Technology. *HortScience* **2017**, *52*, 16–19. [CrossRef]
31. Reeves, S.A.; McBee, G.G. Nutritional Influences on Cold Hardiness of St. Augustinegrass (*Stenotaphrum Secundatum*) [1]. *Agron. J.* **1972**, *64*, 447–450. [CrossRef]
32. Karcher, D.E.; Richardson, M.D. Quantifying Turfgrass Color Using Digital Image Analysis. *Crop Sci.* **2003**, *43*, 943–951. [CrossRef]
33. Richardson, M.D.; Karcher, D.E.; Purcell, L.C. Quantifying Turfgrass Cover Using Digital Image Analysis. *Crop Sci.* **2001**, *41*, 1884–1888. [CrossRef]
34. Morri, K.N.; Shearman, R.C. NTEP Turfgrass Evaluation Guidelines. National Turfgrass Evaluation Program. 2014. Available online: https://www.Ntep.Org/Pdf/Ratings.Pdf (accessed on 18 June 2024).
35. Bunderson, L.D.; Johnson, P.G.; Kopp, K.L.; Dyke, A.V. Tools for Evaluating Native Grasses as Low Maintenance Turf. *HortTechnology* **2009**, *19*, 626–632. [CrossRef]
36. Russell, T.R.; Karcher, D.E.; Richardson, M.D. Daily Light Integral Requirements of Warm-season Turfgrasses for Golf Course Fairways and Investigating in Situ Evaluation Methodology. *Crop Sci.* **2020**, *60*, 3301–3313. [CrossRef]
37. Gu, S. Lethal Temperature Coefficient–a New Parameter for Interpretation of Cold Hardiness. *J. Hortic. Sci. Biotechnol.* **1999**, *74*, 53–59. [CrossRef]
38. Dunn, J.H.; Bughrara, S.S.; Warmund, M.R.; Fresenburg, B.F. Low Temperature Tolerance of Zoysiagrasses. *HortScience* **1999**, *34*, 96–99. [CrossRef]

39. Wang, H.; Cheng, X.; Shi, Q.; Xu, J.; Chen, D.; Luo, C.; Liu, H.; Cao, L.; Huang, C. Cold Tolerance Identification of Nine *Rosa* L. Materials and Expression Patterns of Genes Related to Cold Tolerance in *Rosa hybrida*. *Front. Plant Sci.* **2023**, *14*, 1209134. [CrossRef]
40. Jin, D.; Xu, Y.; Gui, H.; Zhang, H.; Dong, Q.; Sikder, R.K.; Wang, X.; Yang, G.; Song, M. Evaluation of Cotton (*Gossypium hirsutum* L.) Leaf Abscission Sensitivity Triggered by Thidiazuron through Membership Function Value. *Plants* **2020**, *10*, 49. [CrossRef]
41. Singh, A.; Kumar, A.; Kumar, R.; Prakash, J.; Kumar, N.; Verma, A.K. Evaluation of Salt Tolerance in Jamun (*Syzygium cumini* L. Skeels) Using Morpho-Physiological Traits and Membership Function Analysis. *Sci. Hortic.* **2024**, *326*, 112742. [CrossRef]
42. Li, J.; Zhang, Z.; Chong, K.; Xu, Y. Chilling Tolerance in Rice: Past and Present. *J. Plant Physiol.* **2022**, *268*, 153576. [CrossRef] [PubMed]
43. Wingler, A. Comparison of Signaling Interactions Determining Annual and Perennial Plant Growth in Response to Low Temperature. *Front. Plant Sci.* **2015**, *5*, 125650. [CrossRef] [PubMed]
44. Zhang, Q.; Chen, Q.; Wang, S.; Hong, Y.; Wang, Z. Rice and Cold Stress: Methods for Its Evaluation and Summary of Cold Tolerance-Related Quantitative Trait Loci. *Rice* **2014**, *7*, 24. [CrossRef] [PubMed]
45. Yang, X.; Brown, H.E.; Teixeira, E.I.; Moot, D.J. Development of a Lucerne Model in APSIM next Generation: 1 Phenology and Morphology of Genotypes with Different Fall Dormancies. *Eur. J. Agron.* **2021**, *130*, 126372. [CrossRef]
46. Trischuk, R.G.; Schilling, B.S.; Low, N.H.; Gray, G.R.; Gusta, L.V. Cold Acclimation, de-Acclimation and Re-Acclimation of Spring Canola, Winter Canola and Winter Wheat: The Role of Carbohydrates, Cold-Induced Stress Proteins and Vernalization. *Environ. Exp. Bot.* **2014**, *106*, 156–163. [CrossRef]
47. Janská, A.; Maršík, P.; Zelenková, S.; Ovesná, J. Cold Stress and Acclimation—What Is Important for Metabolic Adjustment? *Plant Biol.* **2010**, *12*, 395–405. [CrossRef] [PubMed]
48. Dhont, C.; Castonguay, Y.; Avice, J.-C.; Chalifour, F.-P. VSP Accumulation and Cold-Inducible Gene Expression during Autumn Hardening and Overwintering of Alfalfa. *J. Exp. Bot.* **2006**, *57*, 2325–2337. [CrossRef] [PubMed]
49. Luca, V.D.; de Barreda, D.G. Effect of a Biostimulant on Bermudagrass Fall Color Retention and Spring Green-Up. *Agronomy* **2021**, *11*, 608. [CrossRef]
50. Munshaw, G.C.; Ervin, E.H.; Beasley, J.S.; Shang, C.; Zhang, X.; Parrish, D.J. Effects of Late-season Ethephon Applications on Cold Tolerance Parameters of Four Bermudagrass Cultivars. *Crop Sci.* **2010**, *50*, 1022–1029. [CrossRef]
51. Annicchiarico, P.; Collins, R.P.; Fornasier, F.; Rhodes, I. Variation in Cold Tolerance and Spring Growth among Italian White Clover Populations. *Euphytica* **2001**, *122*, 407–416. [CrossRef]
52. Caturegli, L.; Ramazani, R.; Volterrani, M.; Grossi, N.; Magni, S.; Macolino, S.; Pornaro, C.; Bella, S.L.; Tuttolomondo, T.; Minelli, A. St. Augustinegrass Accessions Planted in Northern, Central and Southern Italy: Growth and Morphological Traits during Establishment. *Ital. J. Agron.* **2018**, *13*, 332–337. [CrossRef]

horticulturae

MDPI

Article

# The Molecular Biology Analysis for the Growing and Development of *Hydrangea macrophylla* 'Endless Summer' under Different Light and Temperature Conditions

**Zheng Li [1,†], Tong Lyu [2,*,†] and Yingmin Lyu [1,*]**

1 Beijing Key Laboratory of Ornamental Plants Germplasm Innovation & Molecular Breeding, China National, Engineering Research Center for Floriculture, College of Landscape Architecture, Beijing Forestry University, Beijing 100083, China
2 Beijing Flower Engineering Technology Research Center, Plant Institute, China National Botanical Garden North Garden, Beijing 100093, China
* Correspondence: chinabjlutong@163.com (T.L.); luyingmin@bjfu.edu.cn (Y.L.)
† These authors contributed equally to this work.

**Abstract:** *Hydrangea macrophylla*, a celebrated ornamental worldwide, thrives in semi-shaded growth environments in its natural habitat. This study utilizes *Hydrangea macrophylla* 'Endless Summer' as the experimental material to delve into its molecular mechanisms for adapting to semi-shaded conditions. Transcriptome analysis was conducted on leaves from four different natural light growth scenarios, showcasing phenotypic variations. From each sample, we obtained over 276,305,940 clean reads. Following de novo assembly and quantitative assessment, 88,575 unigenes were generated, with an average length of 976 bp. Gene ontology analysis of each control group elucidated the terms associated with the suitable environmental conditions for normal growth, development, and flowering, such as "reproductive bud system development" and "signal transduction". The exploration of gene interactions and the identification of key genes with strong connectivity were achieved by constructing a protein–protein interaction (PPI) network. The results indicate that hydrangea grows vigorously and blooms steadily under semi-shaded conditions; the photosynthetic efficiency of hydrangea is stabilized through genes related to photosynthesis, such as *PHYB*, *PSBR*, *FDC*, etc. Hormone signal transduction genes like *PIN3*, *LAX2*, *TIF6B*, and *EIN3* play important roles in responding to environmental stimulation and regulating growth and development, while genes such as *SOC1*, *COL4/5/16*, and *AGL24* promote flowering. The expression of genes such as *BGLUs* and *TPSs* provides additional energy substances to support flowering.

**Keywords:** *Hydrangea macrophylla*; transcriptome (RNA-seq); semi-shaded conditions; protein–protein interaction (PPI) network

## 1. Introduction

As a member of the Hydrangeaceae family, *Hydrangea macrophylla* is a shrub gaining prominence as the most promising ornamental flower species [1]. Due to its large inflorescences and captivating colors, various species and cultivars of *H. macrophylla* are applied as cut flowers and potted plants and in landscaping. Notably, the cultivation of blue varieties of *H. macrophylla* can be achieved through the modification of external conditions. For instance, changing soil pH or introducing exogenous $Al^{3+}$ can result in a color change in certain infertile flowers of *H. macrophylla* [2]. To date, horticulturists have cultivated four categories of hydrangea, including *Hydrangea macrophylla*, *Hydrangea arborescens*, *Hydrangea paniculata*, and *Hydrangea quercifolia*.

*H. macrophylla* can be seamlessly integrated with other plants in environmental greening, significantly elevating the sense of hierarchy and affinity in landscape design. For instance, when strategically combined with trees during landscape configuration, it produces a positive impact. Trees provide shading, and when complemented by hydrangeas amidst green foliage, they contribute to the softening of the environment and the enhancement of the overall aesthetic appeal. Despite *H. macrophylla*'s robust adaptability and preference for semi-shaded environments, its flowering period from May to August coincides with intense sunlight. Direct exposure to sunlight may lead to leaf yellowing and burning, resulting in sunburn that adversely affects the ornamental appeal during the flowering season. Conversely, excessive shading impedes photosynthesis due to insufficient light, causing nutrient deficiencies and impacting flowering. Therefore, when cultivating *H. macrophylla* in open fields, it is advisable to plant them under sparse tree shade or along tree-lined paths. In cases of excessive sunlight, shading measures become necessary. Xu Hui et al. discovered that under 50% light conditions, hydrangea flowers exhibit optimal blooming, with larger inflorescences, flowers, and petals [3].

Light and temperature constantly change under natural conditions, profoundly affecting plant growth and development. Changes in the cultivating environment, especially local lighting and temperature conditions, exerted influences on plants, with plants responding by adjusting developmental programs to adapt to new conditions. Appropriate lighting and temperature promote flowering. Plants perceive light through photoreceptors, such as photosensitive pigment A-E (phyA-E); cryptochrome (CRY); photosensitizers (PHOT), light/oxygen/voltage (LOV), etc. [4]. Temperature sensing relies on a variety of diverse cellular mechanisms, including those encompassing photosensitive pigments activities under warm conditions, the induction of HSPs, and the physical changes in lipid membranes during heat stress. The currently identified photoreceptors appear to work in temperature reactions. For instance, in Arabidopsis (*Arabidopsis thaliana*), the photosensitive pigment B (phyB) signaling not only responds to light but also reacts to heat, exacerbating disruptions in phyB signaling. Additional temperature-sensing mechanisms recently uncovered in Arabidopsis involve early flowering 3 (ELF3) and the bHLH TF plant pigment interaction factor 7 (PIF7) [5]. These prove the existence of crosstalk between light and temperature reactions.

The strategies employed by plants in response to stress vary significantly among species. However, for most species, stress induces osmotic pressure and ion stress simultaneously at the plant and cellular levels. Heat stress exacerbates this by promoting the accumulation of reactive oxygen species (ROS), leading to damage in the photosynthetic apparatus, particularly PSII, and ultimately resulting in photoinhibition. The ROS serves as a signaling molecule, regulating numerous physiological processes. Excessive ROS accumulation can induce cytotoxic conditions, causing oxidative damage to lipids, proteins, and nucleic acids. In response, plants have evolved both non-enzymatic and enzymatic systems to prevent ROS-caused oxidative damage. Non-enzymatic antioxidants, such as carotenoids, tocopherols, glutathione (GSH), and ascorbic acid, act in conjunction with enzymatic systems comprising superoxide dismutase (SOD), catalase (CAT), ascorbic acid peroxidase (APX), and peroxiredoxin (Prxs). Changes in proteomics and gene expression levels are intricately linked to environmental adaptation for maintaining their normal growth [6]. TFs are likely key players in this process. Genes involved in various pathways have been identified, encompassing signaling, regulation of transcription (especially through abscisic acid-dependent or -independent pathways), production of reactive oxygen species and detoxification, membrane transport, and synthesis of osmoprotectors, such as proline [7].

While considerable attention has been devoted to hydrangea research, the understanding of the molecular mechanism of semi-shaded environment adaptation is still in its infancy. There is a noticeable lack of information to quantitatively assess its long-term resilience to light and temperature, as well as the molecular mechanisms facilitating its smooth growth and development under semi-shaded conditions. Preliminary phenotypic observations indicate that 'Endless Summer' tends to wilt during high noon temperatures in summer and does not bloom under forests with high canopy closure, which makes it an ideal subject for the study of the optimal growth of *H. macrophylla* under semi-shaded conditions. Unraveling the molecular basis of the growth and development of *H. macrophylla* under suitable conditions is pivotal for addressing the genetic improvement challenge in developing genotypes tolerant to light and temperature stress. Furthermore, identifying the genes involved in maintaining the hydrangea's normal development and flowering is also a prerequisite for targeting these genes at the biotechnological level (such as through gene enrichment, etc.) to enhance their resistance to light and temperature stress.

## 2. Materials and Methods

### *2.1. Experimental Materials and Procedures*

Two-year potted seedlings of *H. macrophylla* 'Endless Summer' propagated by cuttings were transplanted to an open field for growth in April, during which normal water and fertilizer management was carried out. After two months of growth, leaf samples were collected on 23 June 2022, at noon, under continuously sunny conditions. The maximum and minimum daily temperatures of the experimental site for the month are illustrated in Figure 1. Light intensity and leaf surface temperatures were measured using a CL-500A spectroradiometer (KM, JPN). For treatment WL01, the light intensity was recorded as 351 $\mu mol \cdot m^{-2} \cdot s^{-1}$, with an average leaf surface temperature of 34.5 °C; for WL02, the light intensity was 751 $\mu mol \cdot m^{-2} \cdot s^{-1}$ with an average temperature of 36.3 °C; and for WL03, it was 1503 $\mu mol \cdot m^{-2} \cdot s^{-1}$ with a temperature of 41.7 °C. WL04 was taken after one week of recovery from WL03 to WL02, and the environmental conditions during sampling were the same as for WL02 (Figure 2). From each plant, 4 to 5 mature leaves were selected, previously chosen from the top. The leaves were promptly immersed in liquid nitrogen and preserved at −80 °C for subsequent analyses.

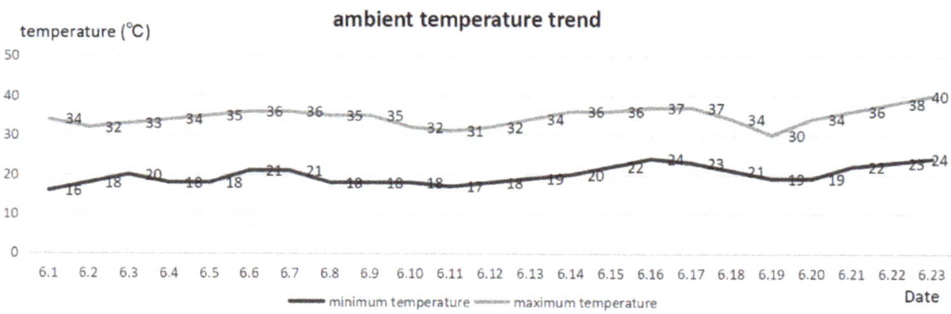


**Figure 1.** Monthly temperature variations up to the day of sampling, https://data.cma.cn (last accessed on 20 November 2023).

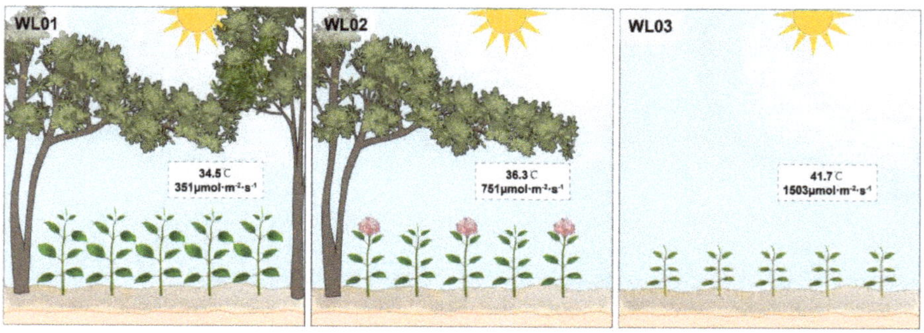


**Figure 2.** Experimental processing simulation diagram. The values inside the white square represent the average leaf surface temperature and light intensity during sampling.

### *2.2. RNA Extraction, cDNA Library Construction, and Sequencing*

Total RNA extraction was carried out using a Trizol reagent kit (Invitrogen, Carlsbad, CA, USA), following professional instructions. The extracted RNA quality was assessed using an Agilent 2100 Bioanalyzer (Agilent Technologies, Palo Alto, CA, USA) and verified through RNase-free agarose gel electrophoresis. Subsequently, eukaryotic mRNA was enriched utilizing Oligo(dT) beads. The enriched mRNA underwent fragmentation into short fragments using a corresponding buffer and was then reversely transcribed into cDNA using the NEBNext Ultra RNA Library Prep Kit for Illumina (NEB #7530, New England Biolabs, Ipswich, MA, USA). The purified double-stranded cDNA fragments underwent end repair, base addition, and ligation to Illumina sequencing adapters. The ligation reaction was purified using AMPure XP beads ($1.0\times$) and subjected to polymerase chain reaction (PCR) amplification. The resulting cDNA library was sequenced using Illumina Novaseq6000 by Gene Denovo Biotechnology Co. (Guangzhou, China).

### *2.3. Data Quality Control and Sequence Alignment Analysis*

For quality control of the raw reads obtained from the sequencing machine, the Fastp tool (version 0.18.0) was employed. Data with low quality, including adapters, unknown nucleotides(N) content exceeding 10%, and bases with a mass value $Q \leq 20$ constituting more than 50% of the entire read were eliminated, resulting in the generation of clean reads [8]. Then, the clean reads were assembled using Trinity to construct a unique consensus sequences for a subsequent information analysis [9].

### *2.4. Gene Expression Quantification and Differential Gene Analysis*

Initially, all unigene sequences were aligned against protein databases, specifically the NCBI Non-Redundant protein database, SwissProt protein database, KEGG pathway database, and KOG database, using blast x with an e-value threshold of less than 0.00001. After alignment, we obtained the highest similarity with a given unigene that contained functional annotation information [10]. Subsequently, all the gene expression levels in each sample were calculated to obtain the per kilobase per million mapped fragments (FPKM) value. EdgeR 4.0.2 software was employed for the analysis and normalization of read counts. *p*-value was computed based on the model, and multiple hypothesis testing correction was performed to derive the false discovery rate (FDR) value, indicating the error detection rate. Following the differential analysis results, genes with FDR < 0.05 and $|\log2FC| > 1$ were chosen as notably differentially expressed candidates. GO and KEGG analyses were then conducted on these differentially expressed genes (DEGs). GO analysis provided all the GO terms significantly enriched in DEGs compared to genome background and filtered the DEGs corresponding to specific biological functions. Pathway-based KEGG pathway analysis was performed to gain further insights into the gene biological functions.

### *2.5. Trend Analysis of Gene Expression Levels*

A heatmap was plotted by https://www.bioinformatics.com.cn (last accessed on 20 February 2024), an online platform for data analysis and visualization. Mfuzz algorithms were used to divide the transcriptome data into different clusters, and the cluster parameter was set to 5 based on the elbow inflection point value calculated by the default parameters. The networks were visualized using Cytoscape v3.9.1 [11].

### *2.6. Protein–Protein Interaction (PPI) Network Construction*

Differential gene–protein interaction networks were analyzed utilizing interaction relationships within the STRING Protein Interaction Database. A subset of DEGs for the included species was extracted from the database, and an interaction network diagram was constructed using Cytoscape 3.9.1. For species not covered in the database, an initial step involved applying blast x alignment to the protein sequences of the reference species included in the STRING database from the target gene set [12]. Subsequently, an interaction network was constructed using protein interaction relationships of reference species identified through the alignment process.

*2.7. Real-Time Quantitative PCR Verification*

RNA extraction from various treated leaf samples was conducted using the Aidlab Easy Spin Plus RNA Extraction Kit. Extracted RNA integrity was verified through agarose gel electrophoresis, and the concentration was determined using the NanoDrop 2000 spectrophotometer. Reverse transcription of the RNA into cDNA was accomplished using the Aidlab PC54-TRUEscript RT kit (+gDNA Eraser). qRT-PCR primers were designed by Primer Premier 5.0 (Table S1), and the TAKARA TB Green® Premix Ex TaqTM II kit was employed for qRT-PCR detection. The reaction system comprised each upstream and downstream primer of 0.4 μL (10 $\mu mol{\cdot}L^{-1}$), cDNA of 1.0 μL, $ddH_2O$ of 3.2 μL, and SYBR Premix Ex Taq of 5.0 μL constituting a total volume of 20 μL. The thermal cycling program involved an initial denaturation at 95 °C for 30 s, followed by 38 cycles of denaturation at 95 °C for 5 s, annealing at 60 °C for 30 s, and extension at 72 °C. The $2^{-\Delta\Delta CT}$ method was employed to normalize the DEG expressions [13], with *RPL34* serving as a reference gene [14]. Three biological replicates were performed.

## 3. Results

*3.1. Growth Status Analysis*

In treatment WL01, the growth of plants seemed to be regular. Compared to WL02, the leaves in WL01 exhibited a larger size while the degree of green color was similar to that in WL02, as illustrated in Figure 3. However, no flowering was observed. The leaves in WL02 were of intermediate size compared to WL01 and WL03 (Figure 3), and the plants maintained a similar height to those in WL01 but exhibited a flowering phenotype. Throughout the WL03 period, the leaves were smaller, prominently displaying yellowing and sunburn, as depicted in Figure 3, and the plants were notably shorter. WL04 had smaller yellow leaves but showed a recovery state compared to WL03.

**Figure 3.** Leaf phenotypes under four shading conditions, as observed under macrography. Scale bars: 2 cm.

*3.2. Sequencing Data Assembly Results*

In this study, transcriptome sequencing was conducted on four samples, yielding a total of 41.32 G of high-quality sequence reads. The GC content of the samples was above 45.07%, and the Q30 score was ≥92.14% (Table 1). A total of 151,939 transcript sequences (transcript number) and 54,935 non-redundant gene sequences (unigene number) were obtained with high assembly completeness. Post-assembly, the N50 length of the overlapping fragments was 1561 nt, with an average length of 1042 nt. Single-gene clusters within the length range of 300 to 500 nt constituted 36.1% of the assembly, totaling 19,834. The second-largest group, with lengths ranging from 500 to 1000 nt, accounted for 28.94% of the total. Sequences longer than 2000 nt constituted 13.33% of the total. The high-quality reads produced in this study were deposited in the NCBI SRA database (accession number: PRJNA1109561). To ensure the trustworthiness of the RNA-Seq data, we chose 12 differentially expressed genes (DEGs) for qRT-PCR analysis. The RT-qPCR findings revealed comparable expression patterns between the RNA-Seq and RT-qPCR data, thereby affirming the sequencing results' precision and reliability (Figure S1).

**Table 1.** Statistics of sequencing data quality.

| Sample | Raw Reads | Raw Bases | Clean Reads | Clean Bases | Q20 | Q30 | GC |
|---|---|---|---|---|---|---|---|
| WL01 | 77.23 M | 11.58 G | 76.77 M | 11.48 G | 97.56% | 93.33% | 45.14% |
| WL02 | 63.58 M | 9.54 G | 63.14 M | 9.44 G | 97.20% | 92.46% | 45.07% |
| WL03 | 71.22 M | 10.68 G | 70.77 M | 10.58 G | 97.47% | 93.15% | 45.27% |
| WL04 | 66.07 M | 9.91 G | 65.61 M | 98.17 G | 97.05% | 92.14% | 45.14% |

### *3.3. Functional Annotation and Expression Analysis of Transcriptome Unigenes*

The unigene sequences underwent comparison against the NR, Swissprot, KOG, and KEGG databases for functional annotation (Figure 4). The annotation results are as follows: 27,947 genes (58.48%) were annotated in the NR database, 21,138 genes (44.23%) in the Swissprot database, 5583 genes (11.68%) in the KEGG database, and 15,970 genes (33.41%) in the KOG database (Figure 4). The reads were aligned to the unigene sequences, achieving an alignment rate ranging from 88.84% to 89.73%.

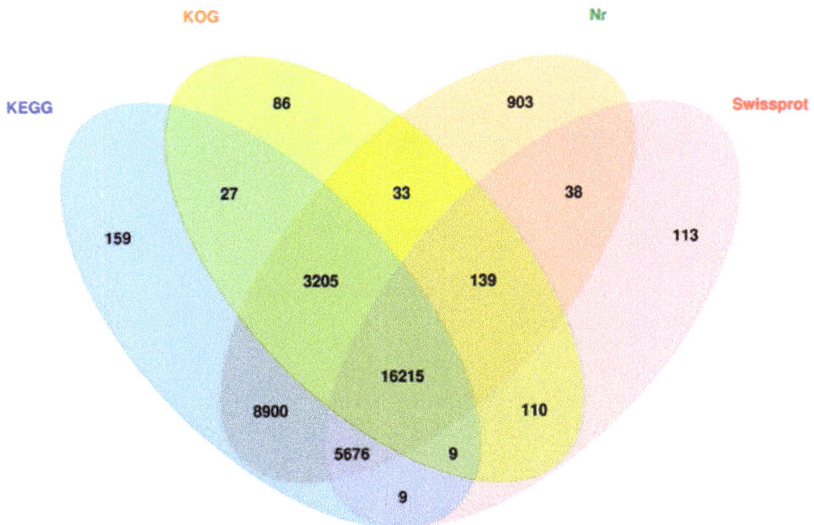


**Figure 4.** Four major databases annotated Venn diagrams.

### *3.4. Gene Ontology Classification of Differentially Expressed Genes*

In WL02, compared to WL01 (Figure 5a), 4271 transcripts were significantly upregulated and 3893 transcripts were significantly downregulated. We firmly established DEGs in the significantly enriched GO terminology, dividing them into three primary categories: molecular functions, cellular components, and biological processes. In the biological process category, the most abundant groups were "Protein phosphorylation" (331 genes), "Phosphorylation" (380 genes), and "Response to auxin" (71 genes). Within the molecular function category, "DNA-binding TF activity" (221 genes), "Protein kinase activity" (328 genes), and "Kinase activity" (398 genes) were the most highly represented groups. Regarding cellular components, "Cell periphery" (511 genes), "Plasma membrane" (680 genes), and "Intrinsic component of membrane" (653 genes) were the most represented groups. In the WL03 vs. WL02 comparison (Figure 5b), "macromolecular modification" (1394 genes), "RNA modification" (342 genes), and "nuclear acid photosensitive bond hydrogenation" (337 genes) were three significantly enriched entries in the biological process module. Within the molecular function category module, "Endonuclease activity" (475 genes), "nuclease activity" (544 genes), and "protein kinase activity" (534 genes) were three significantly enriched entries. However, there were no cellular component-related entries in the top 20 significantly enriched entries. For WL02 vs. WL04 (Figure 5c), cell wall-related terms were significantly enriched.

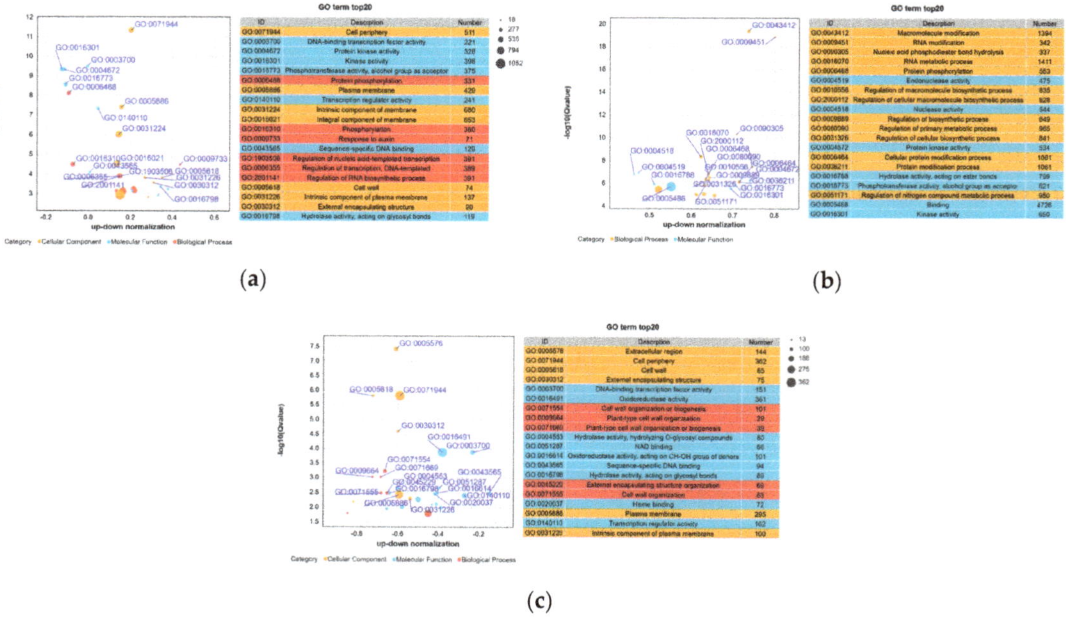

**Figure 5.** Bubble diagram of the top 20 biological process terms enriched in GO. (**a**) WL01 vs. WL02. (**b**) WL03 vs. WL02. (**c**) WL02 vs. WL04.

### 3.5. DEGs Pathway Analysis

Pathway analysis using the KEGG database was employed to identify the DEG-relevant pathways. In WL01 vs. WL02 (Figure 6a), the KEGG results revealed enrichment in secondary metabolites biosynthesis with 259 DEGs, comprising 166 upregulated and 93 downregulated biomarkers. Plant hormone signal transduction exhibited enrichment with 78 genes, of which 61 were upregulated and 17 downregulated. Starch and sucrose metabolism illustrated enrichment with 45 genes, including 34 upregulated and 11 downregulated genes. Phenylpropanoid biosynthesis exhibited enrichment with 36 genes, of which 25 were upregulated and 11 downregulated. In WL03 vs. WL02 (Figure 6b), secondary metabolites biosynthesis demonstrated enrichment with 498 DEGs, consisting of 386 upregulated and 112 downregulated genes. Plant hormone signal transduction exhibited enrichment with 112 DEGs, including 99 upregulated and 13 downregulated genes.

Plant–pathogen interaction showed enrichment with 108 genes, comprising 74 upregulated and 34 downregulated genes. Phenylpropanoid biosynthesis exhibited enrichment with 67 genes, including 55 upregulated and 12 downregulated genes. In WL02 vs. WL04 (Figure 6c), the metabolic pathway enriched up to 409 genes. Biosynthesis of secondary metabolites, plant hormone signal transduction, and phenylpropanoid biosynthesis were found to overlap in the three control groups, indicating that plant hormones and phenylpropanoid work crucially in the *H. macrophylla* process of adaptation to the environment. Additionally, starch and carbohydrates were identified as related to *H. macrophylla* flowering under semi-shaded conditions.

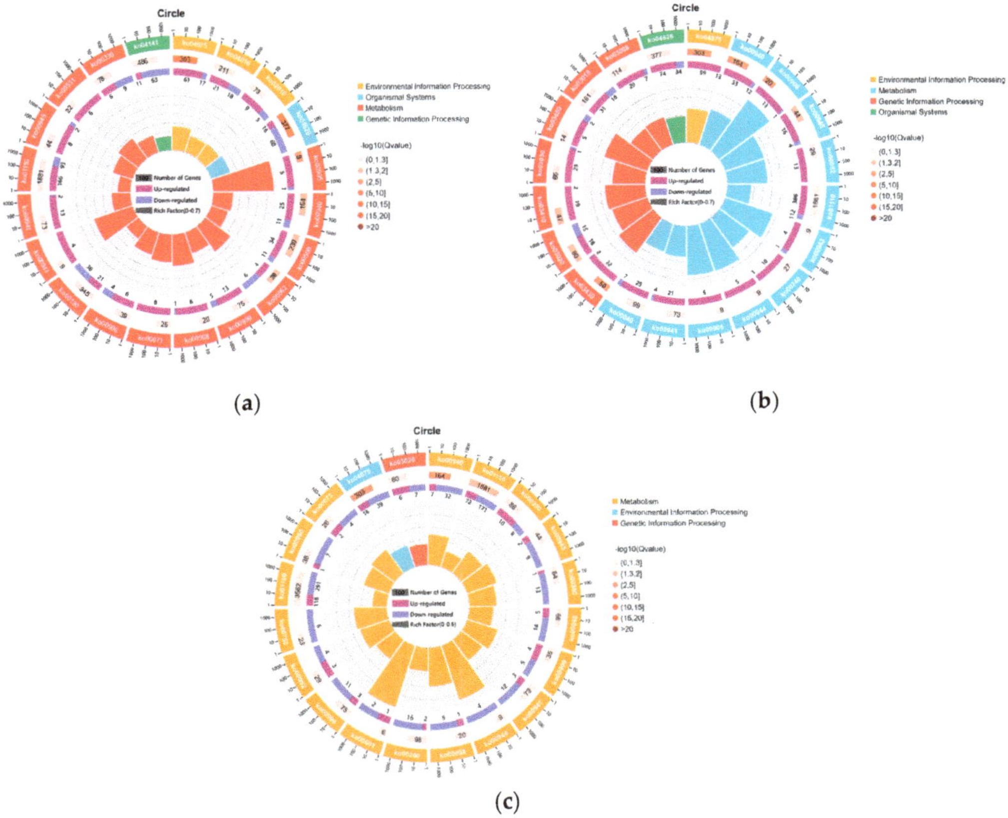


**Figure 6.** Bubble diagram of the top 20 pathways enriched in KEGG. (**a**) WL01 vs. WL02. (**b**) WL03 vs. WL02. (**c**) WL02 vs. WL04. The graphical representation commences with the outermost circle denoting the pathway IDs, where different colors serve to distinguish between various KEGG_A_classes. The second circle illustrates the count of genes from the background gene set that are enriched within each pathway ID, employing a color scheme that represents differing −log10(Qvalue), as elucidated in the legend. The third circle highlights the number of genes from the target gene set that have been enriched in the pathways, utilizing distinct colors to differentiate between upregulated and downregulated genes. The fourth circle presents the gene ratio, computed as the quotient of the number of genes enriched in the pathway from the target gene set divided by the equivalent count from the background gene set for that particular pathway.

### 3.6. Trend Analysis of Gene Expression Levels

By combining the DEGs involved in the process of environmental adaptation in *H. macrophylla* with GO functional enrichment and other genes that exhibit significant differences or high expression, this study identified 344 DEGs that are crucial for the growth and development (Table S2). The functional annotation and previous research findings were used to select these genes. Based on pairwise correlations of gene expression, five modules were identified (marked with different colors in Figure 7a). The most obvious feature of cluster 1 is the high expression level in WL03. The top enriched GO terms for this module were "response to light intensity", "response to reactive oxygen species", and "response to high light intensity". The top three enriched KEGG pathways were "Plant-pathogen interaction", "Phenylpropanoid biosynthesis", and "MAPK signaling pathway—plant". The high expression of these genes may be related to high temperature and high light stress, but it is not conducive to plant growth. The expression levels of most genes in cluster 2 were highest in WL01. The top enriched GO terms were "DNA-binding transcription factor activity", "transcription regulator activity", and "sequence-specific DNA binding", and the top KEGG pathways were "Plant-pathogen interaction", "Phenylpropanoid biosynthesis", and "MAPK signaling pathway—plant". Cluster 3 was associated with the WL02. The expression profile of these DEGs peaked during the WL02 stage. The most abundant GO term indicated that the flavonoids and hormones, especially auxin, played a critical role in controlling the lowering and the good growth of plants. The expression level of cluster 4 was high in WL02 and WL04. The most representative GO terms and KEGG pathways were "response to hormone" and "Plant hormone signal transduction", indicating that the hormone was closely associated with recovery from adversity and maintaining growth. In cluster 5, most of the DEGs maintained high expression levels throughout WL01 and WL02, but significantly decreased in WL03. The expression of these genes may be beneficial for the normal growth of plants (Figures S2 and S3).

The top 20 hub genes were identified by the cytohubba plug-in using the MCC algorithm in Cytoscape 3.9.1 software. In the top 20 hubba genes of cluster 1 (Figure 7b), several genes closely associated with heat tolerance were identified, including heat stress transcription factor (*HSF30, HSFA6b*), heat shock protein (*HSP70-1, HSP70-3, HSP22, HSP83A, HOP*), BAG family molecular chaperone regulator (*BAG5*), along with dehydration-responsive element-binding protein (*DREB2C*). Meanwhile, in the top 20 hubba genes of cluster 2 (Figure 7c), numerous transcription factors were found, including WRKY family members (*WRKY70-1, WRKY70-2, WRKY40, WRKY24, WRKY11*), ERF transcription factor family members (*ERFR118-1, ERF118-2, ERF010, CRF4*), NAC transcription factor family member (*NAC022*), and RAV transcription factor family member (*RAV1*). These transcription factors possess intricate functions, and their high expression may potentially impact the reproductive growth of Hydrangea macrophylla. In cluster 3 (Figure 7d), the hub genes have multiple functions, including *TCP20*, which promotes auxin synthesis [15], *SAUR50*, an auxin response gene [16]; auxin transporter *PIN3* [17]; *PHOTO1*, a photoreceptor; *SCL7*, enhancing resistance to abiotic stress [18]; photosensitive signal transduction and photosynthesis-related *PHYB, PSBY* [19,20], ROS clear genes *PEX5* [21], phenylpropane metabolism, *PAL, PER63* [22,23]; and *YAB2*, which promotes plant growth [24]. The high expression of these genes promotes the growth and development of hydrangea.

### 3.7. PPI Analysis

Biological processes are orchestrated by a complex interplay of proteins and their interactions, often depicted as PPI networks [25]. These PPI networks are constructed using data derived from both wet-lab experiments and computational techniques, with repositories like DIP, STRING, and BioGRID serving as databases. STRING, for instance, amalgamates PPIs from diverse sources, including experimental and computational methods, assigning a comprehensive quality score to each interaction by integrating data from the literature and gene expression profiles [26]. The identification of key protein players and the elucidation of their interaction networks offer invaluable insights into the regulation of plant developmental processes and the intricate interactions between plants and their environment.

Utilizing the differentially expressed genes, 176 genes were chosen for the construction of a PPI network (Table S3). These genes were selected based on their involvement in pathways related to photoperiod, hormone signal transduction, environmental temperature, ROS, flavonoids, and sugar, which are all crucial for the adaptation, nutritional support, and reproductive growth of hydrangea. In the PPI network for WL01 vs. WL02 (Figure 8a), which encompasses 75 nodes and 289 edges, several genes associated with auxin and flavonoid biosynthesis exhibited high connectivity. The top 10 core genes, identified by the degree algorithm, were linked to various biological processes such as flavonoid biosynthesis (*CHS*, *CYP75B2*), porphyrin metabolism (*DVR*, *CHLI*), photosynthesis (VDE1, FDC1), starch and sucrose metabolism (*AGPS1*, *BAM1*), MAPK signaling (*NTF3*), and IAA signaling (*LAX1*, *ABCB1*).

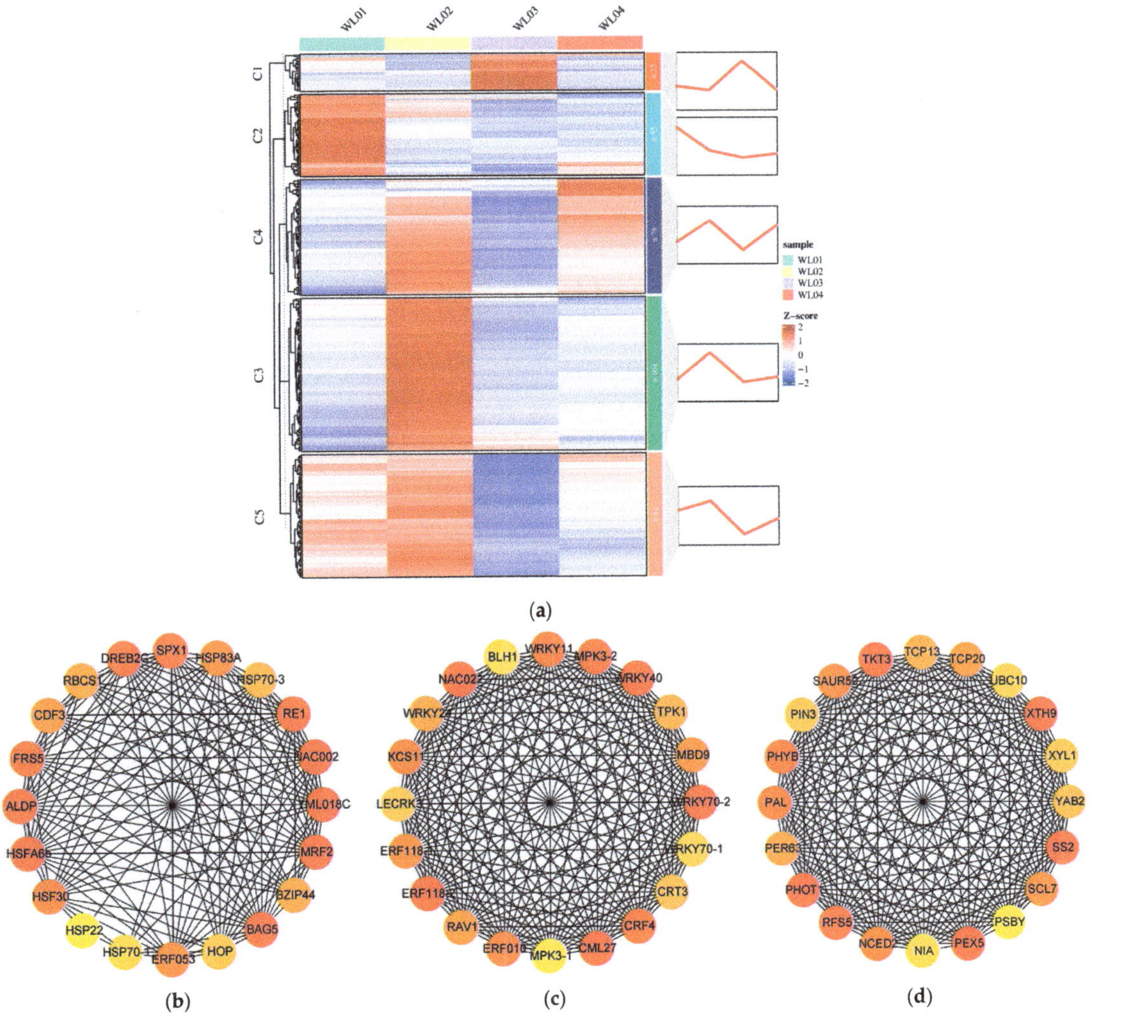


**Figure 7.** Identification of gene clusters using DEGs. (**a**) A heatmap of candidate DEGs' expression trend during natural stress treatment. (**b**) The top 20 genes calculated in cluster 1 using the cytohubba plug-in Cytoscape. The node colors correspond to connectivity values. (**c**) The top 20 genes in cluster 2. (**d**) The top 20 genes in cluster 3.

In the PPI network for WL03 vs. WL02 (Figure 8b), which consists of 123 nodes and 703 edges, the 10 core genes are associated with biological processes, including flavonoid biosynthesis (*CHS*), ROS metabolism (*CAT1*), ethylene and jasmonic acid hormone signal transduction (*EIN2*, *ETR1*, *EIN3*, *OPR3*, *CULR3*, *AOS1*), and photosynthesis (*PHYB*, *GAAPC2*). The results of the PPI network analysis underscore the intricate connections of flavonoid biosynthesis, hormone signal transduction, and photosynthesis with the normal growth and development of hydrangea in semi-shaded environments. The PPI result files are shown in Tables S4 and S5.

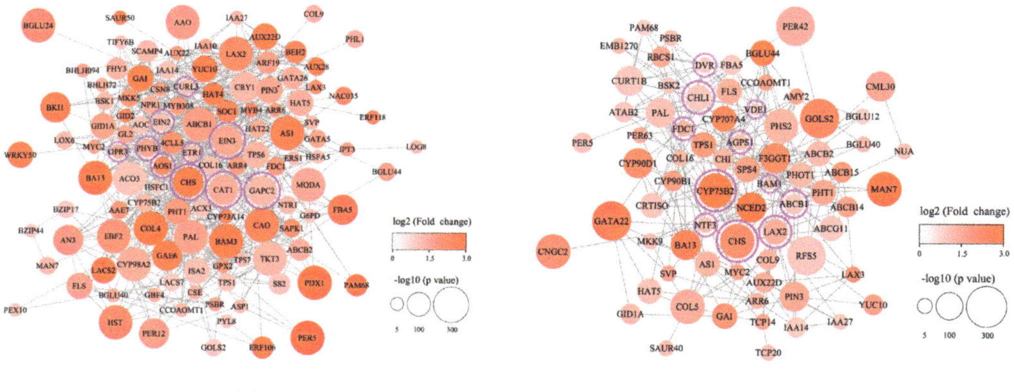


(**a**) (**b**)

**Figure 8.** Protein–protein interaction network of differentially expressed genes. (**a**) WL01 vs. WL02 network, (**b**) WL03 vs. WL02 network. The genes circled in purple are the top ten genes calculated using the degree algorithm.

## 4. Discussion

The normal growth and development of hydrangea under semi-shaded conditions were maintained through complex interactions. Through a comparative analysis of the transcriptome during the WL02 period, marked by robust growth under semi-shaded conditions, with the non-flowering WL01 period under full shade and the less thriving WL03 period under full sunlight, several genes were pinpointed for their pivotal role in the normal growth and flowering of *hydrangea* in semi-shaded environments.

### *4.1. Phenylpropanoid Metabolism in Response to Semi-Shaded Conditions*

Phenylpropanoids are essential bioactive secondary metabolites in plants; they are synthesized from the vital amino acid phenylalanine through PAL enzymatic action (phenylalanine ammonia-lyase) within the shikimate pathway [27,28]. In this study, the phenylpropanoid biosynthesis pathway was found to notably enriched in pivotal KEGG pathways in both comparison pairs. Specifically, several key genes exhibited substantial upregulation in the WL02 group, including *PAL* (Phenylalanine ammonia lyase gene), *CHI* (Chalcone isomerase gene), *F3′H* (Flavanone 3′-hydroxylase gene), *CCOAOMT* (Caffeoyl-CoA O-methyltransferase gene), *PER5* (peroxidase 5-like gene), and *PER42* (peroxidase 42-like gene), exhibited substantial upregulation in WL02. Furthermore, phenylpropanoid metabolism and flavonoid biosynthesis biomarkers displayed more pronounced upregulation in WL03 vs. WL02, such as *CHS* (Chalcone synthase), *CHS4* (coumaroyl triacetic acid synthase), *CYP75B2* (flavonoid-3′-hydroxylase), and *CYP98A2* (cytochrome P450 98A2) [29]. Previous research has demonstrated that flavonoid structural biomarker (e.g., *CHS* and *DFR*) overexpression upregulates flavonol glycosides and anthocyanins levels, preventing ROS accumulation, thus improving tolerance to salt stress in plants like rice (*Oryza sativa*), *Brassica napus* L. 'Hanla', [27] suggesting a crucial role for flavonoids in *H. macrophylla*'s adaptation to semi-shaded conditions.

### 4.2. Photosynthesis and Chlorophyll-Related Genes under Semi-Shaded Conditions

Shading induces reductions in light intensity and alters light quality by modifying the red-to-far-red light ratio. Conversely, high light intensity does not alter light quality but intensifies overall illumination. Noteworthy changes were identified in photosynthesis-related genes in both comparison pairs, featuring distinctive variations in specific genes. In WL01 vs. WL02, the genes associated with chlorophyll synthesis, including *CHLI* (Magnesium chelatase i2 gene), *PSBR* (Photosystem II 10 kDa polypeptide gene), and *DVR* (Divinyl reductase gene), displayed significant upregulation [30]. *PAM68* (Photosynthesis-affected MUTANT68 gene), a photosystem II assembly factor facilitating chlorophyll molecule insertion into the CP47 polypeptide chain [31,32], exhibited an upregulation trend in both comparison pairs. In WL03 vs. WL02, alongside the upregulation of *FDC1* and *PSBR* upregulation, the genes linked to carbon fixation in photosynthesis-related genes, such as *FBA5* (Fructose-bisphosphate aldolase gene), *GAPC2* (Glyceraldehyde-3-phosphate dehydrogenase gene), and *MODA* (NADP-dependent malic enzyme gene), were also upregulated. This supported the importance of photosynthesis in the hydrangea's adaption to semi-shaded environments. Simultaneously, the activation of the response pathway mediated by *SPX1/PHR1* may be related to lower photosynthetic efficiency, potentially influencing the photosynthesis of WL03, due to its high expression level in WL03 [33]. Prior research has indicated that carotenoids present in green leaves play a critical role in facilitating efficient photosynthesis, scavenging diverse reactive oxygen species and safeguarding chlorophyll from photooxidation [34]. In this study, the *VDE*, *CYP707A*, *CRTISO*, and *NCED2* genes related to carotenoid biosynthesis were upregulated in WL01 vs. WL02. This upregulation is indicative of an enhancement in the light-harvesting molecules' efficiency within *H. macrophylla* leaves under semi-shaded conditions, contributing to increased photosynthetic efficacy. Consequently, *H. macrophylla* may elevate gene expressions related to photosystem functionality as an adaptive mechanism for growth under semi-shaded conditions. Additionally, prior research has demonstrated that *COL16* participates in chlorophyll accumulation in morning glory, where elevated *phCOL16* expression correlates with increased chlorophyll levels in the corolla, positively regulating chlorophyll biosynthesis. This study shows that *COL16*'s expression level is significantly increased in WL02. This suggests that *COL16* may also play an important role in the regulation of chlorophyll biosynthesis.

### 4.3. Carbohydrate Metabolism in Response to Semi-Shaded Conditions

Carbohydrates serve crucial functions in plant growth and development, with sugars serving multiple functions in flowering and stress resistance. They act as key floral signals initiating floral induction and participating in non-biological stress response mechanisms simultaneously [35]. BGLUs (Beta-glucosidase), hydrolytic enzyme class members, are involved in glycosidic substance hydrolysis and glycosidic bonds of oligosaccharides, releasing non-reducing glucose residues. This process is implicated in various biological phenomena [36], including stress response phytohormone activation and alpha-hydroxy nitriles release to protect against stresses. In Arabidopsis, *AtBGLU10* has been identified to catalyze free ABA production, thereby enhancing the plant's resistance to drought and salt stress. BGLU genes have also been found to regulate tolerance under dark stress in *Stevia rebaudiana* [37]. In our study, several BGLU family members, including *BGLU20/40/44*, were enriched in both comparative cohorts. These genes likely serve crucial functions in the normal growth and development of *H. macrophylla* under semi-shaded conditions. Beta-amylase (BAM) 1 and *BAM3* belong to the gene family responsible for encoding beta-amylase in plants. These enzymes play a crucial role in plant starch metabolism and energy distribution, influencing key physiological processes such as seed germination and carbon metabolism in the source–sink balance [38]. Consequently, they contribute to plant growth and development modulations. In Arabidopsis, *BAM1* is responsible for breaking down transitory starch to support proline biosynthesis under drought stress conditions [39]. Meanwhile, in citrus, *BAM3* facilitates starch degradation, leading to soluble sugar content increase and improved resistance to abiotic stress [40]. *BAM1* has a notable expression in the WL01 vs. WL02 comparison and *BAM3* in the WL03 vs. WL02 comparison, which emphasizes their crucial roles in these specific conditions.

*4.4. Key Transcription Factors for Normal Growth in Semi-Shade Conditions*

TFs, or transcription factors, act as regulatory proteins positioned at the signal transduction pathway terminus [41]. They function as switches that control the expression of downstream stress response genes. TFs bind to cis-acting elements in the target genes' promotors, thus orchestrating gene expression and sustaining the normal growth and hydrangea development under semi-shaded conditions. The findings suggest that several TF families, including AR2/ERF (75), bHLH (69), ERF (75), and NAC (56), likely play pivotal roles in governing gene expression through hormone signal transduction and ROS signal networks in semi-shaded environments.

The NAC (NAM, no apical meristem, ATAF, and CUC) family stands as one of the largest gene families among plant-specific TFs [42]. NAC TFs work crucially in various plant growth and developmental processes, demonstrating particular significance in bolstering plant resistance against a spectrum of abiotic stresses. NAC TFs displayed distinct variations between the two comparison groups: the majority experienced downregulation in WL01 vs. WL02, while an upregulation trend was observed in WL03 vs. WL02. This suggests noteworthy distinctions in the NAC TFs' expression dynamics under different environmental conditions. Notably, both *NAC002* and *NAC014* exhibited significant downregulation in both control groups. Previous studies have demonstrated that *LlNAC014* overexpression in *Lilium longiflorum* and Arabidopsis enhances heat tolerance but concurrently induces growth defects, aligning with the findings in our study. Although *NAC014* can resist unsuitable environments, it inhibits plant growth, and the specific regulatory mechanism deserves further investigation [43].

BHLH TFs play a pivotal role in regulating various synthetic metabolism and signal transduction processes in plants, influencing crucial aspects of growth and development [44]. These processes include, but are not limited to, seed germination, photomorphogenesis, flowering, leaf senescence, and cell apoptosis. A significant portion of the bHLH TF family members exhibited upregulation in both comparison groups, with up to 95% of bHLH TFs showing upregulation in WL03 vs. WL02. The pronounced upregulation across expression levels underscores its pivotal role. Notably, *BHLH79*, *BHLH51*, *FAMA*, and *GL3* displayed significant upregulation in both comparison groups. Previous studies have associated *BHLH51* and *GL3* with anthocyanin accumulation, while *FAMA* has been implicated in stomatal development [45–47]. Furthermore, *CabHLH79* has been identified as a *CaNAC035* upstream regulator, contributing to cold stress modulation in pepper [48]. These findings underscore the intricate and crucial role of BHLH TFs in the hydrangea's adaptation to semi-shaded environments.

Among the identified flowering-related TFs in this investigation, *SOC1*, characterized as a MADS box TF, serves as an integrator of diverse flowering signals originating from photoperiod, temperature, hormones, and age-related cues [49]. *CONSTANS* (*CO*), functioning as a floral activator, is involved in activating *SOC1*. Additionally, the *SOC1* gene contributes to the developmental process modulation in nutritional organs like leaves and stems in Arabidopsis [50]. *CO* gene family members, strategically positioned between biological clock and flowering integrators, exhibit diverse effects on the induction of plant flowering [51]. *COL3* overexpression in Arabidopsis has been linked to delayed flowering [52], while *RcCO/RcCOL4* expression suppression results in delayed flowering in roses under both short and long photoperiods [53]. Additionally, *AtCOL4* expression is robustly induced by ABA, salt, and osmotic stress. *AtCOL4* plays a pivotal role in regulating plant resistance to abiotic stress, as mutations in atcol4 lead to increased sensitivity to ABA and salt stress during seed germination and cotyledon greening. This underscores *AtCOL4*'s crucial regulatory function in plant responses to abiotic stress conditions [54]. *SiCOL5* has been identified as a positive flowering time modulator [55]. Genes such as *AGL24*, *SOC1*, *COL4*, *COL5*, and *COL9* presently exhibit heightened expression levels in WL02, indicating their significant roles in promoting hydrangea flowering in semi-shaded environments through intricate interactions. In cluster 1, *WRKY70* highlights its central role. Research indicates that *GATA5* and *WRKY70* emerge as crucial candidate genes for soybean pod abscission under shaded conditions [56]. The transcription levels of the flowering time integration gene *FT* and the floral meristem identity genes *APETALA1* (*AP1*) and *LEAFY* (*LFY*) were lower in WRKY7-OE compared to WT [57]. Overexpression of the *BLH1*-like gene *MdBLH14* in Arabidopsis leads to significant dwarfing and delayed flowering phenotypes by inhibiting the accumulation of active GA [58]. The *PHYTOCHROME A SIGNAL TRANSDUCTION 1* (*PAT1*) subfamily is a branch of the GRAS family, and *PAT1* inhibits phyA signal transduction, while participating in the shade avoidance response [59,60]. The interaction between *PAT1* and *CONSTANS-LIKE 13* (*COL13*) negatively regulates flowering [61]. Given their high expression during the WL01 phase, *WRKY70*, *WRKY7*, *BLH1*, and *PAT1* may exert inhibitory effects on the reproductive growth of hydrangea in shaded environments. Overexpression of *DREB2C* in Arabidopsis delays flowering by activating *FLOWERING LOCUS C* (*FLC*) [62], while *FAR1* negatively regulates flowering time by regulating the transcription of multiple genes [63]. The high expression of *DREB2C* and *FRS5/FAR1* in WL03 suggests their potential impact on the reproductive growth of hydrangea under full light conditions.

### *4.5. Plant Hormone and Signal Transduction Play Significant Roles in Hydrangea's Adaptation to Semi-Shaded Conditions*

Plant hormones are essential signaling molecules that play a crucial role in governing various aspects of plant growth, development, and responses to environmental stimuli. They act at multiple levels to mediate the intricate processes that enable plants to respond and adapt to their surroundings [64].

Ethylene is a pivotal plant hormone that governs various physiological processes, including seed germination, root formation, flower differentiation, leaf senescence, and fruit development. Additionally, it functions crucially in responding to abiotic stress. Among plant hormones, the ethylene signaling pathway is well studied. The final step of ethylene biosynthesis is catalyzed by 1-aminocyclopropane-1-carboxylate oxidase 1 (*ACO*) [65]. In WL02, high *ACO* expression facilitates ethylene synthesis. Ethylene perception in Arabidopsis involves five genes, including *ETR1* and *ERS1* [66]. The binding of ethylene to its receptor proteins renders the negative regulator *CTR1* inactive, leading to *EIN2* expression activation [67]. *EIN2*, the ethylene signaling pathway's central component, positively regulates ethylene signal transduction. Upon activation, *EIN2* promotes *EIN3/EIL1* nuclear accumulation; these are crucial nuclear TFs in the ethylene pathway. *EIN3* participates in flower opening, aging [68], and stress resistance modulations [69,70]. It also activates key enzyme gene expressions involved in secondary metabolite biosynthesis [71]. In loquat and kiwifruit, *EIN3/EIL* genes are involved in the fruit cold stress response [72,73]. Notably, *EIN3* significantly regulates numerous photosynthesis-related functional gene expressions, enhancing the light adaptation ability of seedlings after emergence [74]. In WL03 vs. WL02, the high *ETR1*, *ERS1*, *EIN2* expressions and *EIN3* underscore the critical role of ethylene signal transduction in maintaining the hydrangea's normal growth.

GAs are pivotal plant growth hormones that are crucial for stem elongation, seed germination, leaf expansion, and flower development [75]. *GAI*, a DELLA family member, inhibits downstream GA response genes but also plays a positive role in abiotic stress tolerance. DELLA proteins accumulate under high light conditions, enhancing plant survival in salt stress [76]. *GAI* negatively regulates flowering by inhibiting GA signal transduction and interfering with CO protein transcription [77]. However, GA counters this by activating the signaling pathway through *GID1* receptor binding, leading to DELLA protein degradation [78]. *GAI* and *GID1A* showed an increasing trend in this research, prompting further investigation into their impact on the experimental treatments within a dynamic regulatory network. Understanding their roles in semi-shaded environments could unveil valuable insights into hydrangea adaptation mechanisms.

Jasmonates (JAs), crucial oxygenated lipid derivatives (oxylipins), serve as essential plant hormones for growth and environmental adaptation. Genes involved in JA biosynthesis (*LOX6, AOS1, OPR3, AOC*) were upregulated in WL02. The MYC-type bHLH TF family, which is pivotal in the jasmonate response pathway, works crucially in regulating plant resistance to adverse conditions [79]. *MYC2* exhibits both positive and negative regulatory effects as a key member. It can inhibit *CAT2* expression and act as a negative proline biosynthesis regulator, affecting ROS levels and reducing plant resistance to abiotic stress [80,81]. Simultaneously, *MYC2* induces flowering-related gene expressions, like LFY and FT, promoting female flower development [82]. *MYC2* also engages in crosstalk with other hormones, physically interacting with EIN3 and antagonistically regulating wounding-responsive gene expression [83,84]. *MYC2* displayed an increasing trend in both comparison groups, suggesting its potential benefits for flowering but not for stress resistance. JA's complex effects on hydrangea growth and development in semi-shaded environments involve intricate interactions, including both positive and negative feedback loops.

Auxin, which is a pivotal player in plant growth, development, and environmental adaptation, exerts its influence through various genes. The *YUC* gene family, which is integral to auxin synthesis and plant development, and the *PIN-FORMED* (*PIN*) family, comprising key auxin efflux carriers, the *AUXIN1/LIKE-AUX1* (*AUX/LAX*) proteins, serve as prominent auxin influx carriers and are crucial components [85]. This study identified upregulations of auxin synthesis-related gene *YUC10* and auxin transport-related genes *PIN3, LAX2, LAX3,* and *ABCB1* in WL02. The elevated auxin-responsive protein genes *IAA27, IAA14,* and small auxin upregulated RNA gene (*SAUR40*) expressions further underscores auxin's importance in this context. These findings suggest that auxin plays a regulatory role in hydrangea growth processes, with potential implications for its adaptation to semi-shaded environments. Moreover, the intricate crosstalk between auxin and other hormones emerges as a critical factor in shaping plant responses.

In conclusion, this study proposes a gene regulatory network model elucidating the mechanisms underlying the hydrangea's normal growth and development in semi-shaded conditions (Figure 9). The model begins with environmental signal reception by photoreceptors *PHYB* and *CRY1*, which subsequently transmit these signals downstream. *COLs* (*COL4/5/16*) are then regulated by *PHYB* and *CRY1*, positively influencing *SOC1* expression and, consequently, flowering. Genes involved in sucrose and starch metabolism, including *BGLUs* and *TPSs*, also contribute to *SOC1* expression. *EIN3* positively modulates hydrangea growth and development by inhibiting *MYC2* expression. IAA, facilitated by the transport protein genes *ABCB1, PIN3, LAX2,* and *LAX3*, moves within cells to suppress *GAI* expression, promoting the gibberellin pathway and supporting normal plant growth. BR (*CURL3/BR11, BES1/BZR1*) and JA (*LOX6, AOS1, OPR3, AOC, TIF6B*) also regulate growth through their respective biosynthesis, metabolism, and signal transduction genes. Flavonoids contribute positively to these processes. *WRKY70, BLH1, GATA5, PAT1,* and other genes may exert inhibitory effects on the reproductive growth of hydrangea in a full shading environment, while *NAC014, DREB2C, SPX1, FRS5/FAR1* and other genes may induce leaf senescence, reduce photosynthetic efficiency, and inhibit flowering and cause other adverse growth behaviors of hydrangea in a full light environment. However, it is crucial to note that the plant's regulatory network in response to the natural environment is intricate and speculative. While this study postulates the functions and interrelationships of genes, further experimental exploration is necessary to validate and enhance our understanding, considering the potential disparities between mRNA and protein levels and enzyme activity.

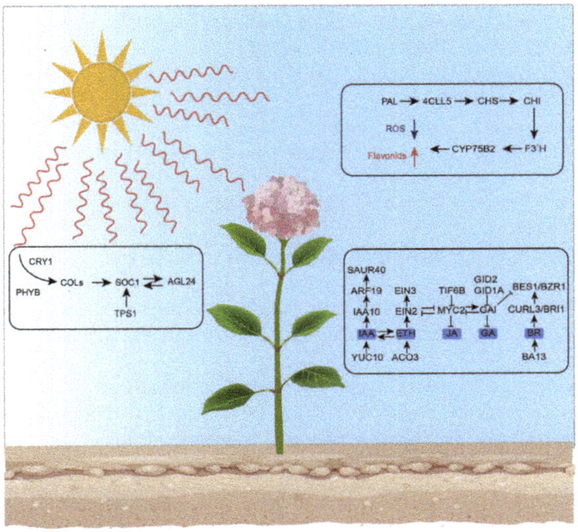


**Figure 9.** The hypothetical model of gene regulatory network for growth under semi-shaded environments in *Hydrangea macrophylla*. The black arrows indicate promotion, the T-shaped arrows indicate inhibition, the blue arrow indicates decrease and the red arrow indicates increase.

**Supplementary Materials:** The following supporting information can be downloaded at: https://www.mdpi.com/article/10.3390/horticulturae10060586/s1.

**Author Contributions:** Z.L., T.L. and Y.L. conceptualized and designed the experiment; Z.L. and T.L. performed the laboratory experiments and statistical analysis; Z.L., T.L. and Y.L. wrote and revised the manuscript; T.L and Y.L. funding acquisition and supervision. All authors have read and agreed to the published version of the manuscript.

**Funding:** This work was supported by the Project supported by China National Botanical Garden North Garden (grant No. BZ202405), Beijing Municipal Park Management Center (grant No. 2019-ZX-13), and the China National Natural Science Foundation (grant No. 31872138).

**Data Availability Statement:** Data are contained within the article.

**Conflicts of Interest:** The authors declare no conflicts of interest.

## References

1. Peng, J.; Dong, X.; Xue, C.; Liu, Z.; Cao, F. Exploring the molecular mechanism of blue flower color formation in *Hydrangea macrophylla* cv. "Forever Summer". *Front. Plant Sci.* **2021**, *12*, 585665. [CrossRef]
2. Rahmati, R.; Hamid, R.; Ghorbanzadeh, Z.; Jacob, F.; Azadi, P.; Zeinalabedini, M.; Farsad, L.K.; Kazemi, M.; Ebrahimi, M.A.; Shahinnia, F.; et al. Comparative transcriptome analysis unveils the molecular mechanism underlying sepal colour changes under acidic pH substratum in *Hydrangea macrophylla*. *Int. J. Mol. Sci.* **2022**, *23*, 15428. [CrossRef]
3. Xu, H.; Liu, C.; Zhong, H.D. The effect of different light intensities on the blooming of *Hygrangea macrophylla*. *Nothern Hortic.* **2014**, *1*, 81–82.
4. Li, X.; Liang, T.; Liu, H. How plants coordinate their development in response to light and temperature signals. *Plant Cell* **2021**, *34*, 955–966. [CrossRef]
5. Chen, Z.L.; Galli, M.; Gallavotti, A. Mechanisms of temperature-regulated growth and thermotolerance in crop species. *Curr. Opin. Plant Biol.* **2022**, *65*, 102134. [CrossRef]
6. Gururani, M.A.; Venkatesh, J.; Tran, L.S.P. Regulation of photosynthesis during abiotic stress-induced photoinhibition. *Mol. Plant* **2015**, *8*, 1304–1320. [CrossRef]
7. Saibo, N.J.M.; Lourenco, T.; Oliveira, M.M. Transcription factors and regulation of photosynthetic and related metabolism under environmental stresses. *Ann. Bot.* **2009**, *103*, 609–623. [CrossRef]
8. Chen, S.F.; Zhou, Y.Q.; Chen, Y.R.; Gu, J. Fastp: An ultra-fast all-in-one FASTQ preprocessor. *Bioinformatics* **2018**, *34*, i884–i890. [CrossRef]
9. Langmead, B.; Salzberg, S.L. Fast gapped-read alignment with Bowtie 2. *Nat. Methods* **2012**, *9*, 357–359. [CrossRef]

10. Pertea, M.; Pertea, G.M.; Antonescu, C.M.; Chang, T.C.; Mendell, J.T.; Salzberg, S.L. String Tie enables improved reconstruction of a transcriptome from RNA-seq reads. *Nat. Biotechnol.* **2015**, *33*, 290–295. [CrossRef]
11. Tang, D.D.; Chen, M.J.; Huang, X.H.; Zhang, G.C.; Zeng, L.; Zhang, G.S.; Wu, S.J.; Wang, Y.W. SR plot: A free online platform for data visualization and graphing. *PLoS ONE* **2023**, *18*, e0294236. [CrossRef]
12. Szklarczyk, D.; Franceschini, A.; Wyder, S.; Forslund, K.; Heller, D.; Huerta-Cepas, J.; Simonovic, M.; Roth, A.; Santos, A.; Tsafou, K.P.; et al. STRING v10: Protein-protein interaction networks, integrated over the tree of life. *Nucleic Acids Res.* **2015**, *43*, D447–D452. [CrossRef]
13. Livak, K.J.; Schmittgen, T.D. Analysis of relative gene expression data using real-time quantitative PCR and the 2(-delta delta C(T)) method. *Methods* **2001**, *25*, 402–408. [CrossRef]
14. Liu, Y.; Lyu, T.; Lyu, Y. Study on the flower induction mechanism of *Hydrangea macrophylla*. *Int. J. Mol. Sci.* **2023**, *24*, 7691. [CrossRef]
15. Krouk, G.; Lacombe, B.; Bielach, A.; Perrine-Walker, F.; Malinska, K.; Mounier, E.; Hoyerova, K.; Tillard, P.; Leon, S.; Ljung, K.; et al. Nitrate-regulated auxin transport by NRT1.1 defines a mechanism for nutrient sensing in plants. *Dev. Cell.* **2010**, *18*, 927–937. [CrossRef]
16. Bao, Y.; Chen, C.; Fu, L.; Chen, Y.Q. Comparative transcriptome analysis of *Rosa chinensis* 'Old Blush' provides insights into the crucial factors and signaling pathways in salt stress response. *Agron. J.* **2021**, *113*, 3031–3050. [CrossRef]
17. Su, N.N.; Zhu, A.Q.; Tao, X.; Ding, Z.J.; Chang, S.H.; Ye, F.; Zhang, Y.; Zhao, C.; Chen, Q.; Wang, J.Q.; et al. Structures and mechanisms of the *Arabidopsis* auxin transporter PIN3. *Nature* **2022**, *610*, 616–621. [CrossRef]
18. Ma, H.S.; Liang, D.; Shuai, P.; Xia, X.L.; Yin, W.L. The salt- and drought-inducible poplar GRAS protein SCL7 confers salt and drought tolerance in *Arabidopsis thaliana*. *J. Exp. Bot.* **2010**, *61*, 4011–4019. [CrossRef]
19. Yuan, L.; Zheng, Y.; Nie, L.; Zhang, L.; Wu, Y.; Zhu, S.; Hou, J.; Shan, G.L.; Liu, T.K.; Chen, G.; et al. Transcriptional profiling reveals changes in gene regulation and signaling transduction pathways during temperature stress in wucai (*Brassica campestris* L.). *BMC Genom.* **2021**, *22*, 687. [CrossRef]
20. Jeon, J.; Rahman, M.M.; Yang, H.W.; Kim, J.; Gam, H.J.; Song, J.Y.; Jeong, S.W.; Kim, J.I.; Choi, M.G.; Shin, D.H.; et al. Modulation of warm temperature-sensitive growth using a phytochrome B dark reversion variant, phyB[G515E], in Arabidopsis and rice. *J. Adv. Res.* 2023; *in press*. [CrossRef]
21. Khan, B.R.; Zolman, B.K. 5 Mutants that differentially disrupt pts1 and pts2 peroxisomal matrix protein import in Arabidopsis. *Plant Physiol.* **2010**, *154*, 1602–1615. [CrossRef]
22. Jie, H.D.; He, P.L.; Zhao, L.; Ma, Y.S.; Jie, Y.C. Molecular mechanisms regulating phenylpropanoid metabolism in exogenously-sprayed ethylene forage ramie based on transcriptomic and metabolomic analyses. *Plants* **2023**, *12*, 3899. [CrossRef]
23. Dong, C.J.; Cao, N.; Zhang, Z.G.; Shang, Q.M. *Phenylalanine ammonia-lyase* gene families in cucurbit species: Structure, evolution, and expression. *J. Integr. Agr.* **2016**, *15*, 1239–1255. [CrossRef]
24. Wang, Q.Q.; Li, Y.Y.; Chen, J.T.; Zhu, M.J.; Liu, X.D.; Zhou, Z.; Zhang, D.Y.; Liu, Z.J.; Lan, S.R. Genome-wide identification of YABBY genes in three *Cymbidium* species and expression patterns in *C. ensifolium* (Orchidaceae). *Front. Plant Sci.* **2022**, *13*, 995734. [CrossRef]
25. Wimalagunasekara, S.S.; Weeraman, J.W.J.K.; Tirimanne, S.; Fernando, P.C. Protein-protein interaction (PPI) network analysis reveals important hub proteins and sub-network modules for root development in rice (*Oryza sativa*). *J. Genet. Eng. Biotechn.* **2023**, *21*, 69. [CrossRef]
26. Struk, S.; Jacobs, A.; Martín-Fontecha, E.S.; Gevaert, K.; Cubas, P.; Goormachtig, S. Exploring the protein-protein interaction landscape in plants. *Plant Cell Environ.* **2019**, *42*, 387–409. [CrossRef]
27. Dong, N.Q.; Lin, H.X. Contribution of phenylpropanoid metabolism to plant development and plant-environment interactions. *J. Integr. Plant Biol.* **2021**, *63*, 180–209. [CrossRef]
28. Rahim, M.A.; Zhang, X.B.; Busatto, N. Editorial: Phenylpropanoid biosynthesis in plants. *Front. Plant Sci.* **2023**, *14*, 1230664. [CrossRef]
29. Jiang, N.; Doseff, A.I.; Grotewold, E. Flavones: From biosynthesis to health benefits. *Plants* **2016**, *5*, 27. [CrossRef]
30. Wang, P.R.; Wan, C.M.; Xu, Z.J.; Wang, P.Y.; Wang, W.M.; Sun, C.H.; Ma, X.Z.; Xiao, Y.H.; Zhu, J.Q.; Gao, X.L.; et al. One divinyl reductase reduces the 8-vinyl groups in various intermediates of chlorophyll biosynthesis in a given higher plant species, but the isozyme differs between species. *Plant Physiol.* **2013**, *161*, 521–534. [CrossRef]
31. Bucinská, L.; Kiss, É.; Koník, P.; Knoppová, J.; Komenda, J.; Sobotka, R. The ribosome-bound protein Pam68 promotes insertion of chlorophyll into the CP47 subunit of photosystem II. *Plant Physiol.* **2018**, *176*, 2931–2942. [CrossRef]
32. Armbruster, U.; Rühle, T.; Kreller, R.; Strotbek, C.; Zühlke, J.; Tadini, L.; Blunder, T.; Hertle, A.P.; Qi, Y.F.; Rengstl, B.; et al. The PHOTOSYNTHESIS AFFECTED MUTANT68-LIKE protein Evolved from a PSII assembly factor to mediate assembly of the chloroplast NAD(P)H dehydrogenase complex in *Arabidopsis*. *Plant Cell* **2013**, *25*, 3926–3943. [CrossRef]
33. Van Rooijen, R.; Harbinson, J.; Aarts, M.G.M. Photosynthetic response to increased irradiance correlates to variation in transcriptional response of lipid-remodeling and heat-shock genes. *Plant Direct* **2018**, *2*, e00069. [CrossRef]
34. An, J.; Wei, X.; Huo, H. Transcriptome analysis reveals the accelerated expression of genes related to photosynthesis and chlorophyll biosynthesis contribution to shade-tolerant in *Phoebe bournei*. *BMC Plant Biol.* **2022**, *22*, 268. [CrossRef]
35. Bolouri Moghaddam, M.R.; Van den Ende, W. Sugars, the clock and transition to flowering. *Front. Plant Sci.* **2013**, *4*, 22. [CrossRef]

36. Wei, J.; Chen, Q.; Lin, J.; Chen, F.; Chen, R.; Liu, H.; Chu, P.; Lu, Z.; Li, S.; Yu, G. Genome-wide identification and expression analysis of tomato glycoside hydrolase family 1 β-glucosidase genes in response to abiotic stresses. *Biotechnol. Biotechnol. Equip.* **2022**, *36*, 268–280. [CrossRef]
37. Yang, Y.; Zhang, T.; Xu, X.; Sun, Y.; Zhang, Y.; Hou, M.; Huang, S.; Yuan, H.; Tong, H. Identification of GH1 gene family fgt members in *Stevia rebaudiana* and their expression when grown in darkness. *Mol. Biol. Rep.* **2020**, *47*, 8739–8746. [CrossRef]
38. Ma, Y.; Han, Y.; Feng, X.; Gao, H.; Cao, B.; Song, L. Genome-wide identification of *BAM* (β-amylase) gene family in jujube (*Ziziphus jujuba* Mill.) and expression in response to abiotic stress. *BMC Genom.* **2022**, *23*, 438. [CrossRef]
39. Zanella, M.; Borghi, G.L.; Pirone, C.; Thalmann, M.; Pazmino, D.; Costa, A.; Santelia, D.; Trost, P.; Sparla, F. β-amylase 1 (BAM1) degrades transitory starch to sustain proline biosynthesis during drought stress. *J. Exp. Bot.* **2016**, *67*, 1819–1826. [CrossRef]
40. Zhang, Y.; Zhu, J.; Khan, M.; Wang, Y.; Xiao, W.; Fang, T.; Qu, J.; Xiao, P.; Li, C.L.; Liu, J.H. Transcription factors ABF4 and ABR1 synergistically regulate amylase-mediated starch catabolism in drought tolerance. *Plant Physiol.* **2023**, *191*, 591–609. [CrossRef]
41. Erpen, L.; Devi, H.S.; Grosser, J.W.; Dutt, M. Potential use of the DREB/ERF, MYB, NAC and WRKY transcription factors to improve abiotic and biotic stress in transgenic plants. *Plant Cell Tiss. Org.* **2018**, *132*, 1–25. [CrossRef]
42. Adem, M.; Gadissa, F.; Zhao, K.; Beyene, D. Transcriptome-wide identification and expression analysis of the NAC gene family in lowland bamboo [Oxytenanthera abyssinica (A.Rich) Munro] under abiotic stresses. *Plant Sci. Today* **2023**, *10*, 250–259. [CrossRef]
43. Wu, Z.; Li, T.; Xiang, J.; Teng, R.D.; Zhang, D.H.; Teng, N.J. A lily membrane-associated NAC transcription factor LlNAC014 is involved in thermotolerance via activation of the DREB2-HSFA3 module. *J. Exp. Bot.* **2023**, *74*, 945–963. [CrossRef] [PubMed]
44. Wang, K.N.; Liu, H.Y.; Mei, Q.L.; Yang, J.; Ma, F.W.; Mao, K. Characteristics of bHLH transcription factors and their roles in the abiotic stress responses of horticultural crops. *Sci. Hortic.* **2023**, *310*, 111710. [CrossRef]
45. Ohashi-Ito, K.; Bergmann, D.C. Arabidopsis FAMA controls the final proliferation/differentiation switch during stomatal development. *Plant Cell* **2006**, *18*, 2493–2505. [CrossRef] [PubMed]
46. Nemie-Feyissa, D.; Heidari, B.; Blaise, M.; Lillo, C. Analysis of interactions between heterologously produced bHLH and MYB proteins that regulate anthocyanin biosynthesis: Quantitative interaction kinetics by microscale thermophoresis. *Phytochemistry* **2015**, *111*, 21–26. [CrossRef] [PubMed]
47. Ning, G.X.; Li, W.F.; Chu, M.Y.; Ma, Z.H.; Wang, P.; Mao, J.; Chen, B.H. *MdbHLH51* plays a positive role in anthocyanin accumulation in 'Red Delicious' apples. *Trees* **2022**, *36*, 1687–1695. [CrossRef]
48. Wang, Z.Y.; Zhang, Y.M.; Hu, H.F.; Chen, L.; Zhang, H.F.; Chen, R.G. CabHLH79 acts upstream of to CaNAC035 regulate cold stress in Pepper. *Int. J. Mol. Sci.* **2022**, *23*, 2537. [CrossRef] [PubMed]
49. Lee, J.; Lee, I. Regulation and function of SOC1, a flowering pathway integrator. *J. Exp. Bot.* **2010**, *61*, 2247–2254. [CrossRef] [PubMed]
50. Wang, S.; Beruto, M.; Xue, J.; Zhu, F.; Liu, C.; Yan, Y.; Zhang, X. Molecular cloning and potential function prediction of homologous SOC1 genes in tree peony. *Plant Cell Rep.* **2015**, *34*, 1459–1471. [CrossRef]
51. Suárez-López, P.; Wheatley, K.; Robson, F.; Onouchi, H.; Valverde, F.; Coupland, G. CONSTANS mediates between the circadian clock and the control of flowering in *Arabidopsis*. *Nature* **2001**, *410*, 1116–1120. [CrossRef]
52. Datta, S.; Hettiarachchi, G.H.C.M.; Holm, D.M. *Arabidopsis* CONSTANS-LIKE is a positive regulator of red light signaling and root growth. *Plant Cell* **2006**, *18*, 70–84. [CrossRef] [PubMed]
53. Lu, J.; Sun, J.J.; Jiang, A.Q.; Bai, M.J.; Fan, C.G.; Liu, J.Y.; Ning, G.G.; Wang, C.Q. Alternate expression of *Constans-Like 4* in short days and *CONSTANS* in long days facilitates day-neutral response in *Rosa chinensis*. *J. Exp. Bot.* **2020**, *71*, 4057–4068. [CrossRef] [PubMed]
54. Min, J.H.; Chung, J.S.; Lee, K.H.; Kim, C.S. The CONSTANS-like 4 transcription factor, AtCOL4, positively regulates abiotic stress tolerance through an abscisic acid-dependent manner in *Arabidopsis*. *J. Integr. Plant Biol.* **2015**, *57*, 313–324. [CrossRef] [PubMed]
55. Li, F.F.; Niu, J.H.; Yu, X.; Kong, Q.H.; Wang, R.F.; Qin, L.; Chen, E.Y.; Yang, Y.B.; Liu, Z.Y.; Lang, L.N.; et al. Isolation and identification of *SiCOL5*, which is involved in photoperiod response, based on the quantitative trait locus mapping of. *Front. Plant Sci.* **2022**, *13*, 969604. [CrossRef] [PubMed]
56. Sun, H.X.; He, D.X.; Wang, N.; Yao, X.D.; Xie, F.T. Transcriptome and metabolome jointly revealed the regulation and pathway of flower and pod abscission caused by shading in soybean (*Glycine max* L.). *Agronomy* **2024**, *14*, 106. [CrossRef]
57. Chen, W.; Hao, W.J.; Xu, Y.X.; Zheng, C.; Ni, D.J.; Yao, M.Z.; Chen, L. Isolation and characterization of CsWRKY7, a subgroup IId WRKY transcription factor from camellia sinensis, linked to development in *Arabidopsis*. *Int. J. Mol. Sci.* **2019**, *20*, 2815. [CrossRef]
58. Jia, P.; Sharif, R.; Li, Y.M.; Sun, T.B.; Li, S.K.; Zhang, X.M.; Dong, Q.L.; Luan, H.A.; Guo, S.P.; Ren, X.L.; et al. The *BELL1-like homeobox* gene *MdBLH14* from apple controls flowering and plant height via repression of *MdGA20ox3*. *Int. J. Biol. Macromol.* **2023**, *242*, 124790. [CrossRef] [PubMed]
59. Cai, H.; Xuan, L.; Xu, L.-A.; Huang, M.-R.; Xu, M. Identification and characterization of nine *PAT1* branch genes in poplar. *Plant Growth Regul.* **2017**, *81*, 355–364. [CrossRef]
60. Bolle, C.; Koncz, C.; Chua, N.H. PAT1, a new member of the GRAS family, is involved in phytochrome A signal transduction. *Genes Dev.* **2000**, *14*, 1269–1278. [CrossRef]
61. Muntha, S.T.; Zhang, L.L.; Zhou, Y.F.; Zhao, X.; Hu, Z.Y.; Yang, J.H.; Zhang, M.F. Phytochrome a signal transduction 1 and CONSTANS-LIKE 13 coordinately orchestrate shoot branching and flowering in leafy *Brassica juncea*. *Plant Biotechnol. J.* **2019**, *17*, 1333–1343. [CrossRef] [PubMed]

62. Song, C.; Baek, J.; Choe, U.; Lim, C.O. Overexpression of DREB2C delays flowering in arabidopsis thaliana via the activation of FLC. *J. Plant Biol.* **2022**, *65*, 133–143. [CrossRef]
63. Lin, R.C.; Wang, H.Y. Arabidopsis *FHY3/FAR1* gene family and distinct roles of its members in light control of arabidopsis development. *Plant Physiol.* **2004**, *136*, 4010–4022. [CrossRef] [PubMed]
64. Paik, I.; Huq, E. Plant photoreceptors: Multi-functional sensory proteins and their signaling networks. *Semin. Cell Dev. Biol.* **2019**, *92*, 114–121. [CrossRef] [PubMed]
65. Amor, M.B.; Guis, M.; Latché, A.; Bouzayen, M.; Pech, J.C.; Roustan, J.P. Expression of an antisense 1-aminocyclopropane-1-carboxylate oxidase gene stimulates shoot regeneration in *Cucumis melo*. *Plant Cell Rep.* **1998**, *17*, 586–589. [CrossRef] [PubMed]
66. Hall, A.E.; Findell, J.L.; Schaller, G.E.; Sisler, E.C.; Bleecker, A.B. Ethylene perception by the ERS1 protein in Arabidopsis. *Plant Physiol.* **2000**, *123*, 1449–1458. [CrossRef] [PubMed]
67. Catala, R.; Lopez-Cobollo, R.; Mar Castellano, M.; Angosto, T.; Alonso, J.M.; Ecker, J.R.; Salinas, J. The Arabidopsis 14-3-3 protein RARE COLD INDUCIBLE 1A links low-temperature response and ethylene biosynthesis to regulate freezing tolerance and cold acclimation. *Plant Cell* **2014**, *26*, 3326–3342. [CrossRef]
68. Jensen, L.; Hegelund, J.N.; Olsen, A.; Lutken, H.; Muller, R. A natural frameshift mutation in *Campanula EIL2* correlates with ethylene insensitivity in flowers. *BMC Plant Biol.* **2016**, *16*, 117. [CrossRef]
69. Liu, C.; Li, J.; Zhu, P.; Yu, J.; Hou, J.; Wang, C.; Long, D.; Yu, M.; Zhao, A. Mulberry EIL3 confers salt and drought tolerances and modulates ethylene biosynthetic gene expression. *PeerJ* **2019**, *7*, e6391. [CrossRef]
70. Dolgikh, V.A.; Pukhovaya, E.M.; Zemlyanskaya, E.V. Shaping ethylene response: The role of EIN3/EIL1 transcription factors. *Front. Plant Sci.* **2019**, *10*, 1030. [CrossRef]
71. Nieuwenhuizen, N.J.; Chen, X.; Wang, M.Y.; Matich, A.J.; Perez, R.L.; Allan, A.C.; Green, S.A.; Atkinson, R.G. Natural ariation in monoterpene synthesis in kiwifruit: Transcriptional regulation of eerpene synthases by NAC and ETHYLENE-INSENSITIVE3-like transcription factors. *Plant Physiol.* **2015**, *167*, 1243–1258. [CrossRef] [PubMed]
72. Yin, X.R.; Allan, A.C.; Chen, K.S.; Ferguson, I.B. Kiwifruit *EIL* and *ERF* genes involved in regulating fruit ripening. *Plant Physiol.* **2010**, *153*, 1280–1292. [CrossRef] [PubMed]
73. Wang, P.; Zhang, B.; Li, X.; Xu, C.; Yin, X.; Shan, L.; Ferguson, I.; Chen, K. Ethylene signal transduction elements involved in chilling injury in non-climacteric loquat fruit. *J. Exp. Bot.* **2010**, *61*, 179–190. [CrossRef] [PubMed]
74. Liu, X.Q.; Liu, R.L.; Li, Y.; Shen, X.; Zhong, S.W.; Shi, H. EIN3 and PIF3 form an interdependent module that represses chloroplast development in buried seedlings. *Plant Cell* **2017**, *29*, 3051–3067. [CrossRef] [PubMed]
75. Fu, X.; Sudhakar, D.; Peng, J.; Richards, D.E.; Christou, P.; Harberd, N.P. Expression of Arabidopsis GAI in transgenic rice represses multiple gibberellin responses. *Plant Cell* **2001**, *13*, 1791–1802. [CrossRef] [PubMed]
76. Arain, S.; Meer, M.; Sajjad, M.; Yasmin, H. Light contributes to salt resistance through GAI protein regulation in *Arabidopsis thaliana*. *Plant Physiol. Bioch.* **2021**, *159*, 1–11. [CrossRef]
77. Wang, H.P.; Pan, J.J.; Li, Y.; Lou, D.J.; Hu, Y.R.; Yu, D.Q. The DELLA-CONSTANS transcription factor cascade integrates gibberellic acid and photoperiod signaling to regulate flowering. *Plant Physiol.* **2016**, *172*, 479–488. [CrossRef] [PubMed]
78. Fukazawa, J.; Ohashi, Y.; Takahashi, R.; Nakai, K.; Takahashi, Y. DELLA degradation by gibberellin promotes flowering via GAF1-TPR-dependent repression of floral repressors in *Arabidopsis*. *Plant Cell* **2021**, *33*, 2258–2272. [CrossRef]
79. Ogawa, S.; Miyamoto, K.; Nemoto, K.; Sawasaki, T.; Yamane, H.; Nojiri, H.; Okada, K. *OsMYC2*, an essential factor for JA-inductive sakuranetin production in rice, interacts with MYC2-like proteins that enhance its transactivation ability. *Sci. Rep.* **2017**, *7*, 40175. [CrossRef]
80. Verma, D.; Jalmi, S.K.; Bhagat, P.K.; Verma, N.; Sinha, A.K. A bHLH transcription factor, MYC2, imparts salt intolerance by regulating proline biosynthesis in *Arabidopsis*. *FEBS J.* **2020**, *287*, 2560–2576. [CrossRef]
81. Song, R.F.; Li, T.T.; Liu, W.C. Jasmonic acid impairs arabidopsis seedling salt stress tolerance through MYC2-mediated repression of expression. *Front. Plant Sci.* **2021**, *12*, 730228. [CrossRef] [PubMed]
82. Cheng, H.; Zha, S.X.; Luo, Y.Y.; Li, L.; Wang, S.Y.; Wu, S.; Cheng, S.Y.; Li, L.L. *JAZ1-3* and *MYC2-1* synergistically regulate the transformation from completely mixed flower buds to female flower buds in *Castanea mollisima*. *Int. J. Mol. Sci.* **2022**, *23*, 6452. [CrossRef] [PubMed]
83. Zheng, Y.Y.; Lan, Y.H.; Shi, T.L.; Zhu, Z.Q. Diverse contributions of MYC2 and EIN3 in the regulation of Arabidopsis jasmonate-responsive gene expression. *Plant Direct* **2017**, *1*, e00015. [CrossRef] [PubMed]
84. Chini, A.; Fonseca, S.; Fernández, G.; Adie, B.; Chico, J.M.; Lorenzo, O.; García-Casado, G.; López-Vidriero, I.; Lozano, F.M.; Ponce, M.R.; et al. The JAZ family of repressors is the missing link in jasmonate signalling. *Nature* **2007**, *448*, 666–671. [CrossRef]
85. Swarup, R.; Péret, B. AUX/LAX family of auxin influx carriers—An overview. *Front. Plant Sci.* **2012**, *3*, 225. [CrossRef]

horticulturae

MDPI

Article

# Exogenous Application of Gamma Aminobutyric Acid Improves the Morpho-Physiological and Biochemical Attributes in *Lavandula dentata* L. under Salinity Stress

**Awad Y. Shala [1,*], Amira N. Aboukamar [2] and Mayank A. Gururani [3,*]**

1 Medicinal and Aromatic Plants Research Department, Horticulture Research Institute, Agricultural Research Center, Giza 12619, Egypt
2 Botanical Gardens Research Department, Horticulture Research Institute, Agricultural Research Center, Giza 12619, Egypt; amera_aboukamar@yahoo.com
3 Biology Department, College of Science, UAE University, Al Ain P.O. Box 15551, United Arab Emirates
* Correspondence: awad.shala@yahoo.com (A.Y.S.); gururani@uaeu.ac.ae (M.A.G.)

**Abstract:** Saline water has been proposed as a solution to partially supply plants with their water requirements due to a lack of fresh water for cultivation in arid and semi-arid sites. Gamma-aminobutyric acid (GABA) is a non-protein amino acid participating in numerous metabolic processes to mitigate the undesirable effects of salinity. A pot experiment was carried out during 2021 and 2022 at Sakha Horticulture Research Station to investigate the effect of foliar application of GABA at 20 and 40 mM on vegetative growth and biochemical changes in French lavender under increasing levels of sea water salinity irrigation treatments (0, 1000, 2000, and 3000 ppm). Results indicated that increasing salinity concentration noticeably decreased plant height, number of branches, herb fresh and dry weight, root length, root fresh and dry weights, photosynthetic pigments, relative water content, and essential oil percentage. On the other hand, accumulation of proline and antioxidant enzymes was increased under increasing salinity concentrations. We conclude that foliar application of GABA acid at 40 mM can alleviate the adverse effects of salinity on the abovementioned French lavender plant characteristics by improving vegetative growth and root characteristics, as well as diminishing chlorophyll degradation, maintaining high leaf relative water content, increasing proline accumulation and antioxidant activity.

**Keywords:** lavender; gamma-aminobutyric acid; antioxidant activity; essential oil; proline

## 1. Introduction

Water availability is a crucial factor that confines crop productivity in Egypt, as the environment is mainly arid and semiarid [1]. Thus, it is imperative in these areas to adopt alternative water sources, such as brackish, reclaimed, and drainage water, to combat water scarcity. Furthermore, the Egyptian water share of the Nile freshwater amount is fixed, and the Egyptian government tends to expand the cultivation of aromatic plants in new reclaimed desert regions that are located in arid zones and recognized by elevated salt levels. Additionally, fresh water will remain a scarce commodity, particularly with the anticipated impacts of global warming in addition to sea level rise that is becoming a serious issue in coastal areas. The availability of water will become a top priority and a great challenge for the Egyptian government in the foreseeable future.

Generally, the influences of salinity on plants are exhibited by a severe-to-moderate contraction in plant development and yield [2–6]. Plants can adapt to the adverse effects of salinity via both enzymatic and nonenzymatic antioxidant protective mechanisms to eliminate provoked ROS to protect plants against unfavorable influences of salinity. Antioxidant enzymes combining peroxidase, catalase, and superoxide dismutase are employed in enzymatic antioxidant defense systems [4,7–10]. Moreover, certain substances like proline, soluble sugars, ascorbic acid, and phenolics are exploited by plant cells to counteract the negative impacts of salinity [4,8,11–13].

One of the most effective adaptation strategies as osmotic regulators is the exogenous application of plant growth compounds to enhance plant adaptation under adverse conditions. A nonprotein amino acid called gamma-aminobutyric acid (GABA) is among these stress-responsive compounds that can rapidly improve the resilience of plants to abiotic stress. The application of GABA alleviated drought stress by enhancing antioxidant activity, osmolytes accumulation, water absorption, chlorophyll, proline accumulation, and soluble sugar content, as well as water efficiency [14–17]. Moreover, GABA could boost saline-alkaline stress tolerance [18] via accumulating soluble sugars and boosting secondary antioxidant activity and lignification of roots. Additionally, it encourages plant adaptation to salinity stress through reinforcing the upregulation of antioxidant potential, photosynthesis, and proline content [7,19–21]. Also, a rising accumulation of endogenous GABA in plant tissue subjected to salt stress and other abiotic stresses was detected [7,15,19,20]. Considering these promising findings, GABA's protective role in revamping yields and quality of medicinal plants under a variety of environmental stresses must be taken into consideration.

French lavender (*Lavandula dentata* L.) is a flowering plant, belonging to the Lamiaceae family, native to the Mediterranean basin and is broadly cultivated around the world. Little shrubs with purple flower spikes begin to grow, highlighting their decorative potential, both as a garden and as a potted plant [22]. The French lavender essential oil is rich in oxygenated monoterpenes, which represent the major oil constituents of linalool, camphor, and borneol. Additionally, the oil has a wide spectrum of biological effects entailing antioxidants, antifungal, and insecticidal potencies [23,24].

Previous research unveiled that French lavender vegetative growth and plant chemical analysis were significantly affected by salinity stress [25]; hence, salinity stress could be a critical limiting factor influencing French lavender oil yield. Under sea water irrigation, however, neither lavender growth nor productivity nor biochemical characteristics have been examined with exogenous GABA application. We hypothesized that saline stress would inhibit the metabolism of lavender plants, but that exogenous GABA might alleviate its negative effects. The current study examined, for the first time, the protective effects of GABA on lavender plants' growth characteristics and biochemical changes under saline water irrigation.

## 2. Materials and Methods

### 2.1. Plant Growth and Experimental Conditions

At Sakha Horticulture Research Station (31°07′ N latitude, 30°05′ E longitude), Kafr El-Sheikh Governorate, North Nile Delta of Egypt, a pot experiment was executed during the two successive seasons of 2021 and 2022 to examine the effect of irrigation with the mixture of fresh water and natural sea water from the white sea (electrical conductivity EC of 35 dS $m^{-1}$), on elevated salinity levels.

Two-month-old rooted transplants with 2–3 true leaves were procured from El-Kanatar El-Khaireya Experimental Farm for Medicinal and Aromatic Plants, Horticulture Research Institute, Agricultural Research Center, Egypt. Then, they were transplanted into plastic pots (30 cm in diameter and 18 cm in height) containing 5 kg clayey soil and subjected to saline water treatments and foliar application of GABA. Physical and chemical soil properties of the used soil were determined according to Page et al. [26], and are displayed in Table 1. The experimental design was a complete randomized blocks design with four replications.

**Table 1.** Some physical and chemical soil properties of the experimental site as mean values of the two growth seasons.

| Soil Depth (cm) | Field Capacity (%) | Wilting Point (%) | Bulk Density ($Mg\ m^{-3}$) | Total Porosity (%) | Sand (%) | Silt (%) | Clay (%) | Texture Class | pH |
|---|---|---|---|---|---|---|---|---|---|
| 0–15 | 45.37 | 22.91 | 1.19 | 55.09 | 19.42 | 24.97 | 55.61 | Clayey | 7.92 |
| | $EC_e$ ($dS\ m^{-1}$) | Inions concentration (meq $L^{-1}$) | | | | Cation concentration (meq $L^{-1}$) | | | |
| | | $CO_3^-$ | $HCO_3^-$ | $Cl^-$ | $SO_4^-$ | $Ca^{++}$ | $Mg^{++}$ | $Na^+$ | $K^+$ |
| 0–15 | 2.74 | --- | 2.52 | 14.94 | 12.28 | 5.02 | 6.84 | 17.22 | 0.66 |

### 2.2. Salinity Stress Treatments

The plants were irrigated with fresh water at 290 ppm (0.45 ds/m) from transplanting to 30 days; then, irrigation with saline water treatments was initiated when lavender plants had 4–5 fully grown leaves and continued until harvesting at the beginning of October for both growing seasons 2021 and 2022. The plants received their water requirements plus 20% as leaching requirements for all treatments during both growing seasons until harvest. The salinity levels 0, 1000 ppm (1.56 dS $m^{-1}$), 2000 ppm (3.12 dS $m^{-1}$), and 3000 ppm (4.69 dS $m^{-1}$) were achieved by adding the proper amount of sea water [EC 35,000 ppm (54.69 dS $m^{-1}$)] to fresh water, which was adjusted via EC meter instrument.

The foliar spray of GABA on plants was performed two times in the morning; the first treatment was performed one month after the initiation of saline irrigation treatments, while the second foliar addition was carried out one month later. Different concentrations of GABA (20 and 40 mM) were prepared in 0.1% (*v*/*v*) Tween-20, while tap water was used as a control. The other agricultural practices were implemented according to the Egyptian Ministry of Agriculture and Land Reclamation recommendations.

### 2.3. Measurement of Growth Parameters

The plants were harvested at the full blooming stage at the beginning of October in both growing seasons; then, the following parameters were recorded: plant height (cm), root length, plant's fresh and dry weight, and root's fresh and dry weight.

### 2.4. Determination of Photosynthetic Pigments

Chlorophyll a, b and carotenoid contents in fresh lavender leaves were extracted by using 0.25 g with 80% acetone according to the method of Lichtenthaler and Buschmann [27]. Absorbances of leaf extract were estimated spectrophotometrically at 663 and 645 nm (for Chl. a and b) and 470 nm (for carotenoids). Finally, pigment concentrations were expressed as mg $g^{-1}$ FW.

### 2.5. Relative Water Content

To determine the leaf relative water content (RWC), fully expanded young leaves were weighed immediately after harvesting, put in vials containing deionized water at room temperature for 24 h, blotted on dry filter paper to obtain the turgid weight, and finally dried in an oven at 70 °C for 48 h. To determine dry weight, RWC was calculated from the following equation: [fresh weight − dry weight/turgid weight − dry weight] × 100 [28].

### 2.6. Measurement of Proline Content

Centrifuging at 10,000× *g* for 10 min was executed after 0.5 g of fresh leaves was combined with 3% sulphosalicylic acid. A mixture of 2 mL of supernatant, glacial acetic acid, and ninhydrin reagent was prepared. For one hour, the reaction mixtures were maintained in a bath of boiling water. Next, the mixture was extracted using 4 mL of toluene after the reaction was stopped in an ice bath. Using spectrophotometry, the absorbance of the organic phase was detected at 520 nm. Proline was quantified in μmol $g^{-1}$ and its concentration was estimated by a standard curve [29].

### 2.7. Essential Oil Extraction

Harvested lavender plants were air-dried, chopped, and subjected to hydrodistillation using Clevenger apparatus for 3 h, conforming to Pharmacopoeia [30]. Essential oil percentage was measured by the following formula: [Volume oil in graduated tube/sample dry weight] × 100, whereas essential oil yield (mL/plant) was calculated by using the dry weight of the plant aerial parts.

### 2.8. Antioxidant Enzyme Activities

To assess the antioxidant enzyme activities, 0.5 g of leaf tissue was homogenized in liquid nitrogen with 3 mL of extraction buffer that consisted of 1 mM ethylenediaminetetraacetic acid (EDTA), 1% (*w*/*v*) polyvinylpyrrolidone, 1 mM phenylmethylsulfonyl fluoride (PMSF), and 0.05% Triton X-100 in 50 mM potassium-phosphate buffer (pH = 7.0) via a prechilled mortar. After carrying out a four-layer cheesecloth filtering process, the homogenate was centrifuged at 12,000 rpm for 20 min at 4 °C. The total soluble enzyme activity assay was performed via a UV-160A spectrophotometer (160A, Shimadzu, Kyoto, Japan) on the supernatant, after it had been centrifuged again at 12,000 rpm for 20 min at 4 °C.

Catalase (CAT) activity was evaluated by the technique described by Aebi [31]. The reaction mixture was composed of 30 mmol $L^{-1}$ $H_2O_2$ in a 50 mmol $L^{-1}$ phosphate buffer (pH 7.0) and 0.1 mL enzyme extract in a total volume of 3 mL. By tracking the consumption of $H_2O_2$ at 240 nm, the activity of catalase was estimated.

Peroxidase activity (EC1.11.1.7) was ascertained in line with the method suggested by Hammerschmidt et al. [32] and expressed as units of peroxidase/mg protein.

Polyphenol oxidase activity (EC 1.10.3.1) was assessed concurring with the procedure claimed by Malik and Singh [33]. The reaction mixture comprised 3.0 mL of buffered catechol solution (0.01 M) newly prepared in 0.1 M phosphate buffer (pH 6.0). 100 mL of the crude enzyme extract was added to initiate the process. Absorbance alterations at 495 nm were noted at 30 s for 3 min. A rise in absorbance $min^{-1}g^{-1}$ fresh weight signified enzyme activity.

### 2.9. Statistical Analysis

All data were tested by analysis of variance (ANOVA) performed by the CoStat 6.45 software program. The comparison of means was evaluated via Duncan's multiple range test in agreement with Snedecor and Cochran [34].

## 3. Results and Discussion

### 3.1. Vegetative Growth Characteristics

Plant vegetative growth traits were negatively influenced under the highest concentration of saline water treatments as compared with the control (Table 2). In this regard, plant height and branch number were significantly cut down in plants exposed to saline irrigation water >2000 ppm, while no significant differences were noted among lower saline water application and control during both growing seasons. Additionally, herb fresh weight as well as herb dry weight gradually diminished with rising saline irrigation water concentration as paralleled to the unstressed plants.

Plant growth cutback is a well-known response to salinity for diverse plant species. This impact has also been reported on other lavender species. In this concern, Szekely-Varga et al. [11] found a substantial curtailment in number of branches, stems, and leaves. The fresh weight of *Lavandula angustifolia* was pronounced in plants subjected to elevated salinity treatments (200 and 300 mM NaCl) with respect to nonstressed control. Also, thyme and lavender plants grown under 50 and 100 mM NaCl significantly reduced the growth as compared with nonsalinized control [2]. Additionally, Wang et al. [7] reported that plant height, leaf freshness, and dry weight of maize seedlings were curtailed under salinity stress (150 mM and 300 mM NaCl concentration). Furthermore, Moghith et al. [3] argued that elevating salinity levels lowered the vegetative growth of *Salvia hispanica* L. plants. In a study conducted in Cyprus, the authors discovered that the plant height of *Lavandula angustifolia* Mill. crumbled in plants exposed to saline conditions >50 mM. Furthermore, all salinity concentrations (25, 50, and 100 mM) trimmed fresh biomass and biomass dry matter [4].

**Table 2.** The effect of saline irrigation water and GABA application on plant height, branch number, herb freshness, and herb dry weight of *Lavandula dentata* L. plants.

| Treatments | Plant Height (cm) | | No. of Branches | | Herb Fresh Weight (g/Plant) | | Herb Dry Weight (g/Plant) | |
|---|---|---|---|---|---|---|---|---|
| | 1st Season | 2nd Season | 1st Season | 2nd Season | 1st Season | 2nd Season | 1st Season | 2nd Season |
| Control (fresh water) | 44.00 a | 47.33 a | 23.67 a | 21.67 abc | 68.09 cd | 78.03 c | 24.51 c | 27.45 b |
| Sea water at 1000 ppm | 42.33 ab | 46.00 ab | 21.33 ab | 19.67 abc | 64.30 d | 70.63 d | 24.57 c | 27.38 b |
| Sea water at 2000 ppm | 39.67 cd | 44.00 bc | 20.33 ab | 18.33 c | 57.13 e | 61.24 f | 20.78 d | 22.61 c |
| Sea water at 3000 ppm | 37.00 e | 38.67 e | 13.00 d | 11.33 d | 50.20 f | 60.73 f | 15.96 e | 18.63 e |
| 1000 ppm sea water + 20 mM GABA | 43.00 a | 47.00 a | 24.00 a | 22.00 ab | 68.65 cd | 77.94 c | 26.27 bc | 29.19 b |
| 1000 ppm sea water + 40 mM GABA | 42.67 ab | 46.33 ab | 22.00 ab | 20.00 abc | 71.27 c | 80.32 c | 26.15 bc | 29.00 b |
| 2000 ppm sea water + 20 mM GABA | 40.33 bc | 44.00 bc | 21.33 ab | 18.67 bc | 83.59 b | 91.43 b | 29.34 ab | 33.29 a |
| 1000 ppm sea water + 40 mM GABA | 42.33 ab | 45.67 ab | 24.33 a | 22.67 a | 90.50 a | 96.02 a | 31.37 a | 34.20 a |
| 3000 ppm sea water + 20 mM GABA | 37.67 de | 41.00 de | 16.00 cd | 13.67 d | 51.91 ef | 67.72 de | 17.93 de | 21.10 cd |
| 3000 ppm sea water + 40 mM GABA | 39.67 cd | 42.00 cd | 18.33 bc | 14.00 d | 55.83 ef | 64.36 ef | 16.10 e | 19.14 de |

Means followed by the same letter at each column are not significantly different at the 5% level according to Duncan's multiple range test.

Considering the effect of GABA on lavender growth, it was observed that the foliar spray of gamma-aminobutyric acid significantly advanced the abovementioned characteristics (Table 1) under saline conditions. This alleviated growth reduction caused by salinity was observed when it was combined with mixed fresh sea water at 1000 and 2000 ppm relative to GABA-untreated plants. Foliar spray with GABA at 20 and 40 mM on plants irrigated with mixed fresh water with concentration at 1000 ppm, as well as plants sprayed with 40 mM GABA and irrigated with mixed saline fresh water at 2000 ppm, noticeably boosted plant height and branches number without significant variations among treatments. Furthermore, applying GABA at 40 mM resulted in the highest values of fresh weight for plants irrigated with 2000 ppm, as well as the heaviest dry weight achieved from plants cotreated with GABA at 20 and 40 mM and irrigated with 2000 ppm without significant differences among them for both growing seasons.

Also, Wang et al. [7] stated that GABA application at 0.5 mM lifted all vegetative growth traits for maize seedlings grown under saline conditions (150 mM and 300 mM NaCl concentration). Similarly, Kalhor et al. [35] noticed that shoot fresh and dry weight of salt-exposed lettuce (40 mM and 80 mM NaCl) were expanded by exogenous supplementation with GABA at 25 μM. Likewise, Ullah et al. [36] reported that exogenous application of GABA at 2 mM inflated shoot length and shoot fresh weight, as well as shoot dry weight of chufa (*Cyperus esculentus*) under salinity stress (0, 100, and 200 mM NaCl and $Na_2SO_4$).

In our case, GABA exogenous application on plants under salinity produced vegetative growth enhancement especially at 1000 and 2000 ppm, respectively. Therefore, the effect of GABA was growth-enhancing at low-stress saline water levels but not at high-saline stress levels which was in accordance with the findings of Kaur and Zhawar [18]. Accordingly, Wang et al. [7] detected that endogenous GABA concentration was accumulated in moderately maize salt-stressed seedlings, while endogenous GABA accumulation was lessened in severely maize salt-stressed seedlings that were generated after exogenous GABA application.

*3.2. Roots Characteristics*

All root attributes were adversely affected by escalating salinity concentrations (Table 3). Additionally, applying GABA as a foliar spray ameliorated the negative effects of salinity on root characteristics, especially at low and medium salinity concentrations (1000 and

2000 ppm). The combination of GABA at both concentrations (20 and 40 mM) with plants irrigated with saline water (1000 and 2000 ppm) noticeably upgraded root length, as well as root fresh weight, without significant variations among these treatments and plants treated with nonsaline water (control). Otherwise, the highest dry weight value was obtained when salinity influences were amended via applying GABA at both concentrations (20 and 40 mM) with plants exposed to saline water at 1000 ppm without significant variations in between for both seasons and plants subjected to saline water at 2000 ppm without significant variation in between for the first season only. The curtailment of root aspects observed in our study was in harmony with the formerly published outcomes by Chrysargyris et al. [4] on *Lavandula angustifolia* Mill. who claimed that root fresh weight and root length declined at 50 and 100 mM NaCl, respectively. Also, Moghith et al. [3] ascertained that escalating salinity levels trimmed root growth characteristics of the *Salvia hispanica* L. plant. Furthermore, Szekely-Varga et al. [11] disclosed that a substantial cutback in root fresh weight of *Lavandula angustifolia* was pronounced in plants subjected to elevated salinity treatments (200 and 300 mM NaCl) with respect to nonstressed controls. Salinity stress (100 and 200 mM NaCl and $Na_2SO_4$) negatively impacted root length and root fresh and root dry weight of *Cyperus esculentus* plants [36].

**Table 3.** Effect of saline irrigation water and GABA application on root length, root fresh weight, and root dry weight of *Lavandula dentata* L. plants during both growing seasons.

| Treatments | Root Length (cm) | | Root Fresh Weight (g/Plant) | | Root Dry Weight (g/Plant) | |
|---|---|---|---|---|---|---|
| | 1st Season | 2nd Season | 1st Season | 2nd Season | 1st Season | 2nd Season |
| Control (fresh water) | 12.33 a | 14.00 ab | 4.04 ab | 5.87 ab | 1.30 b | 1.72 c |
| Sea water at 1000 ppm | 10.67 bc | 13.50 abc | 3.30 b | 5.10 b | 1.30 b | 1.90 bc |
| Sea water at 2000 ppm | 10.33 bc | 12.00 c | 2.37 c | 4.08 c | 1.01 c | 1.54 d |
| Sea water at 3000 ppm | 8.33 d | 10.00 d | 1.90 c | 2.55 d | 0.71 d | 0.90 f |
| 1000 ppm sea water + 20 mM GABA | 11.5 ab | 13.33 abc | 4.09 a | 6.23 a | 1.57 a | 2.27 a |
| 1000 ppm sea water + 40 mM GABA | 11.67 ab | 13.5 abc | 4.20 a | 6.17 a | 1.61 a | 2.23 a |
| 2000 ppm sea water + 20 mM GABA | 12.17 a | 14.00 ab | 3.90 ab | 5.44 ab | 1.49 a | 2.01 b |
| 2000 ppm sea water + 40 mM GABA | 12.5 a | 15.00 a | 3.82 ab | 5.23 ab | 1.45 ab | 1.94 bc |
| 3000 ppm sea water + 20 mM GABA | 10.00 c | 12.50 bc | 2.32 c | 3.13 cd | 1.02 c | 1.22 e |
| 3000 ppm sea water + 40 mM GABA | 9.50 cd | 12.67 bc | 2.12 c | 2.97 d | 0.97 c | 1.21 e |

Means followed by the same letter at each column are not significantly different at the 5% level according to Duncan's multiple range test.

In the present study, many growth parameters were evaluated in lavender salt-stressed plants, involving the reduction of herb, fresh and dry weight of roots, and leaves' relative water content which is probably the major reliable features to identify growth inhibition.

Plants often respond quickly to salinity stress via inhibiting growth which causes osmotic stress in plants and, as a first result, lessens cell expansion as well as cell turgor [37]. However, growth inhibition under stress is eventually linked to the reallocation of plant resources, which are normally utilized for growth and primary metabolism (i.e., biomass accumulation) towards the stimulation of defense mechanisms [37].

Some investigations have highlighted the valuable impacts of GABA on plants subjected to saline conditions as well as unstressed conditions. In this concern, Sheteiwy et al. [19] marked that shoot length and fresh and dry weight of rice seedlings crumbled under salinity conditions. In contrast to unstressed seedlings, the abovementioned physiological parameters were augmented when seeds were primed with 0.5 mM GABA. Likewise, in chufa (*Cyperus esculentus*), Ullah et al. [36] uncovered that exogenous supplementation of GABA at 2 mM enlarged root length and root fresh weight, as well as root dry weight under salinity stress (0, 100, and 200 mM NaCl and $Na_2SO_4$). On lettuce plants, Kalhor et al. [35] outlined that root fresh and dry weight were cut down under NaCl concentrations (40 and 80 mM), but GABA application (25 µM) alleviated the negative influences of salinity on fresh and dry weight. Additionally, Yousef and Nasef [38] declared that the exogenous spray of GABA (1 mM and 2 mM) significantly refined the vegetative growth and yield of garlic genotypes under unstressed conditions which were attributed to rising chlorophyll content, endogenous total phenols, and elements content. Moreover, Feizi et al. [21] divulged that GABA application at 10 mM upgraded root length and root fresh weight of saffron plants under salinity level (15 dS $m^{-1}$). Also, Ramzan et al. [39] recognized that GABA application noticeably amended length as well as fresh and dry biomass of shoots and roots of *Capsicum annuum* L. under salinity stress.

### 3.3. Photosynthetic Pigments

Proliferating mixed fresh saline water from 0 to 3000 ppm led to a gradual decline in Chl. A, Chl. B, and total carotenoid content in *Lavandula dentata* L. leaves under saline water conditions (Figure 1; Table S1). The lowest values in this respect were noted from plants that received the highest concentration of saline water (3000 ppm). On the other hand, foliar spraying of GABA significantly promoted the accumulation of photosynthetic pigments under salinity conditions; hence, the highest content of Chl. a and Chl. b was marked when GABA was exogenously applied at both concentrations (20 and 40 mM) with plants exposed to saline water at 2000 ppm and control without significant differences in between during both growing seasons (Figure 1a–d).

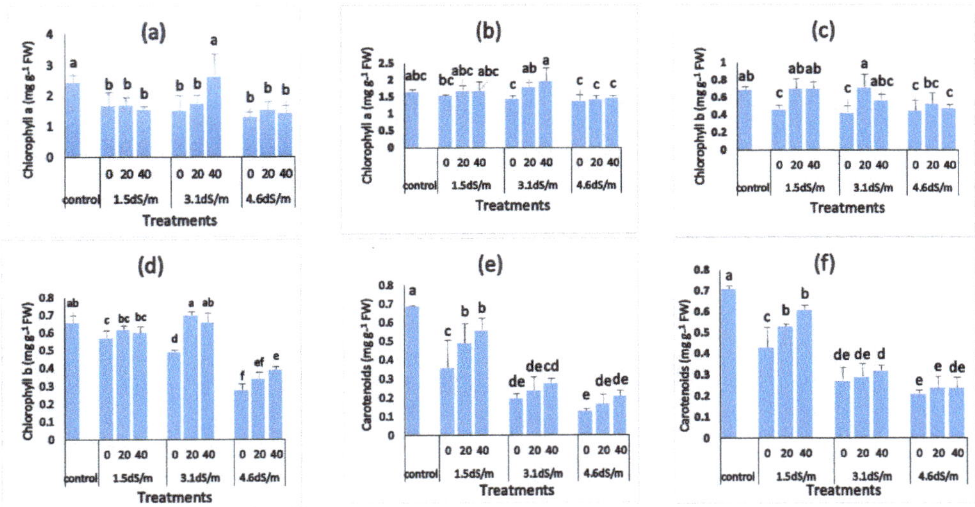


**Figure 1.** Effect of saline irrigation water and GABA application on chlorophyll a: (**a**) first season; (**b**) second season, on chlorophyll b; (**c**) first season; (**d**) second season; and carotenoid content of (**e**) first season and (**f**) second season, of *Lavandula dentata* L. plants (0, 20, 40 (GABA) gamma-aminobutyric acid) and (1.5, 3.1, 4.6 ds/m saline water treatments). Means ± SEs, $n$ = 3; different letters indicate significant differences ($p \leq 0.05$) between the treatments after performing Duncan multiple range test.

Additionally, total carotenoid content was reduced in response to booming concentrations of saline water as observed. The elevated contraction occurred at high salinities (3000 ppm) in both seasons. On the other hand, the highest carotenoid content resulted from plants irrigated with fresh water (control) (Figure 1e,f; Table S1). An earlier published report exhibited that photosynthetic pigment decline was recognized as a useful biochemical stress in plants subjected to salinity conditions. Regarding this, Szekely-Varga et al. [11] unveiled that high salinity treatments (100 and 200 mM NaCl) lowered Chl. a and Chl. B, as well as total carotenoid contents as paralleled with control in *L. angustifolia*. Likewise, Chrysargyris et al. [4] revealed that Chl. a and Chl. b contents were considerably decreased in *Lavandula angustifolia* plants exposed to salinity stress with >50 mM NaCl. Additionally, the promotion effect of GABA on the photosynthetic pigment accumulation has been previously reported by Yousef and Nasef [38]. They discovered that foliar supplementation of GABA at 1 mM and 2 mM significantly heightened total chlorophyll content in garlic genotypes under unstressed conditions and was associated with the enhancement in plant growth as well as garlic yield. Recently, Ullah et al. [36] documented that chlorophyll a and b concentrations were diminished on chufa plants subjected to salinity stress (100 and 200 mM NaCl and $Na_2SO_4$), while these pigments were progressed via exogenous application of GABA at 2 mM. The enhancement role of GABA for growth and photosynthetic pigments in lavender plants may be attributed to the fact that the exogenous application of GABA diminished chlorophyll degradation and developed photosynthetic capacity, as well as growth under salinity conditions [35,36].

### 3.4. Relative Water Content

When the salinity level was aggravated, the relative water content steadily fell (Figure 2a,b; Table S2), which led to a decline in growth rate and a discernible impact on the growth of lavender. The control group had the highest relative water content. On the other hand, foliar application of GABA intensified the relative water content of *Lavandula dentata* L. leaves under medium 2000 ppm diluted sea water followed by low diluted sea water level. Therefore, the greatest relative water content was noted for control and plants that received GABA at 20 and 40 mM under 2000 ppm diluted sea water. This lessened the adverse effects of salinity on leaves' water content without significant differences in between during both growing seasons. A decrease in relative water content under saline conditions is in harmony with the previous results on *Nigella sativa* [40] and *Cassia italica* [41] that significantly declined the growth of these plants. In addition, the GABA-treated plants not only maintained high leaf relative water content, but also ameliorated the negative impacts induced by salinity than non-GABA-treated plants in salt-stressed *Cassia italica*, maize, and rice seedlings. This may be because GABA helps to restore hydration status in addition to protecting cell and organelle membranes from oxidative damage [7,19,41].

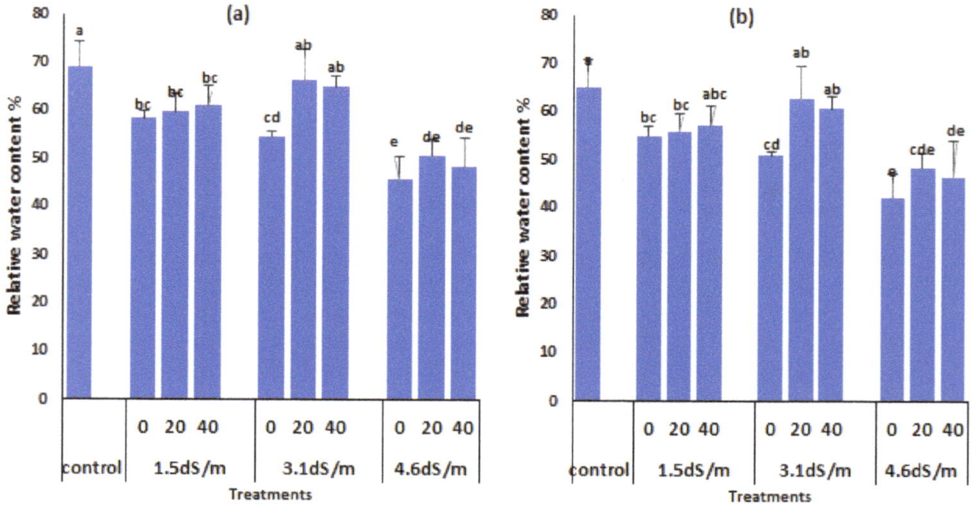


**Figure 2.** Effect of saline irrigation water and GABA application on relative water content of *Lavandula dentata* L. plants (**a**) first season, (**b**) second season, (0, 20, 40 gamma-aminobutyric acid) and (1.5, 3.1, 4.6 ds/m saline water treatments). Means ± SEs, $n = 3$; different letters indicate significant differences ($p \leq 0.05$) between the treatments after performing Duncan multiple range test.

### 3.5. Changes in Proline Accumulation

Proliferating the diluted sea water concentration from 1000 to 3000 ppm gradually inflated proline concentrations (Figure 3a,b; Table S2). Plants subjected to the high saline irrigation water treatment (3000 ppm) and treated with GABA at 20 and 40 mM noticeably expanded proline concentration, followed by plants exposed to diluted sea water at 3000 ppm. The GABA-treated plants at 40 mM combined with 2000 ppm saline irrigation water without significant differences were remarked among treatments; conversely, the control plants reported the least proline concentration in the fresh leaves of *Lavandula dentata* L. The significant increment in leaf proline concentrations in our findings is high enough to have an appropriate osmotic impact in the protective role against elevated salinity.

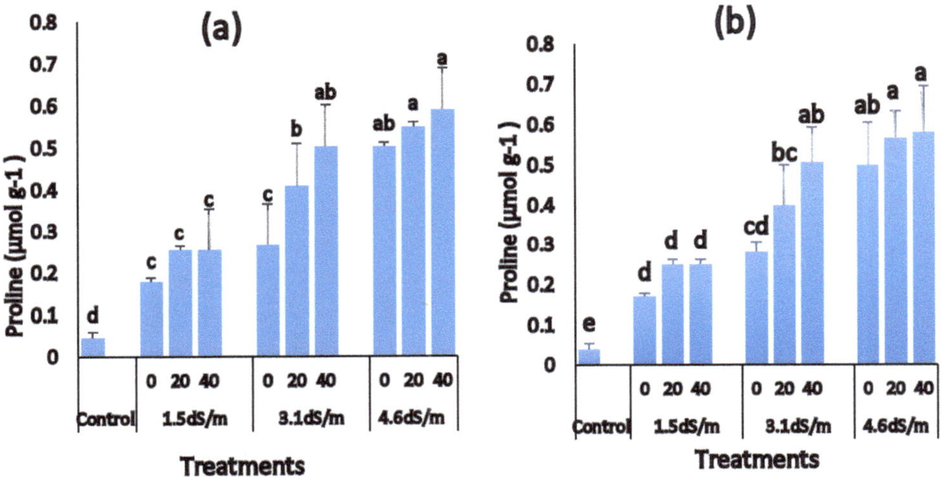


**Figure 3.** Effect of saline irrigation water and GABA application on proline of *Lavandula dentata* L. plants (**a**) first season, (**b**) second season, (0, 20, 40 (GABA) gamma-aminobutyric acid) and (1.5, 3.1, 4.6 ds/m saline water treatments). Means ± SEs, $n = 3$; different letters indicate significant differences ($p \leq 0.05$) between the treatments after performing Duncan multiple range test.

Our results were in accordance with the formerly reported outcomes in which proline concentration was significantly enhanced in the fresh leaves of *Lavandula angustifolia* L. and *Salvia hispanica* L. plants exposed to high salt treatments and high saline irrigation water concentration (4.96 dS $m^{-1}$) [3,4,11].

GABA application significantly progressed proline accumulation in irrigated plants with fresh mixed water at 2000 and 3000 ppm as compared with that of control, which was in agreement with Wang et al. [7], who argued that GABA application at 0.5 mM markedly promoted proline content by 1.2-fold of salt-stressed maize seedlings relative to untreated plants. Lately, Ullah et al. [36] unearthed that proline crumbled in the chufa seedlings with escalating salinity stress (100 and 200 mM NaCl and $Na_2SO_4$); on the contrary, GABA supplementation (2 mM) inflated proline concentrations. Additionally, Ramzan et al. [39] documented that proline decreased in NaCl-stressed *Capsicum annuum* L. seedlings while proliferating in exogenously GABA-applied plants.

On the other hand, Sheteiwy et al. [19] unveiled that proline content significantly developed in rice seedlings subjected to salinity as compared with the unstressed seedlings. Furthermore, proline concentration significantly decreased by priming seeds with 0.5 mM GABA relative to unprimed seeds. Also, proline level was heightened under salt stress in lettuce plants Kalhor et al. [35], but GABA application (at 25 µM) lessened proline content in both saline conditions (40 and 80 mM of NaCl), which may be attributed to the fact that GABA function can partially bypass proline participation in ROS scavenging activities and generate a decrease in proline biosynthesis. Moreover, their results proposed GABA as a plant-boosting antioxidant component, which was able to diminish oxidative damage through enzymatic and nonenzymatic metabolism being activated under salt stress.

The accumulation of stress-protective proline induced by GABA application in plants subjected to salinity stress has been demonstrated. This proved to be efficient in relieving stress as stress-relieving solute due to the role of GABA in inhibiting proline degradation, as well as its participation indirectly in the citric acid cycle under NaCl stress, which will improve the capacity of plants to withstand adverse conditions [7,20].

### *3.6. Changes in Essential Oil Yield*

Lavender essential oil percentage as well as essential oil yield of hydrodistilled aerial parts declined in plants grown under elevated saline irrigation water compared with plants cultivated in nonsaline conditions and low saline irrigation water (Figure 4a–d; Table S2). Interestingly, the highest essential oil percentage and oil yield resulted from plants exposed to medium saline irrigated plants with 2000 ppm and sprayed with GABA at both concentrations without significant differences among them. Therefore, the foliar application of GABA significantly weakened the reduced oil percentage and oil yield noted in medium saline irrigated plants (Table S2).

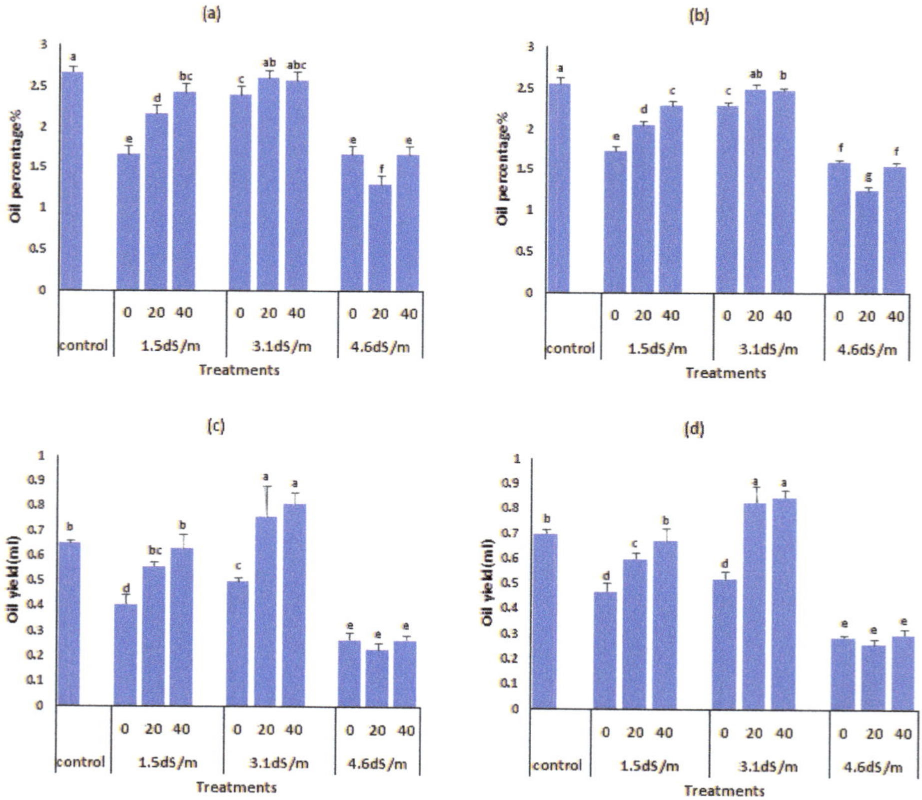


**Figure 4.** Effect of saline irrigation water and GABA application on oil percentage% (**a**) first season, (**b**) second season, and on oil yield (**c**) first season, (**d**) second season of *Lavandula dentata* L. plants (0, 20, 40 (GABA) gamma-aminobutyric acid) and (1.5, 3.1, 4.6 ds/m saline water treatments). Means ± SEs, $n = 3$; different letters indicate significant differences ($p \leq 0.05$) between the treatments after performing Duncan multiple range test.

The extracted essential oil percentage in nonsaline irrigated plants is 2.67 and 2.56% for both seasons, respectively, which is higher than that obtained by Barkaoui et al. [23]. They uncovered that the essential oil percentage of hydrodistilled aerial parts in Morocco was 1.06%. The variation in essential oil yield depended on diverse factors, namely, species, environmental factors, geographical origin, harvest time, and drying method.

The abovementioned outcomes are in harmony with the previous results of Cordovilla et al. [2] who declared that elevated salinity (100 mM) advanced essential oil production of thyme that might be ascribed to the accumulation of secondary metabolites, as a self-defense component versus stress conditions. Also, Chrysargyris et al. [4] mentioned that high salinity (100 mM) trimmed oil yield of *Lavandula angustifolia* as compared with control and low saline levels (25–50 mM).

The stimulation of French lavender essential oil percentage in response to medium saline conditions may be ascribed to an increment in glandular hair number and their densities, as well as the role of secondary metabolites involving essential oil as a self-defense component versus stress conditions [2]. The rise in oil content noticed in plants under salt stress may be attributed to the cutback of the primary metabolites as a result of salinity impacts which, in turn, releases intermediary compounds that facilitate the synthesis of secondary metabolites.

Additionally, the changes in essential oil yield may also be due to herb dry weight variations in plants exposed to irrigation with saline water. Furthermore, the role of GABA application on essential oil yield may be due to photosynthetic pigment enhancement, which produced more metabolites being available to be transformed into essential oil.

### 3.7. *Changes in Antioxidant Activity*

Antioxidant enzyme activities (peroxidase, catalase, and polyphenol) were significantly surged in moderately salt-stressed (1000 and 2000 ppm) and declined in severely salt-stressed (3000 ppm) plants as compared with control (Figure 5a–f; Table S3). Additionally, the application of external GABA to plants markedly boosted the efficacy of selected enzymes in plants exposed to salinity conditions. In plants exposed to a combination of 2000 ppm saline water and 20 or 40 mM GABA, there was a noticeable amplification in the activities of antioxidant enzymes, such as polyphenol, peroxidase, and catalase. Our findings of augmented antioxidant enzymes potential under saline conditions concur with those of Alqarawi et al. [41] for *Cassia italica*, Wang et al. [7] for *Zea mays* L., Chrysargyris et al. [4] for *Lavandula angustifolia*, Sheteiwy et al. [19] for *Oryza sativa* L., Szekely-varga et al. [42] for *Lavandula angustifolia*, and Ullah et al. [36] for *Cyperus esculentus*.

Salinity induces the formation of reactive oxygen species which can be eliminated via antioxidant enzyme activity (peroxidase, catalase, and polyphenol) and play a crucial role in protecting plants from cellular damage resulting from oxidative stress. In our results, GABA-treated plants displayed elevated antioxidant activity and hence mitigated the detrimental impacts of salinity which is repeatedly described in previous studies. For instance, in *Cassia italica* Alqarawi et al. [41], reported that antioxidant enzyme activities (catalase, ascorbate peroxidase, peroxidase, and polyphenol oxidase) were remarkably intensified in plants subjected to 250 mM NaCl + 50 mM GABA. As claimed by Wang et al. [7], in maize plants, all enzyme activities were significantly boosted in GABA-treated plants under salinity stress conditions. It has been demonstrated that the application of GABA to lettuce plants provoked an advancement in their ability to withstand salt stress. This advancement is attributed to the increased activities of catalase and ascorbate peroxidase enzymes. Thus, this finding is significant, as it suggests that GABA plays a vital role in regulating the redox state of the plants and preventing the excessive accumulation of ROS [35]. Additionally, Sheteiwy et al. [19] showed that priming rice seeds with 0.5 mM GABA upgraded the efficacy of catalase, polyphenol oxidase, and ascorbic peroxidase enzymes as compared to the unprimed rice seeds. Also, the authors assumed that the application of 0.5 mM GABA could regulate the antioxidant enzymes efficacy, which played a profound role in scavenging $H_2O_2$ and helping to minimize excessive ROS levels in the stressed plants.

Hence, it boosts a defense mechanism against oxidative stress via upregulation of phenylalanine ammonia lyase, polyphenol oxidase, and shikimate dehydrogenase activity in rice seedlings exposed to salinity. In saffron plants, the foliar application of GABA substantially fortified salt resistance by triggering antioxidant defense mechanisms [21]. Also, it was stated that GABA supplementation at 50 mmol/L alleviated the damage symptoms of *Morus multicaulis* seedlings under salt stress and lowered oxidative damage, as well as antioxidant activity [43].

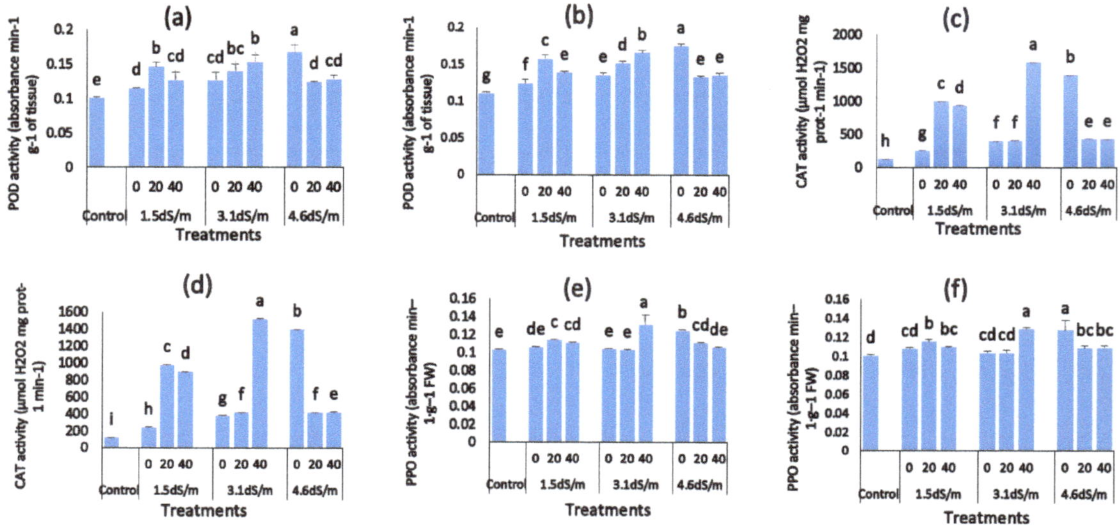


**Figure 5.** Effect of saline irrigation water and GABA application on peroxidase (POD) activity (**a**) first season, (**b**) second season, catalase (CAT) activity (**c**) first season, (**d**) second season, and (0, 20, 40 gamma-aminobutyric acid) and polyphenol (PPO) activity (**e**) first season, (**f**) second season (1.5, 3.1, 4.6 ds/m saline water treatments) of *Lavandula dentata* L. plants. Means ± SEs, $n = 3$; different letters indicate significant differences ($p \leq 0.05$) between the treatments after performing Duncan multiple range test.

Lately, Ullah et al. [36] proclaimed that the GABA application (2 mM) to the chufa seedlings magnified peroxidase activity exposed to salinity levels (100 and 200 mM NaCl and $Na_2SO_4$). On the other hand, GABA supplementation did not foster catalase activity under salinity conditions. Furthermore, Ramzan et al. [39] divulged that GABA application promoted salt tolerance of *C. annuum* L. by reinforcing peroxidase and catalase enzyme activities.

## 4. Conclusions

Our results showed that GABA supplementation positively adjusted various physiological and biochemical mechanisms which could alleviate the detrimental impacts of salinity on growth and biomass. It could magnify vegetative root growth characters, photosynthetic pigment content, relative water content, proline concentration, and antioxidant enzyme activity. Based on our outcomes, GABA is an effective growth regulator that alters growth as well as biochemical responses of lavender plants. We thus suggest that the exogenous application of GABA can help the production of lavender plants in salinity-affected soils in arid and semiarid regions through alleviation of the undesirable impacts of saline irrigation water, which is only available from ground water sources. Thus, we suggest that GABA application at 40 mM could be a viable strategy to alleviate the adverse effects of salinity when French lavender plants are irrigated with saline water at 2000 ppm. Potential studies should delve deeper into the molecular processes underlying GABA's protective benefits in medicinal plants.

**Supplementary Materials:** The following supporting information can be downloaded at: https://www.mdpi.com/article/10.3390/horticulturae10040410/s1, Table S1. The effect of saline irrigation water and GABA application on chlorophyll (a and b) mg/g F.W and carotenoid content mg/g F.W of *Lavandula dentata* L. leaves during both growing seasons. Table S2. The effect of saline irrigation water and GABA application on relative water content, proline, essential oil percentage, and oil yield of *Lavandula dentata* L. plants. Table S3. The effect of saline irrigation water and GABA application to peroxidase, catalase, and polyphenol activity of *Lavandula dentata* L. plants.

**Author Contributions:** A.Y.S. and A.N.A. conceived the study, designed, and performed the experiments, and wrote the manuscript. M.A.G. contributed to data analysis, compilation and writing of the manuscript. All authors have read and agreed to the published version of the manuscript.

**Funding:** This research received no external funding. The APC was funded by UAEU-AUA grant [12R172] received by MAG.

**Data Availability Statement:** The original contributions presented in the study are included in the article and Supplementary Materials, further inquiries can be directed to the corresponding authors.

**Acknowledgments:** Authors gratefully acknowledge the El-Kanatar El-Khaireya Experimental Farm for Medicinal and Aromatic Plants, Horticulture Research Institute, Agricultural Research Center, Egypt for providing French lavender transplants.

**Conflicts of Interest:** The authors declare no competing interests.

## References

1. Abou El Fadl, I.A.; Abd-Ella, M.K.; Hussein, E.H. Effect of saline irrigation water on the growth and some principal compounds of peppermint and spearmint in two types of soil. *J. Agric. Res. Tanta Univ.* **1990**, *16*, 276–295.
2. Cordovilla, M.P.; Bueno, M.; Aparicio, C.; Urrestarazu, M. Effects of salinity and the interaction between *Thymus vulgaris* and *Lavandula angustifolia* on growth, ethylene production and essential oil contents. *J Plant Nutr.* **2014**, *37*, 875–888. [CrossRef]
3. Moghith, W.M.A.; Youssef, A.S.M.; El-Wahab, M.A.A.; Mohamed, Y.F.Y.; Eman, M.A.E.-G. Effect of saline water stress in the presence of silicon foliar application on growth, productivity and chemical constituents of chia (*Salvia hispanica* L.) under Egyptian conditions. *Asian Plant Res. J.* **2020**, *4*, 28–45. [CrossRef]
4. Chrysargyris, A.; Michailidi, E.; Tzortzakis, N. Physiological and biochemical responses of *Lavandula angustifolia* to salinity under mineral foliar application. *Front. Plant Sci.* **2018**, *9*, 489. [CrossRef] [PubMed]
5. Gururani, M.A. In vivo assessment of salinity stress tolerance in transgenic Arabidopsis plants expressing *Solanum tuberosum* D200 gene. *Biol. Plant* **2022**, *66*, 123–131. [CrossRef]
6. Kappachery, S.; Sasi, S.; Alyammahi, O.; Alyassi, A.; Venkatesh, J.; Gururani, M.A. Overexpression of cytoplasmic *Solanum tuberosum* glyceraldehyde 3-phosphate dehydrogenase (GAPDH) gene improves PSII efficiency and alleviates salinity stress in Arabidopsis. *J. Plant Interact.* **2021**, *16*, 398–410. [CrossRef]
7. Wang, Y.; Gu, W.; Meng, Y.; Xie, T.; Li, L.; Li, J.; Wei, S. γ-Aminobutyric acid imparts partial protection from salt stress injury to maize seedlings by improving photosynthesis and upregulating osmoprotectants and antioxidants. *Sci. Rep.* **2017**, *7*, 43609. [CrossRef] [PubMed]
8. Golnari, S.; Vafaee, Y.; Nazari, F.; Ghaderi, N. Gamma-aminobutyric acid (GABA) and salinity impacts antioxidative response and expression of stress-related genes in strawberry cv. *Aromas. Braz. J. Bot.* **2021**, *44*, 639–651. [CrossRef]
9. Gururani, M.A.; Venkatesh, J.; Ghosh, R.; Strasser, R.J.; Ponpandian, L.N.; Bae, H. Chlorophyll-a fluorescence evaluation of PEG-induced osmotic stress on PSII activity in *Arabidopsis* plants expressing *SIP1*. *Plant Biosyst. Int. J. Deal. All Asp. Plant Biol.* **2017**, *3504*, 945–952. [CrossRef]
10. Akilan, S.; Halima, T.H.; Sasi, S.; Kappachery, S.; Baniekal-Hiremath, G.; Venkatesh, J.; Gururani, M.A. Evaluation of osmotic stress tolerance in transgenic Arabidopsis plants expressing *Solanum tuberosum* D200 gene. *J. Plant Interact.* **2019**, *14*, 79–86. [CrossRef]
11. Szekely-Varga, Z.; González-Orenga, S.; Cantor, M.; Jucan, D.; Boscaiu, M.; Vicente, O. Effects of drought and salinity on two commercial varieties of *Lavandula angustifolia* mill. *Plants* **2020**, *9*, 673. [CrossRef] [PubMed]
12. Eisa, E.A.; Honfi, P.; Tilly-Mándy, A.; Gururani, M.A. Exogenous application of melatonin alleviates drought stress in *Ranunculus asiaticus* by improving its morphophysiological and biochemical attributes. *Horticulturae* **2023**, *9*, 262. [CrossRef]
13. Gururani, M.A.; Mohanta, T.K.; Bae, H. Current understanding of the interplay between phytohormones and photosynthesis under environmental stress. *Int. J. Mol. Sci.* **2015**, *16*, 19055–19085. [CrossRef] [PubMed]

14. Ghahremani, Z.; Mikaealzadeh, M.; Barzegar, T.; Ranjbar, M.E. Foliar application of ascorbic acid and gamma aminobutyric acid can improve important properties of deficit irrigated cucumber plants (*Cucumis sativus* cv. Us). *Gesunde Pflanz.* **2021**, *73*, 77–84. [CrossRef]
15. Vijayakumari, K.; Puthur, J.T. γ-Aminobutyric acid (GABA) priming enhances the osmotic stress tolerance in *Piper nigrum* Linn. plants subjected to PEG-induced stress. *Plant Growth Regul.* **2016**, *78*, 57–67. [CrossRef]
16. Rezaei-Chiyaneh, E.; Seyyedi, S.M.; Ebrahimian, E.; Moghaddam, S.S.; Damalas, C.A. Exogenous application of gamma-aminobutyric acid (GABA) alleviates the effect of water deficit stress in black cumin (*Nigella sativa* L.). *Ind. Crops Prod.* **2018**, *112*, 741–748. [CrossRef]
17. Jamshideyni, M.; Behdani, M.A.; Parsa, S.; Khoramdel, S. Evaluation of yield and water use efficiency of Quinoa under irrigation regimes, gamma aminobutyric acid, and vermicompost application. *Acta Agric. Slov.* **2023**, *119*, 1–15. [CrossRef]
18. Kaur, R.; Zhawar, V.K. Regulation of secondary antioxidants and carbohydrates by gamma-aminobutyric acid under salinity–alkalinity stress in rice (*Oryza sativa* L.). *Biol. Futur.* **2021**, *72*, 229–239. [CrossRef]
19. Sheteiwy, M.S.; Shao, H.; Qi, W.; Hamoud, Y.A.; Shaghaleh, H.; Khan, N.U.; Yang, R.; Tang, B. GABA-alleviated oxidative injury induced by salinity, osmotic stress and their combination by regulating cellular and molecular signals in rice. *Int. J. Mol. Sci.* **2019**, *20*, 5709. [CrossRef]
20. Ma, Y.; Wang, P.; Chen, Z.; Gu, Z.; Yang, R. GABA enhances physio-biochemical metabolism and antioxidant capacity of germinated hulless barley under NaCl stress. *J. Plant Physiol.* **2018**, *231*, 192–201. [CrossRef]
21. Feizi, H.; Moradi, R.; Pourghasemian, N.; Sahabi, H. Assessing saffron response to salinity stress and alleviating potential of gamma amino butyric acid, salicylic acid and vermicompost extract on salt damage. *S. Afr. J. Bot.* **2021**, *141*, 330–343. [CrossRef]
22. Peter, K.V.; Shylaja, M.R. Introduction to herbs and spices: Definitions, trade and applications. In *Handbook of Herbs and Spices*; Woodhead Publishing: Cambridge, UK, 2012; pp. 1–24.
23. Barkaoui, H.; Chafik, Z.; Benabbas, R.; Chetouani, M.; El Mimouni, M.; Hariri, E. Antifungal activity of the essential oils of *Rosmarinus officinalis*, *Salvia officinalis*, *Lavandula dentata* and *Cymbopogon citratus* against the mycelial growth of *Fusarium oxysporum* f.sp.Albedinis. *Arab. J. Med. Aromat. Plants* **2022**, *8*, 108–133. [CrossRef]
24. El Abdali, Y.; Agour, A.; Allali, A.; Bourhia, M.; El Moussaoui, A.; Eloutassi, N.; Salamatullah, A.M.; Alzahrani, A.; Ouahmane, L.; Aboul-Soud, M.A.; et al. *Lavandula dentata* L.: Phytochemical analysis, antioxidant, antifungal and insecticidal activities of its essential oil. *Plants* **2022**, *11*, 311. [CrossRef]
25. Hammam, K.A.; Awadalla, S.S.S. Mitigation of saline water stress on French lavender (*Lavandula dentata* L.) plants. *J. Hortic. Sci. Ornam. Plants* **2020**, *12*, 8–16.
26. Page, A.L. *Methods of Soil Analysis: Part 2 Chemical and Microbiological Properties*; American Society of Agronomy, Inc.: Madison, WI, USA; Soil Science Society of America, Inc.: Madison, WI, USA, 1983. [CrossRef]
27. Lichtenthaler, H.K.; Buschmann, C. Chlorophylls and carotenoids: Measurement and characterization by UV-VIS spectroscopy. In *Current Protocols in Food Analytical Chemistry*; John Wiley & Sons: Hoboken, NJ, USA, 2001; pp. F421–F426.
28. Whetherley, P.E. Studies in the water relations of cotton plants. I. The field measurement of water deficit in leaves. *New Phytol.* **1950**, *49*, 81–87. [CrossRef]
29. Bates, L.S.; Waldren, R.P.; Teare, I.D. Rapid determination of free proline for water-stress studies. *Plant Soil* **1973**, *39*, 205–207. [CrossRef]
30. Pharmacopoeia, B. *Determination of Volatile Oil in Drugs*; The Pharmaceutical Press: London, UK, 1963.
31. Aebi, H. Catalase in vitro. *Methods Enzymol.* **1984**, *105*, 121–126.
32. Hammerschmidt, R.; Nuckles, E.M.; Kuć, J. Association of enhanced peroxidase activity with induced systemic resistance of cucumber to *Colletotrichum lagenarium*. *Physiol. Plant Pathol.* **1982**, *20*, 73–82. [CrossRef]
33. Malik, C.P.; Singh, M.B. *Plant Enzymology and Histoenzymology. A Text Manual*; Kalyani Publishers: New Delhi, India, 1980; p. 286.
34. Snedecor, G.W.; Cochran, W.G. *Statistical Methods*, 7th ed.; The Iowa State University Press: Ames, IA, USA, 1980.
35. Kalhor, M.S.; Aliniaeifard, S.; Seif, M.; Asayesh, E.J.; Bernard, F.; Hassani, B.; Li, T. Enhanced salt tolerance and photosynthetic performance: Implication of ɣ-amino butyric acid application in salt-exposed lettuce (*Lactuca sativa* L.) plants. *Plant Physiol. Biochem.* **2018**, *130*, 157–172. [CrossRef] [PubMed]
36. Ullah, A.; Tariq, A.; Zeng, F.; Noor, J.; Sardans, J.; Asghar, M.A.; Zhang, Z.; Peñuelas, J. Application of GABA (γ-aminobutyric acid) to improve saline stress tolerance of chufa (*Cyperus esculentus*, L. var. sativus Boeck) plants by regulating their antioxidant potential and nitrogen assimilation. *S. Afr. J. Bot.* **2023**, *157*, 540–552. [CrossRef]
37. Munns, R.; Tester, M. Mechanisms of salinity tolerance. *Annu. Rev. Plant Biol.* **2008**, *59*, 651–681. [CrossRef] [PubMed]
38. Yousef, E.A.A.; Nasef, I.N. Growth and yield response of garlic genotypes to foliar application of γ-aminobutyric acid. *Hortsci. J. Suez Canal Univ.* **2019**, *8*, 35–43. [CrossRef]
39. Ramzan, M.; Shah, A.A.; Ahmed, M.Z.; Bukhari, M.A.; Ali, L.; Casini, R.; Elansary, H.O. Exogenous application of glutathione and gamma amino-butyric acid alleviates salt stress through improvement in antioxidative defense system and modulation of CaXTHs stress-related genes. *S. Afr. J. Bot.* **2023**, *157*, 266–273. [CrossRef]
40. Rashed, N.M.; Shala Awad, Y.; Mahmoud, M.A. Alleviation of salt stress in *Nigella sativa*, L. by gibberellic acid and rhizobacteria. *Alex. Sci. Exch. J.* **2017**, *38*, 785–799. [CrossRef]
41. Alqarawi, A.A.; Hashem, A.; Abd Allah, E.F.; Al-Huqail, A.A.; Alshahrani, T.S.; Alshalawi, S.R.; Egamberdieva, D. Protective role of gamma amminobutyric acid on *Cassia italica* Mill under salt stress. *Legume Res.* **2016**, *39*, 396–404. [CrossRef]

42. Szekely-varga, Z.; González-orenga, S.; Cantor, M.; Boscaiu, M.; Vicente, O. Antioxidant responses to drought and salinity in *Lavandula angustifolia* Mill. *Not. Bot. Horti Agrobot. Cluj-Napoca* **2020**, *48*, 1980–1992. [CrossRef]
43. Zhang, M.; Liu, Z.; Fan, Y.; Liu, C.; Wang, H.; Li, Y.; Xin, Y.; Gai, Y.; Ji, X. Characterization of GABA-transaminase gene from mulberry (*Morus multicaulis*) and its role in salt stress tolerance. *Genes* **2022**, *13*, 501. [CrossRef]

*horticulturae*

*Article*

# Vegetation and Dormancy States Identification in Coniferous Plants Based on Hyperspectral Imaging Data

**Pavel A. Dmitriev *, Boris L. Kozlovsky and Anastasiya A. Dmitrieva**

Botanical Garden, Academy of Biology and Biotechnologies, Southern Federal University, Rostov-on-Don 344006, Russia; blk@sfedu.ru (B.L.K.); admit@sfedu.ru (A.A.D.)
* Correspondence: pdmitriev@sfedu.ru

**Abstract:** Conifers are a common type of plant used in ornamental horticulture. The prompt diagnosis of the phenological state of coniferous plants using remote sensing is crucial for forecasting the consequences of extreme weather events. This is the first study to identify the "Vegetation" and "Dormancy" states in coniferous plants by analyzing their annual time series of spectral characteristics. The study analyzed *Platycladus orientalis*, *Thuja occidentalis* and *T. plicata* using time series values of 81 vegetation indices and 125 spectral bands. Linear discriminant analysis (LDA) was used to identify "Vegetation" and "Dormancy" states. The model contained three to four independent variables and achieved a high level of correctness (92.3 to 96.1%) and test accuracy (92.1 to 96.0%). The LDA model assigns the highest weight to vegetation indices that are sensitive to photosynthetic pigments, such as the photochemical reflectance index (PRI), normalized PRI (PRI_norm), the ratio of PRI to coloration index 2 (PRI/CI2), and derivative index 2 (D2). The random forest method also diagnoses the "Vegetation" and "Dormancy" states with high accuracy (97.3%). The vegetation indices chlorophyll/carotenoid index (CCI), PRI, PRI_norm and PRI/CI2 contribute the most to the mean decrease accuracy and mean decrease Gini. Diagnosing the phenological state of conifers throughout the annual cycle will allow for the effective planning of management measures in conifer plantations.

**Keywords:** vegetation indices; photochemical reflectance index; *Platycladus orientalis*; acclimatization; deacclimatization

## 1. Introduction

A precise and objective evaluation of the condition of woody plants during their annual development cycle is crucial for anticipating the effects of exposure to extreme temperatures, including untimely negative and positive temperatures. This task is especially relevant in the context of climate change, where the winter temperature regime has become increasingly unstable [1,2]. Winter temperature fluctuations significantly affect the cultivation of woody ornamental plants [3].

The annual cycle of the development of woody plants in the temperate climate zone consists of two main stages: vegetation and dormancy. The transition from vegetation to dormancy is carried out via acclimatization, and the reverse process is carried out by deacclimatization. Plants undergo complex morphological, physiological, biochemical, and genetic changes during these processes [4–7]. Coniferous plants undergo changes in the structure of their photosynthetic apparatus, as well as during photoinhibition and other related processes [8–10]. Significant differences in the state of the photosynthetic apparatus of coniferous plants during the periods of vegetation and dormancy are a prerequisite for the identification of these states via spectral reflections using remote sensing.

Acclimatization is a two-step process [11]. The first stage of this process is cold acclimation, which is initiated by short daylight hours and proceeds at low positive temperatures. This stage leads to the emergence of plant resistance to cold and frost. The second phase of acclimatization, which involves gaining cold hardiness, occurs at negative temperatures that do not freeze plants. This process enables a particular species to achieve its maximum frost resistance. This is a period of deep dormancy (endo-dormancy) during which the plant's growth cannot resume even under favorable conditions. In the latter half of winter, deep dormancy transitions into a state of forced dormancy (eco-dormancy), during which the plant's growth is only restrained by unfavorable weather conditions.

The process of transition from the state of eco-dormancy to the state of vegetation (deacclimatization) in plants has been studied less than the process of acclimatization [1]. This process occurs at the end of winter or in spring under the influence of positive temperatures. After deacclimatization, the plants resume vegetation.

The completion of the processes of acclimatization and deacclimatization is not manifested in external signs; therefore, it cannot be established via phenological methods. For example, the phenological phases of "bud break" and "beginning of shoot growth", which are taken as the beginning of the plant's vegetation, occur later than the completion of the deacclimatization process [12–14]. Therefore, in accordance with phenological characteristics, a plant that is at rest may already be vulnerable to recurrent frosts. According to phenological characteristics, it is also impossible to determine that the plant has passed the second stage of acclimatization (gaining cold hardiness) and has reached maximum frost resistance.

An objective assessment of the frost resistance of woody plants can be obtained via the methods of electrolyte leakage, chlorophyll fluorometry, differential thermal analysis, electrical impedance spectroscopy, and others [15,16]. However, these methods of diagnosing frost resistance are quite laborious and slow. Therefore, it is necessary to develop prompt and relatively simple methods for assessing frost resistance, as well as diagnosing the states of vegetation and dormancy of plants. Such contact methods include an analysis of the amplitude–kinetic characteristics of chlorophyll fluorescence [17,18]. The development of technology for monitoring the solar-induced fluorescence (SIF) of chlorophyll using satellites is currently only at the stage of exploratory research [19,20]. Passive methods of remote sensing have been developed to a much greater extent. Using spectral sensors, it is possible to remotely diagnose the physiological and biochemical characteristics of plants—the content of chlorophylls [21], macroelements [22–25], water [26,27], and nitrogen [28,29]. Spectral sensors proved to be effective in diagnosing stress in plants under the influence of high [30–34] and low temperatures [35].

Vegetation indices (VIs) are utilized to describe plant phenology through remote sensing. Other metrics, such as the leaf area index (LAI) [36] and maximum quantum yield of photosystem II (PSII) [37], are used less frequently. Among the VIs, the normalized difference vegetation index (NDVI) and the extended vegetation index (EVI) [38–41] are the most commonly used.

Whereas the phenology of deciduous plants can be described fairly accurately via a time series of chlorophyll-sensitive VIs, tracking the phenology of evergreens is much more difficult because they retain leaves and photosynthetic pigments throughout the year. At the same time, the LAI of conifers varies little with season [42]. The content of photosynthetic pigments, especially xanthophyll cycle pigments, changes to a greater extent. Therefore, the "photosynthetic phenology" of conifers is well described by the photochemical reflectance index (PRI) and chlorophyll/carotenoid index (CCI), which are related to the carotenoid content and the proportion of chlorophylls and carotenoids [37,43–46].

Further accumulation of factual material on a wide range of species, climatic conditions, and VIs is required for the development of remote sensing phenology of evergreen plants. The possibility of diagnosing "Vegetation" and "Dormancy" states of conifers using remote sensing remains an open question.

Objectives of the study: to correlate spectral responses and climatic adaptations of species of the genus Thuja and Platycladus and to develop hyperspectral remote sensing tools to assess species adaptations to climatic stress.

## 2. Conditions, Objects, and Methods

### *2.1. Study Region*

The study was conducted at the Botanical Garden of Southern Federal University in Rostov-on-Don, Russia (47°13′ N; 39°39′ E). The study utilized plant specimens from the collection of holosemal plants at the Botanical Garden.

### *2.2. Meteorological Characteristics*

Rostov-on-Don has a moderately continental and arid climate, with moderately mild winters and hot summers (Climate of Rostov-on-Don, 1987). The average annual air temperature is +8.9 °C, with an average temperature of −5.7 °C in January and +23 °C in July. The average annual rainfall is 548 mm.

The growing season in 2022 (the period from the transition of the average daily air temperature above +5 °C to the decrease in the average daily air temperature below +5 °C) began on 29 March and ended on 1 November. In 2023, the growing season began on 1 April.

The periods of acclimatization and deacclimatization were determined by indirect indicators, such as the length of the day and the daily temperature fluctuations.

Figure 1 presents the seasonal dynamics of daily air temperatures from 22 January 2022 to 23 May 2023.

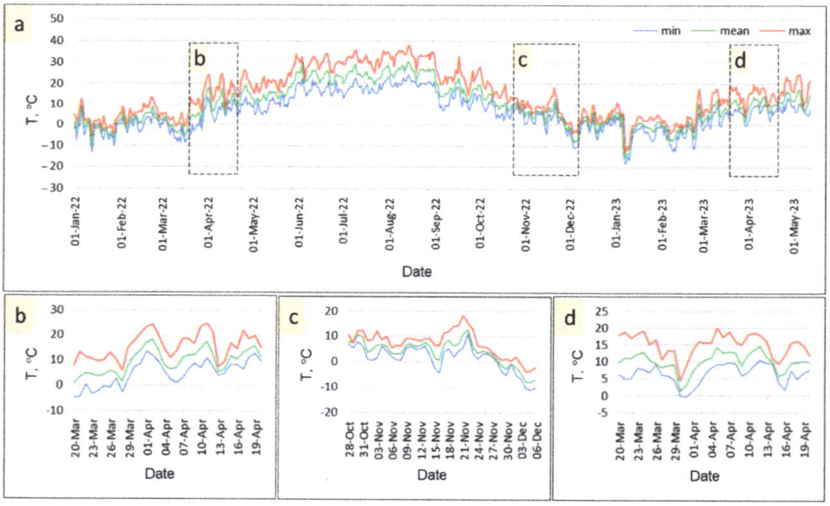


**Figure 1.** Seasonal dynamics of air temperature. (**a**): The entire study period. (**b**): Deacclimatization period in spring 2022. (**c**): Acclimatization period in autumn 2022. (**d**): Deacclimatization period in spring 2023. Min—minimum temperature per day; mean—average temperature per day; max—maximum temperature per day.

Deacclimatization period in spring 2022. Daylight hours began to exceed 12 h after 18 March. On 29 March, the minimum daily air temperature steadily exceeded 0 °C (further on, the minimum daily temperature did not fall below 0 °C, which excludes the reacclimatization process), and on 15 April, the average daily temperature steadily exceeded +10 °C.

Acclimatization period in autumn 2022. Daylight hours began to be less than 12 h after 25 September. The average daily temperature dropped below +10 °C since 1 November. The first nighttime negative temperatures (−2.7 °C) were recorded on 30 November. A sharp decrease in the average daily temperature to −8.0 °C occurred on 4 December and lasted for 4 days. Such a temperature regime is lethal for non-acclimatized woody plants. Therefore, it is believed that the acclimatization of woody plants occurred between 1 November and 30 November 2022. By 4 December, the plants had already entered dormancy.

Deacclimatization period in spring 2023. On 24 March, the minimum daily air temperature steadily exceeded 0 °C (further on, the minimum daily temperature did not fall below 0 °C, which excludes the reacclimatization process), and on 17 April, the average daily temperature steadily exceeded +5 °C.

According to the periods of acclimatization and deacclimatization, the following time frames for the dormancy and vegetation of experimental coniferous plants were established:

— Winter dormancy for 2021–2022 ended on 29 March 2022.
— Vegetation started on 15 April 2022 and ended on 1 November 2022.
— The plants entered dormancy on 30 November 2022 and left it on 24 March 2023.
— Vegetation started again on 17 April 2023.

### 2.3. *Objects of Study*

Three plant species from the genera Thuja and Platycladus were the objects of the study.

*Thuja occidentalis* L.—evergreen tree 12–20 m high. The natural range is located in the southeastern part of Canada and the northern part of the United States. This plant is highly frost-resistant—USDA (United States Department of Agriculture) hardiness zone 6a. *T. occidentalis* is a hygrophilous plant. Within its natural range, it has a positive growth response to humidity conditions and a neutral response to temperature [47]. In the study area, *T. occidentalis* suffers from drought and high summer temperatures [48].

*T. plicata* Donn ex D. Don—tall (up to 60 m) evergreen tree. The natural range of the species is located in the northwest of North America. Ecologically, *T. plicata* is similar to *T. occidentalis*. USDA zone—6b.

*Platycladus orientalis* (L.) Franco—evergreen low (up to 12 m) tree. The natural range is in China and locally in South Korea. The plant is highly drought tolerant and heat tolerant. The frost tolerance of this tree is not as high as *T. occidentalis* (USDA frost tolerance zone—7a). Thermal factors play a more important role in *P. orientalis* culture expansion than humidity factors [49].

Thus, at the study site, the critical season in the annual development cycle for *P. orientalis* is winter, and for *T. occidentalis* and *T. plicata*, summer.

Each species was represented in the experiment by three specimens. All plant samples grew under the same conditions. Seven shoots were taken from each plant sample and transported to the laboratory within an hour. The shoots contained first and second year growths.

### 2.4. *Hyperspectral Imaging Technique*

The study presented here used a time series of 81 VIs and 125 spectral bands (SBs) to cover two periods of deacclimatisation, one period of acclimatisation, and one period of vegetation and dormancy of three coniferous plant species: *Thuja occidentalis*, *T. plicata*, and *Platycladus orientalis*. The most informative VIs and SB for describing the phenological cycle of coniferous plants were determined, and their "Vegetation" and "Dormancy" states were identified.

Hyperspectral imaging (HSI) was carried out in laboratory conditions with an interval of 7–10 days from 2 February 2022 to 10 May 2023. There were 57 HSI in total.

For HSI, a Cubert UHD-185 hyperspectral camera (Cubert GmbH, Ulm, Germany) was used [50,51]. Hyperspectral imaging was conducted using artificial illuminants that had a spectral range overlapping with that of the hyperspectral camera. During the HSI, the camera lens was positioned at 40 cm from the object and directed perpendicular to it. Plant shoots were stacked (Figure 2a). Before each HSI, the shoots in the stack were moved from bottom to top. Seven HSI were made for each stack.

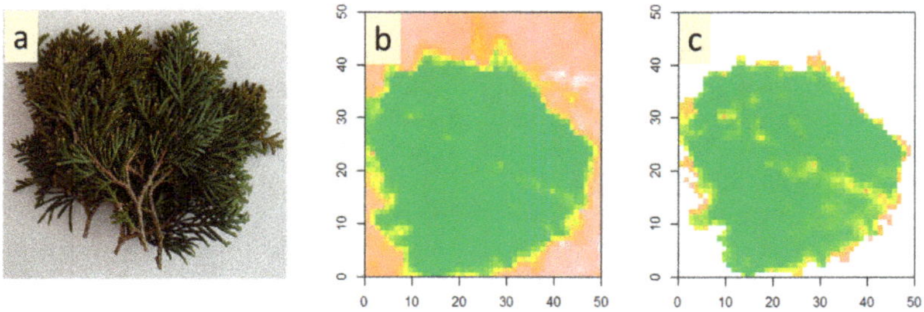


**Figure 2.** Selection ROI. Stack of shoots of *T. occidentalis* (**a**), hyperspectral image before (**b**) and after (**c**) threshold setting "Carter5 > 1.4".

### 2.5. Hyperspectral Imagery Data Preprocessing

During the preprocessing stage, noise was removed from the spectra by using a Savitsky–Golay filter with a length of 15 nm. To select the region of interest (ROI), the threshold of the vegetation index (VI) Carter5 was used with a value of more than 1.4 [52] (Figure 2b,c).

The selection of spectral profiles (pixels) from the image was achieved through automated repeated random selection.

### 2.6. Calculation of Vegetation Indices

Hyperspectral imaging data were used to calculate PRI, CCI, and NDVI for plant phenology analysis. It should be noted that these VIs have traditionally been used to describe plant growth and development. The formulas for calculating these VIs are provided below [45,53,54].

$$\mathrm{NDVI} = \frac{R_{900} - R_{680}}{R_{900} + R_{680}}$$

$$\mathrm{PRI} = \frac{R_{528} - R_{570}}{R_{528} + R_{570}}$$

$$\mathrm{CCI} = \frac{R_{528} - R_{645}}{R_{528} + R_{645}}$$

where $R_{xxx}$: reflectance at the wavelength "xxx".

Additionally, 78 more VIs were calculated. Details regarding these VIs can be found in Dmitriev et al. [55].

### 2.7. Hyperspectral Imagery Data Processing

The study employed the non-parametric Spearman correlation coefficient to identify the most informative VIs for describing conifer phenology. To identify the "Vegetation" and "Dormancy" states of conifers, linear discriminant analysis (LDA) and random forest (RF) were used. The statistical calculations were performed using the R environment (R Core Team, Vienna, Austria).

## 3. Results

### 3.1. Correlation Analysis between Spectral and Climate Data

The processes of acclimatization and deacclimatization and the state of plants during vegetation and dormancy are closely related to temperature and daylight hours. The dynamics of daily temperature allows us to indirectly divide the "Vegetation" and "Dormancy" periods in the annual development cycle of woody plants. Therefore, an assessment was made of the closeness of the relationship between the values of VIs, SB, and the average daily temperature, as well as the duration of daylight hours. For this, the nonparametric Spearman correlation coefficient ($r$) was used. In the annual cycle of plant development, only 4 out of 81 VIs had a high correlation strength with the average daily temperature

($r > 0.7$ according to the Chaddock scale) simultaneously for *P. orientalis*, *T. occidentalis*, and *T. plicata* (Table 1). These were CCI, PRI, ratio of PRI to coloration index 2 (PRI/CI2), and normalized PRI (PRI_norm). These VIs also had a high and statistically significant correlation with daylight hours.

**Table 1.** The value of the Spearman correlation coefficient between the values of VI and the average daily temperature, as well as between the values of VI and the length of daylight hours in the annual cycle of plant development.

| *P. orientalis* | | | *T. occidentalis* | | | *T. plicata* | | |
|---|---|---|---|---|---|---|---|---|
| **VI** | ***r*** | ***p*-Value** | **VI** | ***r*** | ***p*-Value** | **VI** | ***r*** | ***p*-Value** |
| Average daily temperature, °C | | | | | | | | |
| PRI_norm | −0.8 | 0.001 | PRI/CI2 | 0.85 | 0.001 | PRI/CI2 | 0.83 | 0.001 |
| PRI | 0.84 | 0.001 | PRI | 0.84 | 0.001 | PRI | 0.81 | 0.001 |
| PRI/CI2 | 0.82 | 0.001 | PRI_norm | −0.8 | 0.001 | PRI_norm | −0.8 | 0.001 |
| CCI | 0.8 | 0.001 | CCI | 0.77 | 0.001 | CCI | 0.72 | 0.001 |
| DPI | 0.79 | 0.001 | DPI | 0.72 | 0.001 | DPI | 0.65 | 0.001 |
| Vogelmann3 | 0.76 | 0.001 | RARS | −0.6 | 0.001 | RARS | −0.6 | 0.001 |
| D2 | −0.8 | 0.001 | CRI2 | −0.6 | 0.001 | CRI2 | −0.6 | 0.001 |
| MTCI | 0.72 | 0.001 | CRI1 | −0.6 | 0.001 | CRI1 | −0.6 | 0.001 |
| D1 | 0.72 | 0.001 | Gitelson2 | 0.56 | 0.001 | Gitelson2 | 0.51 | 0.001 |
| NDVI | 0.28 | 0.03 | NDVI | 0.09 | 0.50 | NDVI | 0.07 | 0.58 |
| Day length, day | | | | | | | | |
| CCI | 0.78 | 0.001 | PRI | 0.75 | 0.001 | PRI_norm | −0.7 | 0.001 |
| PRI_norm | −0.7 | 0.001 | PRI/CI2 | 0.73 | 0.001 | PRI/CI2 | 0.65 | 0.001 |
| PRI | 0.72 | 0.001 | PRI_norm | −0.7 | 0.001 | CCI | 0.65 | 0.001 |
| PRI/CI2 | 0.71 | 0.001 | CCI | 0.69 | 0.001 | PRI | 0.65 | 0.001 |
| DWSI4 | 0.7 | 0.001 | DPI | 0.64 | 0.001 | RARS | −0.6 | 0.001 |
| NDVI3 | −0.7 | 0.001 | D1 | 0.56 | 0.001 | CRI2 | −0.5 | 0.001 |
| GI | 0.69 | 0.001 | Vogelmann2 | −0.5 | 0.001 | DPI | 0.52 | 0.001 |
| Datt5 | −0.7 | 0.001 | Vogelmann4 | −0.5 | 0.001 | Gitelson2 | 0.47 | 0.001 |
| GMI1 | −0.6 | 0.001 | D2 | −0.5 | 0.001 | CRI1 | −0.5 | 0.001 |
| NDVI | 0.31 | 0.02 | NDVI | 0.24 | 0.08 | NDVI | 0.2 | 0.14 |

Note: VI—vegetation index. *r*—correlation coefficient. *p*-value—level of significance.

The time series of the CCI, double peak index (DPI), PRI, PRI/CI2, and PRI_norm values, in contrast to NDVI, have a well-defined seasonal character (Figure 3).

Preliminarily, it can be stated that the "Dormancy" state of experimental evergreens is characterized by the following VIs values: CCI < 0.15; DPI < 0.075; PRI < −0.05; PRI/CI2 < −0.04; and PRI_norm > 0.003.

The strength of association of the SB with the mean daily temperature and day length is much lower than that of VIs. The highest value of the correlation coefficient ($0.42 < r < 0.57$, at $p < 0.001$) with the average daily temperature has an SB in the range from 514 to 542 nm (Table 2).

### 3.2. Linear Discriminant Analysis of States "Dormancy" and "Vegetation"

In accordance with the periods of "Vegetation" and "Dormancy" identified by climatic characteristics, the HSI data were divided into two classes. "Vegetation" class—HSI data obtained in the interval from 15 April 2022 to 1 November 2022. "Dormancy" class—HSI data obtained in the interval from 30 November 2022 to 24 March 2023.

Each class was divided into two subsets: 70% (training set) spectral data for model building and the remaining 30% (test set) spectral data used for validation. The LDA for

81 Vis gave good results. The testing accuracy for *P. orientalis*, *T. occidentalis*, and *T. plicata* was 99.8, 96.9, and 96.0%, respectively (Figure 4a,b).

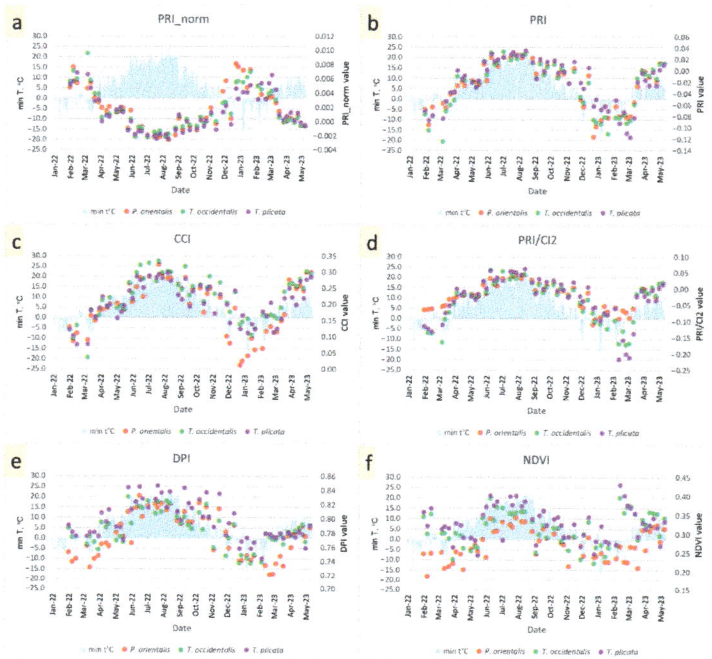


**Figure 3.** Seasonal dynamics of PRI_norm (**a**), PRI (**b**), CCI (**c**), PRI/CI2 (**d**), DPI (**e**), and NDVI (**f**) values compared to the dynamics of the daily temperature.

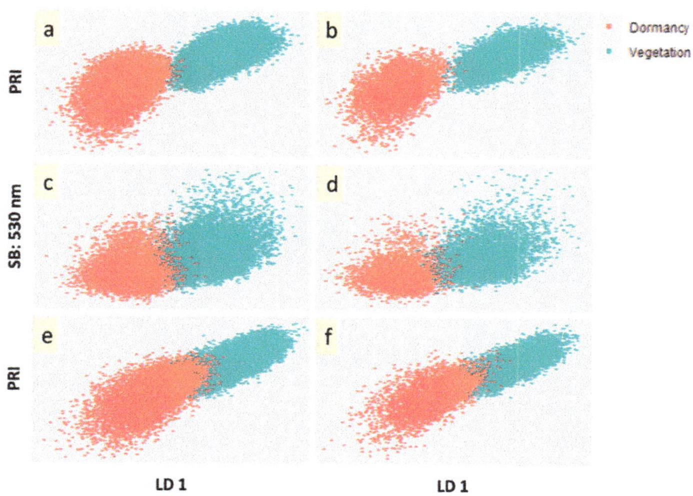


**Figure 4.** Linear discriminant analysis of two states of *P. orientalis*: "Vegetation" and "Dormancy". (**a**): Using 81 VIs (training set); (**b**): using 81 VIs (test set); (**c**): using 125 SB (training set); (**d**): using 125 SB (test set); (**e**): using LDA model LD1 = 0.82 PRI − 0.73 PRI_norm − 0.82 D2 (training set); (**f**) using LDA model LD1 = 0.82 PRI − 0.73 PRI_norm − 0.82 D2 (test set).

A similar result was obtained for 125 SB (Figure 4c,d). The testing accuracy for *P. orientalis*, *T. occidentalis*, and *T. plicata* was 97.7, 92.9, and 92.5, respectively.

The stepwise classification results, using a 10-fold cross-validated correctness rate of the method LDA, provided that adding the next factor does not increase the accuracy by more than 1%, are presented in Table 3.

**Table 2.** The value of the Spearman correlation coefficient between the values of the SB and the average daily temperature and daylight hours in the annual cycle of plant development.

| *P. orientalis* | | | *T. occidentalis* | | | *T. plicata* | | |
|---|---|---|---|---|---|---|---|---|
| **SB** | ***r*** | ***p*-Value** | **SB** | ***r*** | ***p*-Value** | **SB** | ***r*** | ***p*-Value** |
| Average daily temperature, °C | | | | | | | | |
| 518 | 0.46 | 0.001 | 526 | 0.57 | 0.001 | 526 | 0.46 | 0.001 |
| 522 | 0.45 | 0.001 | 530 | 0.57 | 0.001 | 522 | 0.45 | 0.001 |
| 526 | 0.45 | 0.001 | 522 | 0.56 | 0.001 | 530 | 0.44 | 0.001 |
| 514 | 0.44 | 0.001 | 534 | 0.55 | 0.001 | 518 | 0.43 | 0.001 |
| 530 | 0.43 | 0.001 | 518 | 0.54 | 0.001 | 534 | 0.42 | 0.001 |
| 698 | −0.43 | 0.001 | 538 | 0.53 | 0.001 | 514 | 0.39 | 0.001 |
| 694 | −0.43 | 0.001 | 514 | 0.51 | 0.001 | 538 | 0.38 | 0.001 |
| 690 | −0.42 | 0.001 | 542 | 0.51 | 0.001 | 510 | 0.36 | 0.010 |
| 534 | 0.42 | 0.001 | 510 | 0.47 | 0.001 | 542 | 0.36 | 0.010 |
| 702 | −0.42 | 0.001 | 546 | 0.47 | 0.001 | 546 | 0.32 | 0.020 |
| Day length, day | | | | | | | | |
| 526 | 0.44 | 0.001 | 526 | 0.44 | 0.001 | 526 | 0.45 | 0.001 |
| 530 | 0.44 | 0.001 | 522 | 0.44 | 0.001 | 522 | 0.45 | 0.001 |
| 534 | 0.44 | 0.001 | 530 | 0.44 | 0.001 | 530 | 0.45 | 0.001 |
| 538 | 0.43 | 0.001 | 518 | 0.43 | 0.001 | 534 | 0.44 | 0.001 |
| 522 | 0.43 | 0.001 | 534 | 0.41 | 0.001 | 518 | 0.43 | 0.001 |
| 518 | 0.43 | 0.001 | 514 | 0.41 | 0.001 | 514 | 0.40 | 0.001 |
| 542 | 0.42 | 0.001 | 538 | 0.39 | 0.001 | 538 | 0.40 | 0.001 |
| 546 | 0.41 | 0.001 | 510 | 0.38 | 0.001 | 926 | 0.38 | 0.001 |
| 514 | 0.41 | 0.001 | 918 | 0.38 | 0.001 | 542 | 0.38 | 0.001 |
| 550 | 0.40 | 0.001 | 922 | 0.37 | 0.010 | 922 | 0.38 | 0.001 |

Note: SB—spectral band. *r*—correlation coefficient. *p*-value—level of significance.

All LDA models have high model correctness rates and testing accuracy (Table 3, Figure 4e,f).

It should be noted that the LDA models based on the SBs have lower values of the model correctness rate and testing accuracy than the LDA models based on the VIs (Table 3). To improve the accuracy of testing the LDA model based on the SBs, it will be necessary to use many SBs, which contribute little to the accuracy of the model. From a practical point of view, this is a problem since remote sensing will have to use expensive multichannel spectral sensors. Thus, the use of VIs to identify the states of "Vegetation" and "Dormancy" of evergreens gives better results than the use of SBs. The basis of LDA models is the indices of the "PRI group".

### 3.3. Random Forest Pixel-Based Testing of States "Dormancy" and "Vegetation"

RF pixel-based test was used to identify the states of "Vegetation" and "Dormancy" in experimental plants.

The following dataset was used for the training set (Figure 5): "Vegetation" class—HSI data obtained in the interval from 15 April 2022 to 1 November 2022. "Dormancy" class—HSI data obtained in the interval from 30 November 2022 to 24 March 2023.

**Table 3.** LDA models for the two target classes "Vegetation" and "Dormancy".

| Species | Final Model | Model Correctness Rate, % | Testing Accuracy, % |
|---|---|---|---|
| | VIs | | |
| *P. orientalis* | LD1 = 0.82 PRI − 0.73 PRI_norm − 0.82 D2 | 97.20 | 97.01 |
| *T. occidentalis* | LD1 = 0.77 PRI − 0.72 PRI_norm + 0.57 PRI/CI2 | 94.45 | 93.97 |
| *T. plicata* | LD1 = 0.68 PRI − 0.53 PRI_norm + 0.43 PRI/CI2 + 0.39 D1 | 92.31 | 92.10 |
| All species | LD1 = 0.75 PRI − 0.69 PRI_norm + 4.44 PRI/CI2 − 0.44 D2 | 96.09 | 96.03 |
| | SB | | |
| *P. orientalis* | LD1 = $-0.65 R_{450} + 0.37 R_{522} + 0.37 R_{526} + 0.38 R_{530} + 0.37 R_{534} - 0.39 R_{686}$ | 88.63 | 88.14 |
| *T. occidentalis* | LD1 = $-0.52 R_{450} + 0.38 R_{518} + 0.41 R_{522} + 0.40 R_{526} + 0.37 R_{530}$ | 87.40 | 87.67 |
| *T. plicata* | LD1 = $-0.63 R_{450} + 0.31 R_{522} + 0.32 R_{526} + 0.31 R_{530} + 0.36 R_{906} + 0.34 R_{910}$ | 83.72 | 83.79 |
| All species | LD1 = $R_{450} + R_{518} + R_{522} + R_{526} + R_{530}$ | 86.56 | 87.56 |

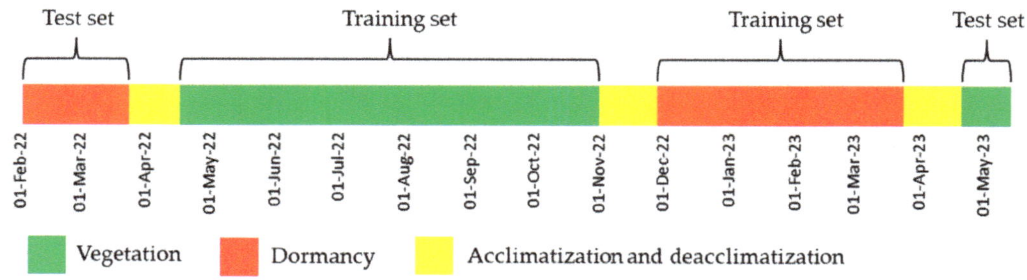


**Figure 5.** Classes for dataset in random forest classification.

For the test set, the HSI data were also divided into two classes (Figure 5): "Vegetation" class—HSI data obtained in the following calendar dates: 19 April, 26 April, 2 May, and 10 May 2023. "Dormancy" class—HSI data obtained in the interval from 2 February to 23 March 2022.

The RF used number of trees 100 (Figure 6), the number of variables tried at each split: 8. The confusion matrix obtained from the RF pixel-based classification of 81 VIs for the two target classes "Vegetation" and "Dormancy" for *P. orientalis* is presented in Table 4. The out-of-bag (OOB) estimate of the error rate was only 0.16%.

**Table 4.** Confusion matrix obtained from the RF pixel-based classification of 81 VIs for two target classes "Vegetation" and "Dormancy" for *P. orientalis*.

| State | Dormancy | Vegetation | Error, % |
|---|---|---|---|
| | Training set | | |
| Dormancy | 11,966 | 34 | 0.28 |
| Vegetation | 27 | 22,973 | 0.12 |
| | Test set | | |
| Dormancy | 6000 | 0 | 0 |
| Vegetation | 274 | 3726 | 6.85 |

The largest contributors to the mean decrease accuracy and mean decrease Gini values are CCI, PRI, and PRI_norm (Figure 7).

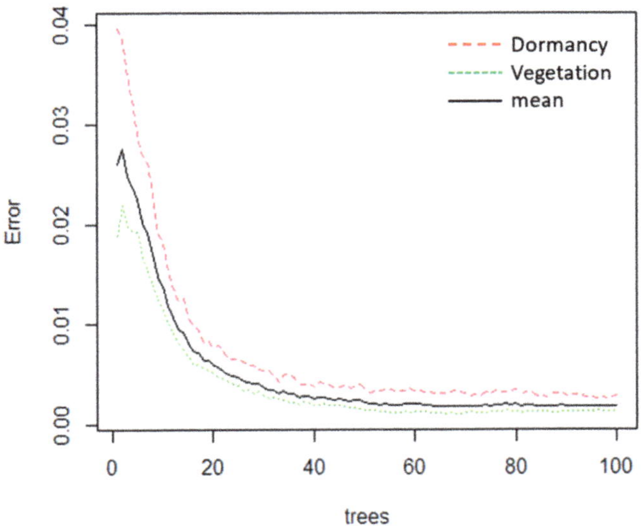


**Figure 6.** Effect of the number of trees on OOB error rate estimation for RF classification of "Vegetation" and "Dormancy" states.

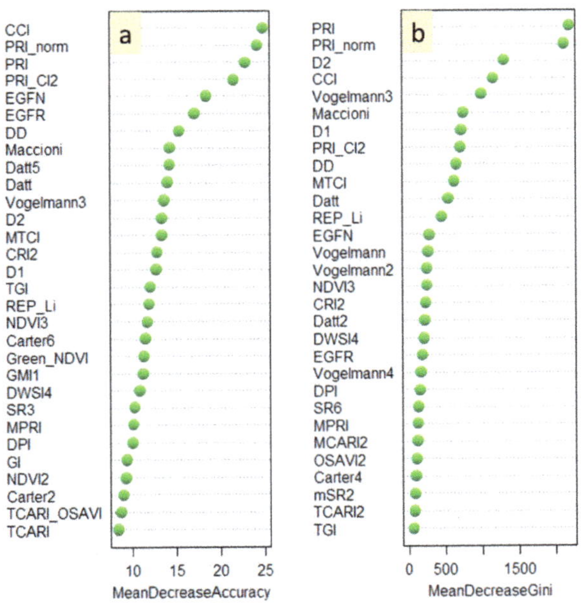


**Figure 7.** Ranking of importance of VIs for differences between the two target classes "Vegetation" and "Dormancy". Vegetation indices are ranked from top to bottom from the most important to the least important by contribution to mean decrease accuracy (**a**) and mean decrease Gini (**b**).

The overall testing accuracy was 97.26%. The "Dormancy" state was tested without error, and the "Vegetation" state was tested with an error of 6.85% (Table 4).

According to the same scheme, RF testing was carried out using the most informative VIs (PRI, PRI_norm, PRI/CI2, and D2) selected based on the results of the LDA (Table 5).

**Table 5.** Confusion matrix obtained from the RF pixel-based classification of four VIs (PRI, PRI_norm, PRI/CI2, and D2) for two target classes "Vegetation" and "Dormancy" for *P. orientalis*.

| State | Dormancy | Vegetation | Error, % |
|---|---|---|---|
| | Training set | | |
| Dormancy | 11,637 | 363 | 3.03 |
| Vegetation | 368 | 22,632 | 1.60 |
| | Test set | | |
| Dormancy | 5873 | 127 | 2.12 |
| Vegetation | 212 | 3788 | 5.30 |

The error of testing the state "Dormancy" was 2.12%, and the error of testing the state "Vegetation" was 5.30%. Thus, the number of VIs for identifying the states of vegetation and dormancy in conifers can be significantly reduced. Similar RF test results are also obtained for *T. occidentalis* and *T. plicata*.

## 4. Discussion

The photosynthetic phenology of plants reflects the seasonal variation in photosynthetic activity, pigment concentrations, and the ratio of their pools using VIs time series and spectral channels [56–58]. Unlike classical phenology, photosynthetic phenology can describe the complete annual development cycle of coniferous plants. It can record both qualitative and quantitative changes in plant states and determine the rate of plant development and senescence [39,59]. Additionally, photosynthetic phenology provides crucial parameters for gross primary productivity models [58]. The PRI and CCI indices, which are sensitive to carotenoids, show potential as metrics for the phenological process in conifers [37,43–46,60]. The use of photosynthetic phenology methods to identify the "Vegetation" and "Dormancy" states in coniferous plants, as well as the transitions between these states (acclimatization and deacclimatization), is of great scientific and practical interest. This study is the first to identify vegetation and winter dormancy states in conifers by analyzing their annual time series of SBs and VIs. The classification was performed using machine learning methods, which are commonly used to predict photosynthetic pigment content [61,62]. The study found that VIs were better metrics for describing annual dynamics in *T. occidentalis*, *T. plicata*, and *P. orientalis* and diagnosing "Vegetation" and "Dormancy" states than SBs.

The correlation analysis revealed that 4 out of 81 vegetation indices (CCI, PRI, PRI/CI2, and PRI_norm) have a strong relationship ($r > 0.7$) with both annual temperature dynamics and daylight hours for all investigated coniferous species. Another group of vegetation indices (CRI1, CRI2, D2, Datt5, DPI, GMI, Gitelson2, RARS, Vogelman, etc.) showed a moderate association ($0.5 < r < 0.7$) with meteorological characteristics. The statistical analysis revealed that there was no significant correlation between NDVI values and meteorological characteristics. Additionally, spectral bands showed a weaker correlation with these meteorological characteristics. The LDA models used to identify the "Vegetation" and "Dormancy" states of *P. orientalis* by VIs contained three to four independent variables, resulting in a high degree of model correctness (ranging from 92.31% to 96.09%) and high test accuracy (ranging from 92.10% to 96.03%). The LDA model identified PRI, PRI_norm, PRI/CI2, and D2 as having the highest weight. The accuracy of LDA models for spectral channels was below 90%. The RF method can identify "Vegetation" and "Dormancy" states with high accuracy. For *P. orientalis*, 81 VIs had a testing accuracy of 97.26%. The most significant contributors to mean decrease accuracy and mean decrease Gini values are CCI, PRI, PRI_norm, and PRI/CI2.

Thus, four carotenoid-sensitive Vis–PRI, PRI_norm, PRI/CI2, and CCI were found to be the most effective in classifying phenological states of experimental plants. The vegetative index PRI is sensitive to rapid changes (within a day) in the state of de-epoxidation, which is a signal of the mutual transformation of xanthophylls. This vegetative index can also be used to track long-term changes in the ratio of chlorophyll and carotenoid pools [45,54,63–66]. PRI indirectly reflects the water content of vegetation [67–70]. Moreover, PRI is an effective tool for diagnosing plant stress [43,71–74]. This effectiveness can be attributed to the fact that the ratio of chlorophyll and carotenoid pools serves as an indicator of the seasonal regulation of photosynthesis and gross primary productivity [75]. During the transition from vegetation to dormancy in coniferous plants, there is a change in the proportions of photosynthetic pigments and the structure of the photosynthetic apparatus. Additionally, there is a steady decrease in the efficiency of PSII [9,10]. For instance, in common pine, the winter suppression of photosynthesis is accompanied by a loss of chlorophylls and a twofold increase in xanthophyll cycle pigments due to light stress [8]. During the transition to winter dormancy, woody plants experience a decrease in tissue water content [76]. The seasonal dynamics of conifers are well described by PRI. However, PRI has a disadvantage of being highly sensitive to light level [77]. On the other hand, CCI records only stable long-term changes in pigments and responds synchronously to seasonal changes in chlorophyll and carotenoid ratios as well as photosynthetic activity [37,45]. It is important to note that the CCI formula has been developed at the leaf level [78,79]. Difficulties may arise when applying this vegetation index at crown level [80]. Generally, CCI and PRI are reliable indices for describing conifer phenology at the shoot level under laboratory conditions. However, NDVI, which is commonly used to describe the phenology of deciduous plants, is not suitable for measuring the phenology of *T. occidentalis*, *T. plicata*, and *P. orientalis* due to the saturation phenomenon [81,82].

In practice, the development of the method of remote diagnostics of phenological states of coniferous plants will make it possible to predict the effect of adverse climatic factors, diagnose stress states of plants, and plan management measures in coniferous plantations.

The results of the study have the following limitations—the study was conducted in laboratory conditions on the shoots of coniferous plants via proximal hyperspectral imaging. The established regularities can be used in remote sensing with certain assumptions.

A potential avenue for future research on the photosynthetic phenology of coniferous plants is the advancement of techniques for the remote diagnosis of the "Vegetation", "Dormancy", "Acclimatisation", and "Deacclimatisation" states. Additionally, it is crucial to promptly evaluate the level of frost tolerance of conifers during specific time periods. Currently, multispectral and hyperspectral data are used to determine the degree of damage to plants caused by negative temperatures [83,84] but not their frost tolerance—their ability to tolerate certain negative temperatures. The ability to solve this issue depends on the correlation between negative temperatures and the values of PRI and CCI. Additionally, during the frost period, highly frost-tolerant *T. occidentalis* and *T. plicata* exhibit significantly different levels of these VIs compared to weakly frost-tolerant *P. orientalis*.

## 5. Conclusions

The operational remote and proximal assessment of conifer plants' (plantation) condition throughout the annual cycle is important for predicting development, diagnosing stress conditions and planning agronomic measures. Photosynthetic phenology, using vegetation indices and spectral bands as metrics, can provide such opportunities. The study analyzed the time series of 81 vegetation index values and 125 spectral bands obtained from hyperspectral imagery for *T. occidentalis*, *T. plicata*, and *P. orientalis*. The time series of carotenoid sensitive vegetation indices (PRI, PRI_norm, PRI/CI2, and CCI) were found to have a more pronounced seasonal character than the time series of chlorophyll sensitive vegetation indices and spectral bands. Using these vegetation indices, "Vegetation" and "Dormancy" states were identified with 97.3% accuracy. The development of this research can be directed toward the development of methods for proximal and remote real-time assessments of frost tolerance in conifers.

**Author Contributions:** Conceptualization, P.A.D. and B.L.K.; data curation, B.L.K. and A.A.D.; formal analysis, A.A.D.; investigation, P.A.D. and A.A.D.; methodology, P.A.D.; project administration, P.A.D.; software, A.A.D.; writing—original draft, P.A.D. and B.L.K.; writing—review and editing, P.A.D. and B.L.K. All authors have read and agreed to the published version of the manuscript.

**Funding:** The project was supported by the Russian Science Foundation under grant No. 23-24-00390, https://rscf.ru/project/23-24-00390/ (accessed on 8 February 2024) and performed in Southern Federal University (Rostov-on-Don, Russian Federation).

**Data Availability Statement:** Data are contained within the article.

**Conflicts of Interest:** The authors declare no conflict of interest.

## References

1. Vyse, K.; Pagter, M.; Zuther, E.; Hincha, D.K. Deacclimation after cold acclimation—A crucial, but widely neglected part of plant winter survival. *J. Exp. Bot.* **2019**, *70*, 4595–4604. [CrossRef]
2. North, M.; Workmaster, B.A.; Atucha, A. Effects of chill unit accumulation and temperature on woody plant deacclimation kinetics. *Physiol. Plant.* **2022**, *174*, e13717. [CrossRef]
3. Hu, X.-G.; Mao, J.-F.; El-Kassaby, Y.A.; Jia, K.-H.; Jiao, S.-Q.; Zhou, S.-S.; Li, Y.; Coops, N.C.; Wang, T. Local Adaptation and Response of *Platycladus orientalis* (L.) Franco Populations to Climate Change. *Forests* **2019**, *10*, 622. [CrossRef]
4. Wisniewski, M.; Bassett, C.; Gusta, L.V. An Overview of Cold Hardiness in Woody Plants: Seeing the Forest through the Trees. *HortScience* **2003**, *38*, 952–959. [CrossRef]
5. Strimbeck, G.R.; Schaberg, P.G.; Fossdal, C.G.; Schröder, W.P.; Kjellsen, T.D. Extreme low temperature tolerance in woody plants. *Front. Plant Sci.* **2015**, *6*, 884. [CrossRef]
6. Li, C.; Junttila, O.; Palva, E.T. Environmental regulation and physiological basis of freezing tolerance in woody plants. *Acta Physiol. Plant* **2004**, *26*, 213–222. [CrossRef]
7. Beck, E.H.; Heim, R.; Hansen, J. Plant resistance to cold stress: Mechanisms and environmental signals triggering frost hardening and dehardening. *J. Biosci.* **2004**, *29*, 449–459. [CrossRef] [PubMed]
8. Ensminger, I.; Sveshnikov, D.; Campbell, D.A.; Funk, C.; Jansson, S.; Lloyd, J.; Shibistova, O.; Öquist, G. Intermittent low temperatures constrain spring recovery of photosynthesis in boreal Scots pine forests. *Glob. Chang. Biol.* **2004**, *10*, 995–1008. [CrossRef]
9. Adams, W.W., III; Demmig-Adams, B. Carotenoid composition and down regulation of photosystem II in three conifer species during the winter. *Physiol. Plant.* **1994**, *92*, 451–458. [CrossRef]
10. Chang, C.Y.Y.; Bräutigam, K.; Hüner, N.P.A.; Ensminger, I. Champions of Winter Survival: Cold acclimation and Molecular Regulation of Cold Hardiness in Evergreen Conifers. *New Phytol.* **2020**, *229*, 675–691. [CrossRef] [PubMed]
11. Weiser, C.J. Cold Resistance and Injury in Woody Plants. *Science* **1970**, *169*, 1269–1278. [CrossRef]
12. Ferguson, J.; Moyer, M.M.; Mills, L.J.; Hoogenboom, G.; Keller, M. Modeling Dormant Bud Cold Hardiness and Budbreak in Twenty-Three Vitis Genotypes Reveals Variation by Region of Origin. *Am. J. Enol. Vitic.* **2013**, *65*, 59–71. [CrossRef]
13. Kovaleski, A.l.; Londo, J. Tempo of gene regulation in wild and cultivated Vitis species shows coordination between cold deacclimation and budbreak. *Plant Sci.* **2019**, *287*, 0168–9452. [CrossRef]
14. Salazar-Gutierrez, M.R.; Chaves, B.; Anothai, J.; Whiting, M.; Hoogenboom, G. Variation in cold hardiness of sweet cherry flower buds through different phenological stages. *Sci. Hortic.* **2014**, *172*, 161–167. [CrossRef]
15. Burr, K.; Hawkins, C.D.B.; L'Hirondelle, S.J.; Binder, W.D.; George, M.F.; Repo, T. Methods for Measuring Cold Hardiness of Conifers. *Tree Physiol.* **2001**, *1*, 369–401. [CrossRef]
16. Atucha Zamkova, A.A.; Steele, K.A.; Smith, A.R. Methods for Measuring Frost Tolerance of Conifers: A Systematic Map. *Forests* **2021**, *12*, 1094. [CrossRef]
17. Hawkins, C.; Lister, G. Invivo chlorophyll fluorescence as a possible indicator of the dormancy stage in Douglas-fir seedlings. *Can. J. For. Res.* **2011**, *15*, 607–612. [CrossRef]
18. Linkosalo, T.; Heikkinen, J.; Pulkkinen, P.; Mäkipää, R. Fluorescence measurements show stronger cold inhibition of photosynthetic light reactions in Scots pine compared to Norway spruce as well as during spring compared to autumn. *Front. Plant Sci.* **2014**, *5*, 264. [CrossRef]
19. Zhang, C.; Atherton, J.; Peñuelas, J.; Filella, I.; Kolari, P.; Aalto, J.; Ruhanen, H.; Bäck, J.; Porcar-Castell, A. Do all chlorophyll fluorescence emission wavelengths capture the spring recovery of photosynthesis in boreal evergreen foliage? *Plant Cell Environ.* **2019**, *42*, 3264–3279. [CrossRef] [PubMed]
20. Du, S.; Liu, L.; Liu, X.; Zhang, X.; Gao, X.; Wang, W. The Solar-Induced Chlorophyll Fluorescence Imaging Spectrometer (SIFIS) Onboard the First Terrestrial Ecosystem Carbon Inventory Satellite (TECIS-1): Specifications and Prospects. *Sensors* **2020**, *20*, 815. [CrossRef]
21. Lang, Q.; Zhiyong, Z.; Longsheng, C.; Hong, S.; Minzan, L.; Li, L.; Junyong, M. Detection of Chlorophyll Content in Maize Canopy from UAV Imagery. *IFAC-PapersOnLine* **2019**, *52*, 330–335. [CrossRef]

22. Noguera, M.; Aquino, A.; Ponce, J.M.; Cordeiro, A.; Silvestre, J.; Arias-Calderón, R.; Marcelo, M.; Jordão, P.; Andújar, J.M. Nutritional status assessment of olive crops by means of the analysis and modelling of multispectral images taken with UAVs. *Biosyst. Eng.* **2021**, *211*, 1–18. [CrossRef]
23. Perry, E.M.; Goodwin, I.; Cornwall, D. Remote Sensing Using Canopy and Leaf Reflectance for Estimating Nitrogen Status in Red-blush Pears. *HortScience* **2018**, *53*, 78–83. [CrossRef]
24. He, C.; Sun, J.; Chen, Y.; Wang, L.; Shi, S.; Qiu, F.; Wang, S.; Tagesson, T. A new vegetation index combination for leaf carotenoid-to-chlorophyll ratio: Minimizing the effect of their correlation. *Int. J. Digit. Earth* **2023**, *16*, 272–288. [CrossRef]
25. Haboudane, D.; Miller, J.R.; Tremblay, N.; Zarco-Tejada, P.J.; Dextraze, L. Integrated narrow-band vegetation indices for prediction of crop chlorophyll content for application to precision agriculture. *Remote Sens. Environ.* **2002**, *81*, 416–426. [CrossRef]
26. Antoniuk, V.; Zhang, X.; Andersen, M.N.; Kørup, K.; Manevski, K. Spatiotemporal Winter Wheat Water Status Assessment Improvement Using a Water Deficit Index Derived from an Unmanned Aerial System in the North China Plain. *Sensors* **2023**, *23*, 1903. [CrossRef]
27. Pôças, I.; Rodrigues, A.; Gonçalves, S.; Costa, P.M.; Gonçalves, I.; Pereira, L.S.; Cunha, M. Predicting Grapevine Water Status Based on Hyperspectral Reflectance Vegetation Indices. *Remote Sens.* **2015**, *7*, 16460–16479. [CrossRef]
28. Hansen, P.M.; Schjoerring, J.K. Reflectance measurement of canopy biomass and nitrogen status in wheat crops using normalized difference vegetation indices and partial least squares regression. *Remote Sens. Environ.* **2003**, *86*, 542–553. [CrossRef]
29. Lapaz Olveira, A.; Saínz Rozas, H.; Castro-Franco, M.; Carciochi, W.; Nieto, L.; Balzarini, M.; Ciampitti, I.; Reussi Calvo, N. Monitoring Corn Nitrogen Concentration from Radar (C-SAR), Optical, and Sensor Satellite Data Fusion. *Remote Sens.* **2023**, *15*, 824. [CrossRef]
30. Ma, H.; Cui, T.; Cao, L. Monitoring of Drought Stress in Chinese Forests Based on Satellite Solar-Induced Chlorophyll Fluorescence and Multi-Source Remote Sensing Indices. *Remote Sens.* **2023**, *15*, 879. [CrossRef]
31. Faqeerzada, M.A.; Park, E.; Kim, T.; Kim, M.S.; Baek, I.; Joshi, R.; Kim, J.; Cho, B.-K. Fluorescence Hyperspectral Imaging for Early Diagnosis of Heat-Stressed Ginseng Plants. *Appl. Sci.* **2023**, *13*, 31. [CrossRef]
32. Burnett, A.C.; Serbin, S.P.; Davidson, K.J.; Ely, K.S.; Rogers, A. Detection of the metabolic response to drought stress using hyperspectral reflectance. *J. Exp. Bot.* **2021**, *72*, 6474–6489. [CrossRef] [PubMed]
33. Bahe, M.M.; Murphy, R.L.; Russell, M.B.; Knight, J.F.; Johnson, G.R. Suitability of a single imager multispectral sensor for tree health analysis. *Urban For. Urban Green.* **2021**, *63*, 127187. [CrossRef]
34. Buitrago, M.F.; Groen, T.A.; Hecker, C.A.; Skidmore, A.K. Changes in thermal infrared spectra of plants caused by temperature and water stress. *ISPRS J. Photogramm. Remote Sens.* **2016**, *111*, 22–31. [CrossRef]
35. Xie, C.; Yang, C.; Moghimi, A. Detection of cold stressed maize seedlings for high throughput phenotyping using hyperspectral imagery. In Proceedings of the Hyperspectral Imaging Sensors: Innovative Applications and Sensor Standards 2017, Anaheim, CA, USA, 28 April 2017; Volume 10213, p. 1021305. [CrossRef]
36. Qiao, K.; Zhu, W.; Xie, Z.; Li, P. Estimating the Seasonal Dynamics of the Leaf Area Index Using Piecewise LAI-VI Relationships Based on Phenophases. *Remote Sens.* **2019**, *11*, 689. [CrossRef]
37. D'Odorico, P.; Besik, A.; Wong, C.Y.S.; Isabel, N.; Ensminger, I. High-throughput drone-based remote sensing reliably tracks phenology in thousands of conifer seedlings. *New Phytol.* **2020**, *226*, 1667–1681. [CrossRef]
38. Kowalski, K.; Senf, C.; Hostert, P.; Pflugmacher, D. Characterizing spring phenology of temperate broadleaf forests using Landsat and Sentinel-2 time series. *Int. J. Appl. Earth Obs. Geoinf.* **2020**, *92*, 102172. [CrossRef]
39. Wang, X.; Zhou, Y.; Wen, R.; Zhou, C.; Xu, L.; Xi, X. Mapping Spatiotemporal Changes in Vegetation Growth Peak and the Response to Climate and Spring Phenology over Northeast China. *Remote Sens.* **2020**, *12*, 3977. [CrossRef]
40. Venkatappa, M.; Anantsuksomsri, S.; Castillo, J.A.; Smith, B.; Sasaki, N. Mapping the Natural Distribution of Bamboo and Related Carbon Stocks in the Tropics Using Google Earth Engine, Phenological Behavior, Landsat 8, and Sentinel-2. *Remote Sens.* **2020**, *12*, 3109. [CrossRef]
41. Thapa, S.; Garcia Millan, V.E.; Eklundh, L. Assessing Forest Phenology: A Multi-Scale Comparison of Near-Surface (UAV, Spectral Reflectance Sensor, PhenoCam) and Satellite (MODIS, Sentinel-2) Remote Sensing. *Remote Sens.* **2021**, *13*, 1597. [CrossRef]
42. Liu, Z.; Jin, G.; Qi, Y. Estimate of Leaf Area Index in an Old-Growth Mixed Broadleaved-Korean Pine Forest in Northeastern China. *PLoS ONE* **2012**, *7*, e32155. [CrossRef]
43. D'Odorico, P.; Schönbeck, L.; Vitali, V.; Meusburger, K.; Schaub, M.; Ginzler, C.; Zweifel, R.; Velasco, V.M.E.; Gisler, J.; Gessler, A.; et al. Drone-based physiological index reveals long-term acclimation and drought stress responses in trees. *Plant Cell Environ.* **2021**, *44*, 3552–3570. [CrossRef]
44. Wong, C.Y.S.; D'Odorico, P.; Bhathena, Y.; Arain, M.A.; Ensminger, I. Carotenoid based vegetation indices for accurate monitoring of the phenology of photosynthesis at the leaf-scale in deciduous and evergreen trees. *Remote Sens. Environ.* **2019**, *233*, e111407. [CrossRef]
45. Gamon, J.A.; Huemmrich, K.F.; Wong, C.Y.S.; Ensminger, I.; Garrity, S.; Hollinger, D.Y.; Noormets, A.; Peñuelas, J. A remotely sensed pigment index reveals photosynthetic phenology in evergreen conifers. *Proc. Natl. Acad. Sci. USA* **2016**, *113*, 13087–13092. [CrossRef]
46. Springer, K.R.; Wang, R.; Gamon, J.A. Parallel Seasonal Patterns of Photosynthesis, Fluorescence, and Reflectance Indices in Boreal Trees. *Remote Sens.* **2017**, *9*, 691. [CrossRef]

47. Kincaid, J.A. Structure and dendroecology of *Thuja occidentalis* in disjunct stands south of its contiguous range in the central Appalachian Mountains. *USA For. Ecosyst.* **2016**, *3*, 25. [CrossRef]
48. Kozlovsky, B.L.; Ogorodnikova, T.K.; Kuropyatnikov, M.V.; Fedorinova, O.I. *Assortment of Woody Plants for Green Building in the Rostov Region*; Southern Federal University: Rostov-on-Don, Russia, 2009; p. 416, ISBN 978-5-9275-0674-3.
49. Li, G.; Du, S.; Wen, Z. Mapping the climatic suitable habitat of oriental arborvitae (*Platycladus orientalis*) for introduction and cultivation at a global scale. *Sci. Rep.* **2016**, *6*, 30009. [CrossRef] [PubMed]
50. Bareth, G.; Aasen, H.; Bendig, J.; Gnyp, M.L.; Bolten, A.; Jung, A.; Michels, R.; Soukkamäki, J. Low-weight and UAV-based hyperspectral full-frame cameras for monitoring crops: Spectral comparison with portable spectroradiometer measurements. *Photogramm. Fernerkund. Geoinf.* **2015**, *103*, 69–79. [CrossRef]
51. Aasen, H.; Burkart, A.; Bolten, A.; Bareth, G. Generating 3D hyperspectral information with lightweight UAV snapshot cameras for vegetation monitoring: From camera calibration to quality assurance. *JPRS* **2015**, *108*, 245–259. [CrossRef]
52. Dmitriev, P.A.; Kozlovsky, B.L.; Dmitrieva, A.A.; Varduni, T.V. Classification of invasive tree species based on the seasonal dynamics of the spectral characteristics of their leaves. *Earth Sci. Inform.* **2023**, *16*, 3729–3743. [CrossRef]
53. Tucker, C.J. Red and photographic infrared linear combinations for monitoring vegetation. *Remote Sens. Environ.* **1979**, *8*, 127–150. [CrossRef]
54. Gamon, J.A.; Gamon, J.A.; Peñuelas, J.; Field, C.B. A narrow-waveband spectral index that tracks diurnal changes in photosynthetic efficiency. *Remote Sens. Environ.* **1992**, *41*, 35–44. [CrossRef]
55. Dmitriev, P.A.; Kozlovsky, B.L.; Kupriushkin, D.P.; Dmitrieva, A.A.; Rajput, V.D.; Chokheli, V.A.; Tarik, E.P.; Kapralova, O.A.; Tokhtar, V.K.; Minkina, T.M.; et al. Assessment of Invasive and Weed Species by Hyperspectral Imagery in Agrocenoses Ecosystem. *Remote Sens.* **2022**, *14*, 2442. [CrossRef]
56. Blackburn, G.A. Hyperspectral remote sensing of plant pigments. *J. Exp. Bot.* **2007**, *58*, 855–867. [CrossRef] [PubMed]
57. Ustin, S.L.; Gitelson, A.A.; Jacquemoud, S.; Schaepman, M.; Asner, G.P.; Gamon, J.A.; Zarco-Tejada, P. Retrieval of foliar information about plant pigment systems from high resolution spectroscopy. *Remote Sens. Environ.* **2009**, *113*, S67–S77. [CrossRef]
58. Wong, C.Y.S. Tracking the phenology of photosynthesis using carotenoid-sensitive and near-infrared reflectance vegetation indices in a temperate evergreen and mixed deciduous forest. *New Phytol.* **2020**, *226*, 1682–1695. [CrossRef] [PubMed]
59. Lebrini, Y.; Boudhar, A.; Laamrani, A.; Htitiou, A.; Lionboui, H.; Salhi, A.; Chehbouni, A.; Benabdelouahab, T. Mapping and Characterization of Phenological Changes over Various Farming Systems in an Arid and Semi-Arid Region Using Multitemporal Moderate Spatial Resolution Data. *Remote Sens.* **2021**, *13*, 578. [CrossRef]
60. Wong, C.Y.S.; Gamon, J.A. The photochemical reflectance index provides an optical indicator of spring photosynthetic activation in evergreen conifers. *New Phytol.* **2015**, *206*, 196–208. [CrossRef]
61. Narmilan, A.; Gonzalez, F.; Salgadoe, A.S.A.; Kumarasiri, U.W.L.M.; Weerasinghe, H.A.S.; Kulasekara, B.R. Predicting Canopy Chlorophyll Content in Sugarcane Crops Using Machine Learning Algorithms and Spectral Vegetation Indices Derived from UAV Multispectral Imagery. *Remote Sens.* **2022**, *14*, 1140. [CrossRef]
62. Yue, J.; Feng, H.; Tian, Q.; Zhou, C. A robust spectral angle index for remotely assessing soybean canopy chlorophyll content in different growing stages. *Plant Methods* **2020**, *16*, 104. [CrossRef]
63. Sims, D.A.; Gamon, J.A. Relationships between leaf pigment content and spectral reflectance across a wide range of species, leaf structures and developmental stages. *Remote Sens. Environ.* **2002**, *81*, 337–354. [CrossRef]
64. Garrity, S.R.; Porcar-Castell, A.; Munné-Bosch, S.; Bäck, J.; Garbulsky, M.F.; Peñuelas, J. Disentangling the relationships between plant pigments and the Photochemical Reflectance Index reveals a new approach for remote estimation of carotenoid content. *Remote Sens. Environ.* **2011**, *115*, 628–635. [CrossRef]
65. Gamon, J.A.; Berry, J.A. Facultative and constitutive pigment effects on the Photochemical Reflectance Index (PRI) in sun and shade conifer needles. *Isr. J. Plant Sci.* **2012**, *60*, 85–95. [CrossRef]
66. Filella, I.; Porcar-Castell, A.; Munné-Bosch, S.; Bäck, J.; Garbulsky, M.F.; Peñuelas, J. PRI assessment of long-term changes in carotenoids/chlorophyll ratio and short-term changes in de-epoxidation state of the xanthophyll cycle. *Int. J. Remote Sens.* **2009**, *30*, 4443–4455. [CrossRef]
67. Asner, G.P.; Nepstad, D.; Cardinot, G.; Ray, D. Drought stress and carbon uptake in an Amazon forest measured with spaceborne imaging spectroscopy. *Proc. Natl. Acad. Sci. USA* **2004**, *101*, 6039–6044. [CrossRef]
68. Peguero-Pina, J.J.; Morales, F.; Flexas, J.; Gil-Pelegrín, E.; Moya, I. Photochemistry, remotely sensed physiological Reflectance index and deepoxidation state of the xanthophyll cycle in *Quercus coccifera* under intense drought. *Oecologia* **2008**, *156*, 1–11. [CrossRef]
69. Suarez, L.; Zarco-Tejada, P.J.; Sepulcre-Cantó, G.; Pérez-Priego, O.; Miller, J.R.; Jiménez-Muñoz, J.C.; Sobrino, J. Assessing canopy PRI for water stress detection with diurnal airborne imagery. *Remote Sens. Environ.* **2008**, *112*, 560–575. [CrossRef]
70. Damm, A.; Cogliati, S.; Colombo, R.; Fritsche, L.; Genangeli, A.; Genesio, L.; Hanus, J.; Peressotti, A.; Rademske, P.; Rascher, U.; et al. Response times of remote sensing measured sun-induced chlorophyll fluorescence, surface temperature and vegetation indices to evolving soil water limitation in a crop canopy. *Remote Sens. Environ.* **2022**, *273*, 112957. [CrossRef]
71. Kohzuma, K.; Tamaki, M.; Hikosaka, K. Corrected photochemical reflectance index (PRI) is an effective tool for detecting environmental stresses in agricultural crops under light conditions. *J. Plant Res.* **2021**, *134*, 683–694. [CrossRef]
72. Berni, J.A.J.; Zarco-Tejada, P.J.; Suarez, L.; Fereres, E. Thermal and Narrowband Multispectral Remote Sensing for Vegetation Monitoring from an Unmanned Aerial Vehicle. *IEEE Trans. Geosci. Remote Sens.* **2009**, *47*, 722–738. [CrossRef]

73. Zarco-Tejada, P.J.; González-Dugo, V.; Berni, J.A.J. Fluorescence, temperature and narrow-band indices acquired from a UAV platform for water stress detection using a micro-hyperspectral imager and a thermal camera. *Remote Sens. Environ.* **2012**, *117*, 322–337. [CrossRef]
74. Biriukova, K.; Pacheco-Labrador, J.; Migliavacca, M.; Mahecha, M.D.; Gonzalez-Cascon, R.; Martin, M.P.; Rossini, M. Performance of singular spectrum analysis in separating seasonal and fast physiological dynamics of solar-induced chlorophyll fluorescence and PRI optical signals. *J. Geophys. Res. Biogeosci.* **2021**, *126*, e2020JG006158. [CrossRef]
75. Wong, C.Y.S.; Gamon, J.A. Three causes of variation in the Photochemical Reflectance Index (PRI) in evergreen conifers. *New Phytol.* **2015**, *206*, 187–195. [CrossRef]
76. Pramsohler, M.; Neuner, G. Dehydration and osmotic adjustment in apple stem tissue during winter as it relates to the frost resistance of buds. *Tree Physiol.* **2013**, *33*, 807–816. [CrossRef]
77. Gamon, J.; Wang, R.; Russo, S.E. Contrasting photoprotective responses of forest trees revealed using PRI light responses sampled with airborne imaging spectrometry. *New Phytol.* **2023**, *238*, 1318–1332. [CrossRef]
78. Gitelson, A.A.; Zur, Y.; Chivkunova, O.; Merzlyak, M. Assessing carotenoid content in plant leaves with reflectance spectroscopy. *Photochem. Photobiol.* **2002**, *75*, 272–281. [CrossRef] [PubMed]
79. Gitelson, A.A.; Keydan, G.P.; Merzlyak, M.N. Three-band model for noninvasive estimation of chlorophyll, carotenoids, and anthocyanin contents in higher plant leaves. *Geophys. Res. Lett.* **2006**, *33*, L11402. [CrossRef]
80. Zarco-Tejada, P.J.; Guillén-Climent, M.L.; Hernández-Clemente, R.; Catalina, A.; González, M.R.; Martín, P. Estimating leaf carotenoid content in vineyards using high resolution hyperspectral imagery acquired from an unmanned aerial vehicle (UAV). *Agric. For. Meteorol.* **2013**, *171–172*, 281–294. [CrossRef]
81. Huang, S.; Tang, L.; Hupy, J.P.; Wang, Y.; Shao, G. A commentary review on the use of normalized difference vegetation index (NDVI) in the era of popular remote sensing. *J. For. Res.* **2021**, *32*, 1–6. [CrossRef]
82. Fawcett, D.; Bennie, J.; Anderson, K. Monitoring spring phenology of individual tree crowns using drone-acquired NDVI data. *Remote Sens. Ecol. Conserv.* **2020**, *7*, 227–244. [CrossRef]
83. Gabbrielli, M.; Corti, M.; Perfetto, M.; Fassa, V.; Bechini, L. Satellite-Based Frost Damage Detection in Support of Winter Cover Crops Management: A Case Study on White Mustard. *Agronomy* **2022**, *12*, 2025. [CrossRef]
84. Shammi, S.; Sohel, F.; Diepeveen, D.; Zander, S.; Jones, M.G.K. A survey of image-based computational learning techniques for frost detection in plants. *Inf. Process. Agric.* **2023**, *10*, 164–191. [CrossRef]

*horticulturae*

MDPI

*Article*

# Melatonin Application Induced Physiological and Molecular Changes in Carnation (*Dianthus caryophyllus* L.) under Heat Stress

**Mohamed S. Elmongy * and Mohaned M. Abd El-Baset**

Department of Vegetable and Floriculture, Faculty of Agriculture, Mansoura University, Mansoura 35516, Egypt; mohaned2005@mans.edu.eg
* Correspondence: mohamedelmongy@mans.edu.eg

**Abstract:** Carnation is one of the most important ornamental plants worldwide; however, heat stress is a problem, which affects carnation cultivation. The harmful effects of heat stress include impaired vegetative development and reduced floral induction. In this study, to enhance carnation growth under conditions of heat stress, various concentrations of melatonin were added to in vitro culture media. The mechanism by which melatonin reduced heat stress damage was then studied by taking measurements of morphological parameters, levels of reactive oxygen species (ROS), antioxidant enzymes, and malondialdehyde (MDA), as well as differential gene expression, in carnation plants during in vitro culture. These data revealed that untreated carnation plants were more harmed by conditions of heat stress than plants treated with melatonin. Melatonin at concentrations of 5 and 10 mM increased chlorophyll content, fresh weight, and plant height to a greater extent than other concentrations. Melatonin may, thus, be used to alleviate damage to carnations caused by heat stress. The application of melatonin was also found to reduce oxidative damage and enhance antioxidant defense mechanisms. In addition, the expression of heat-related genes was found to be upregulated; in melatonin-treated plants, an upregulation was recorded in the expression of *GAPDH, DcPOD1, DcPOD2, DcPOD3, Gols1, MBF1c, HSF30, HSP101, HSP70,* and *sHSP* (MT) genes. In short, we found that melatonin treatment increased heat tolerance in carnation plants. The data presented here may serve as a reference for those seeking to enhance the growth of plants in conditions of heat stress.

**Keywords:** antioxidant enzymes; carnation; gene expression; heat stress; in vitro; melatonin

## 1. Introduction

Carnation (*Dianthus caryophyllus* L.) is one of the most important cultivated plants; it is grown worldwide for use as a cut flower [1]. Carnation is a perennial herbaceous plant that is relatively well adapted to cold seasons but is very sensitive to high temperatures [2]. The best temperature ranges for growing carnations are 13–15 °C in summer and 10–11 °C in winter [3,4]. In recent years, the problem of climate change has made it necessary to better understand the effects of heat stress on plant growth and how any negative effects might be alleviated [5]. Indeed, this is now a critical issue worldwide, because heat stress is known to have an inhibitory effect on the growth of most crops [6], including lower productivity levels in the cultivation of ornamental plants [3,7].

Melatonin (N-acetyl-5-methoxytryptamine) has been shown to enhance plant production under conditions of heat stress [8]. Melatonin participates in many biological processes, including root development, shoot differentiation, leaf senescence, cell elongation, and regulation of the process of photosynthesis [9,10]. The ability of melatonin to enhance the tolerance of plants to abiotic stresses, such as high temperature, salinity, and cold stress, has also been noted [11,12]. The antioxidant effect of melatonin appears to be a direct result of the activation of antioxidant enzymes and scavenging ROS [13,14]. The authors of [15] reported that the mechanism by which melatonin achieved such effects was by increasing the efficiency of mitochondrial electron transport. Researchers have also found that melatonin stimulates molecular pathways in several plants [8,13,14,16]. It has also been reported that, in some ornamental plants, melatonin plays a key role during heat stress through its effect on genes, which are directly related to heat tolerance [7,17].

In the floriculture industry, the search for a protocol to minimize the damage caused to carnations by high temperatures is a pressing issue [17]. The treatment of plants with substances under in vitro and ex vitro conditions is one of the most important methods used for reducing heat stress in plants [18]. Previous studies of carnations have shown that heat stress transcription factors (Hsfs) and heat shock proteins (Hsps) both enhance the tolerance of plants to high temperatures [19–21]. Considering the achievement of such effects using exogenous chemicals, researchers have shown that melatonin helps plants recover from heat stress [22]; however, it remains uncertain whether heat tolerance in carnations may be enhanced by melatonin. Heat stress factors (Hsfs) are important regulatory variables that can directly activate the transcription of downstream Hsps. They are essential for delivering heat stress information and enhancing heat tolerance in plants [23].

In the present study, we evaluated the effects of different levels of melatonin on carnations subjected to heat stress. The relationships between the amounts of melatonin added to in vitro media and physiological parameters, such as antioxidant enzymes, MDA, and ROS, were studied. In addition, at the molecular level, differential expressions of related genes were also investigated. This is the first study conducted to determine the best level of melatonin to be added to in vitro culture media for the cultivation of carnations.

## 2. Materials and Methods

### 2.1. Plant Material and Culture Conditions

Carnation (*Dianthus caryophyllus* L.) seeds were imported from Japan. The experiments were conducted at the Laboratory of Tissue Culture of Vegetable and Ornamental Plants, Horticulture Division, Mansoura University, Mansoura, Egypt. Sodium hypochlorite (2%) was used for sterilizing seed surfaces. Seeds were then rinsed three times with double-distilled sterile water. Subsequently, glass jars (250 mL) containing MS media (30 g $L^{-1}$ sucrose and 8 g $L^{-1}$ agar (Difco Bacto TM Agar, Carolina, Burlington, NC, USA)) were used for seeds without adding any growth regulators to the media [24]. In vitro cultures were grown at 26 $\pm$ 1 °C under a long-day photoperiod (16/8 h light/dark) provided by lamps with a light intensity of 2500 lux.

### 2.2. Heat Application and Morphological Traits

Seedlings of height 4–7 cm were used for heat treatments. Absolute ethanol (99.99%) was used for dissolving the melatonin (Sigma-Aldrich, St. Louis, MO, USA); stock solutions at a concentration of 100 mM were then prepared. To test the melatonin (MT) application under heat stress, all seedlings (40 days old) were assigned to new jars containing MS medium fortified with MT at concentrations of 0 (control), 1, 5, and 10 mM. Four days after transfer to media, seedlings were moved to a growth chamber with a temperature of 42 °C for 1, 3, 6, 12, and 24 h. Samples were directly taken at each time point. Seedling microshoots were harvested, then immersed in liquid nitrogen and stored at −80 °C for subsequent physiological and molecular analyses. To obtain measurements of fresh weight and plant height, the carnation seedlings were first moved to a growth chamber with a temperature of 26 $\pm$ 1 °C. The indicated parameters were then obtained after a recovery period of 3 days.

### 2.3. Photosynthetic Chlorophyll Quantification

Chlorophyll (Chl) content was determined according the method of Frank et al. [25] with some modifications. Briefly, a 100 mg of carnation leaves was ground with a sterile pestle. An amount of 2 mL of ethanol (97%) was added to the ground leaves, and the mixture was allowed to incubate at 4 °C for 2 days. The mixture was then centrifuged at 12,000× *g* at 4 °C for 1 min, and the OD value of the supernatant was assayed by spectrophotometer (Shimadzu, Kyoto, Japan) at 665 nm and 649 nm wavelengths.

### 2.4. Reactive Oxygen Species

Samples of carnation plants (approximately 0.5 g = 3–4 explants) were taken for the purpose of detecting hydrogen peroxide ($H_2O_2$) levels [26]. Shoots were ground in liquid nitrogen, then dissolved in 5.0 mL of 0.1% trichloroacetic acid (TCA). Ground samples were then centrifuged at 10,000× *g* for 15 min. The supernatant was mixed with 10 mM potassium phosphate buffer. $H_2O_2$ levels were detected using a standard curve. Explant samples of 0.5 g weight were also used to determine levels of hydroxyl radical (OH) content. Samples were mixed with 15 mM 2-deoxy-D-ribose at 37–38 °C for 3 h (pH 7.5). Samples of the resulting mixture (0.7–0.8 mL) were then mixed with 0.5% (*w*/*v*) thiobarbituric acid (TBA at concentration 1%, dissolved in 10 mM NaOH) and 1 mL glacial acetic acid, then placed in a water bath (100 °C) (Fisher Scientific Isotemp Digital-Control Water Bath, Kuala Lumpur, Malaysia) for 45 min. Samples were then immediately cooled at 4 °C for 15 min and levels of hydroxyl radicals were detected according to the method of Halliwell et al. [27]. Superoxide radical ($O_2^-$) concentrations were detected according to the method of Elstner and Heupel [28]. Following a pH 7.7 adjustment, 0.5 g of each microshoot sample was combined with potassium phosphate buffer, and the mixture was centrifuged at 4000× *g* for 12 min. The reaction was then maintained at 25 °C for 24 h, and the supernatant's absorbance was measured at 530 nm.

### 2.5. Detection and Quantification of ROS

Levels of accumulated ROS in carnation leaves after heat stress were determined according to the method of Fukao et al. [29]. After the 42 °C heat treatment, samples were taken from plants subjected to different levels of melatonin application for the purpose of detecting hydrogen peroxide. Excised aerial vegetative samples were treated with 1 mg/mL DAB (3,3′diaminobenzidine tetra-hydrochlorride) in 50 mM Tris Acetate buffer, pH 5.0, and then incubated at 25 °C for 24 h under darkness in order to detect hydrogen peroxide. After staining, each plant's uppermost leaf was boiled in 95% *v*/*v* ethanol for 20 min to eliminate chlorophyll and then rehydrated by incubation in 40% *v*/*v* glycerol for 16 h at 25 °C. At least seven distinct plants were used in each experiment's replication, and representative photos are provided.

### 2.6. Antioxidant Enzyme Extraction and Malondialdehide Estimation

Amounts of approximately 0.5 g of fully fresh tissues were collected every 7 days during the whole period of transfer to the acclimatization treatments. The plant material was homogenized in 4 mL of 0.1 M phosphate buffer (pH 7.0, contained 2 mM EDTA + 1% PVP at 4 °C). Then, it was centrifuged at 12,000× *g* for 10 min at a temperature of 4 °C. The supernatant was stored at 4 °C for the purpose of measuring enzyme activity. Three biological replicates were used to measure all enzymes.

Peroxidase (POD, EC 1.11.1.7) was detected using guaiacol (Sigma-Aldrich, Burlington, MA, USA) in a mixture of 3 mL [30] consisting of 2.7 mL phosphate buffer (25 µM, pH 7.0) with 0.1 mL $H_2O_2$ (0.4%), 0.1 mL guaiacol (1.5%), and 0.1 mL of enzyme extract. The absorbance was calculated at 470 nm. By measuring how quickly 1 M of guaiacol oxidized $g^{-1}$ FW $min^{-1}$ at 25 ± 2 °C, POD enzyme activity was determined.

Catalase (CAT, EC. 1.11.1.6) activity was measured following the method of Góth [31]. The total volume of the reaction mixture was 3 mL; this consisted of 0.1 mL enzyme extract and 0.1 mL $H_2O_2$ (0.4%), in addition to 2.8 mL phosphate buffer (25 mM, pH 7.0). The reduction in absorbance was measured at 240 nm.

Superoxide dismutase (SOD, EC 1.15.1.1) enzyme activity was checked by measuring the inhibition of the amount of nitro blue tetrazolium (NBT) photochemical reduction (Sigma-Aldrich, Burlington, MA, USA) following the method of Sheteiwy et al. [32]. The total volume of the reaction mixture was 3.1 mL, consisting of 0.1 mL of enzyme extract and 3 mL NBT solution. Reaction tubes were placed under 15 W fluorescent lamps for 15 min after the addition of 2 μmol $L^{-1}$ riboflavin. The control treatment was the reaction mixture without any enzyme extract. One unit of SOD was determined to be the volume of extract that caused 50% inhibition of NBT reduction. The photoreduction of NBT was measured at 560 nm.

Malondialdehyde (MDA) concentration was estimated using TBA reactive metabolites, according to the method of Heath and Packer [33], so that 1.5 mL of extract solution was added to 2.5 mL of 5% TBA formed in 5% TCA, then subjected to a temperature of 95 °C for 15 min before rapid cooling on ice. The MDA was measured at 532 nm after centrifugation of the supernatant at 5000× *g* for 10 min. Correction of nonspecific turbidity was measured at 600 nm.

### 2.7. Real-Time Quantitative PCR Analysis

Using Trizol reagent (Invitrogen, Carlsbad, CA, USA), total RNA was isolated from carnation microshoots and then processed with RNase-free DNase (Promega, Madison, WI, USA). Using reverse transcriptase (TOYOBO, Otsu, Japan), five micrograms of DNA-free total RNA (about 500 ng/μL) was reverse-transcribed into first-strand cDNA. Total RNA was used to synthesize the first-strand cDNA using the PrimeScript RT Reagent Kit with gDNA Eraser (Perfect Real Time) (Takara, Otsu, Japan). PCR was performed with TB Green® Premix Ex Taq (Takara, Otsu, Japan) on a CFX Connect Real-Time System. The thermal cycle program was as follows: 95 °C for 5 min, 45 cycles of 95 °C for 10 s and 60 °C for 30 s; 95 °C for 5 min, 65 °C for 5 s, 95 °C for 1 min program. Analysis of each sample was conducted with three biological replicates. The relative expression levels of genes were obtained by the $2^{-\Delta\Delta CT}$ method [34]. Primers for qRT-PCR are shown in Table 1.

**Table 1.** Sequences of oligonucleotide primers used in RT-PCR.

| Gene Name | Accession Number | Forward Primer | Reverse Primer | Product Length |
|---|---|---|---|---|
| *GAPDH* | Dc49995 | CACTCCATCACAGCCACACAA | CACGGAAAGCCATACCAGTCA | 190 |
| *DcPOD1* | DT214806 | GTGTAGTCTCGTGTGCCGAT | CTTCGGGGATTTTGCCTTGC | 144 |
| *DcPOD2* | DT214807 | AGCAACCCTTTACCAGCAAC | TCGTCTTCCAACCCAGTGGA | 170 |
| *DCPOD3* | CF259499 | CTGAACGGTAAAGGGTTGCTG | AACAAAACCATGGCCCTAGC | 131 |
| *Gols1* | Dc14879 | GGGGTCAAAGCCGTGGAGAT | CTCTAAAGGGCTCCAGTTTCGT | 155 |
| *MBF1c* | Dc_15682 | TGAATGCCCGGAAACTCGAC | GACCGCCTTTCCATTCTCGT | 179 |
| *HSF30* | Dc83420 | CCGGAGCTAGACAGGCTAATG | TCGATTTCTGGGGGCATTGA | 128 |
| *HSP101* | Dc_70612 | AGGTGGTGACTGAACTGTCG | GATCGACCATCCCTCCGTTT | 126 |
| *HSP70* | Dc_37984 | CAGGCGAAGAGAGAAGCCAT | CTGAGTCACCCCGGTTTCAA | 168 |
| *sHSP (MT)* | Dc_89619 | TCTCCGGCAGTAATGTCGTC | GTTCCTCTCAGAGCGGTCG | 137 |

### 2.8. Statistical Analyses

The experiments were prepared in a completely randomized design. For each treatment, about 30 explants were used, and all experiments were repeated twice. The data were subjected to analysis of variance (ANOVA) using SPSS version 16 (SPSS Inc., Chicago, IL, USA). Duncan's multiple range test was used to test the significance of differences between means (<0.05). The results were presented as mean ± SD.

## 3. Results

### 3.1. Effects of Melatonin Application on Chlorophyll, Plant Height, and Fresh Weight

It can be seen from the data in Figure 1 that heat treatments reduced the levels of chlorophyll in carnation microshoots; however, higher levels of chlorophyll were recorded in media supplemented with melatonin, compared with control treatments (Figure 1A). In addition, when treated with melatonin at a concentration of 5 mM, carnation microshoots exhibited levels of chlorophyll, which were significantly higher than for any other concentration. After 6–24 h of heat treatment, the heights of carnation plants were lower, compared with earlier time points, but plants treated with melatonin exhibited significant differences compared with control plants (Figure 1B), with the best values recorded for melatonin concentrations of 5 and 10 mM. At all time points, there were no significant differences between melatonin treatments and untreated plants in terms of fresh weight measurements, but weights were higher at all times in melatonin-treated plants compared with controls (Figure 1C). These data indicated that melatonin can alleviate the harmful effects of high temperature in carnations (Figure 2).

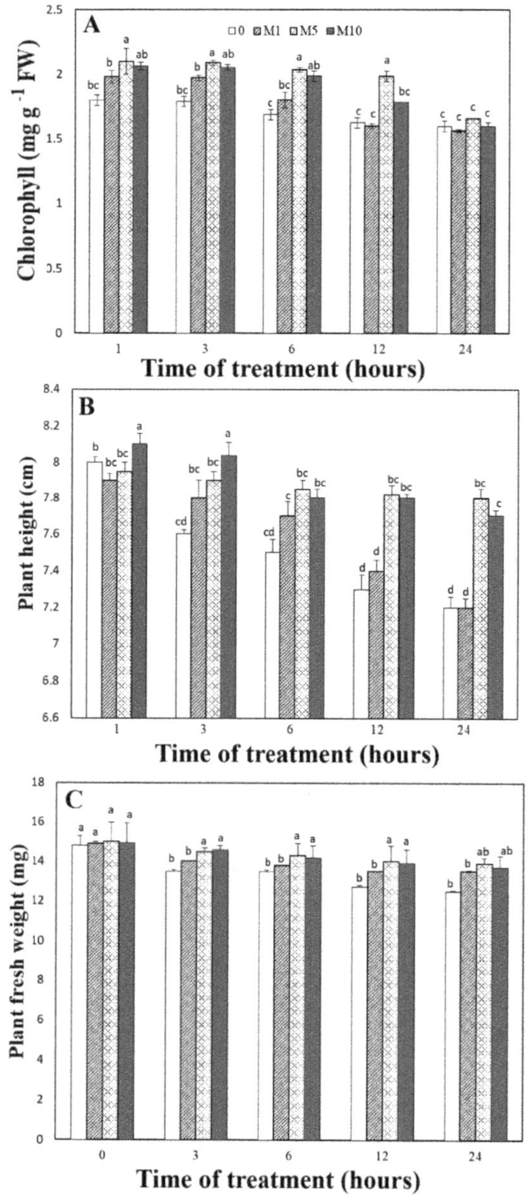


**Figure 1.** Effect of melatonin at (0, 1, 5, and 10 mM) on the carnation physiological parameters (**A**) chlorophyll content, (**B**) Plant height, and plant fresh weight (**C**). The plants were treated with in vitro heat stress on the growth chamber with 42 °C temperature for 1, 3, 6, 12, and 24 h. Different lowercase letters represent significant differences according to Duncan's tests ($p < 0.05$).

**Figure 2.** Effects of melatonin application on the morphological traits of in vitro carnation plants under heat stress condition after 24 h. Microshoots of carnation plants subjected to free medium (control) (**A**). The microshoots' morphological traits under melatonin application at concentration of 1 (**B**), 5 (**C**), and 10 mM (**D**) after 24 h of culture.

### *3.2. Effects of Melatonin Application on Accumulation of Reactive Oxygen Species (ROS) in Carnation Plants under Heat Stress*

Melatonin was found to reduce the levels of $H_2O_2$ content in carnation plants at all testing times (Figure 3A). However, the levels of $H_2O_2$ in carnation plants declined more notably during the later periods of the experiment. Furthermore, after three hours of heat treatment, control treatment samples showed the highest $H_2O_2$ levels. The $O_2$ levels were assessed in order to determine how melatonin influences ROS metabolisms in carnations during heat stress. DAB solution was used to identify the accumulation of ROS ($H_2O_2$) in carnation leaves (Figure 4). The observations demonstrated that ROS were formed as a heat shock reaction; however, the application of melatonin at a concentration of 5 mM reduced the buildup of ROS to a greater degree than any other melatonin concentration, including controls. In control samples, levels of superoxide content significantly increased under heat stress after 3 h (Figure 3B). However, after 24 h, carnation plants treated with melatonin exhibited lower levels of superoxide content than control plants. These support the idea that melatonin can positively affect superoxide generation under high-temperature conditions. The data in Figure 3 also showed that application of melatonin increased OH levels to a greater degree than control treatments (Figure 3C).

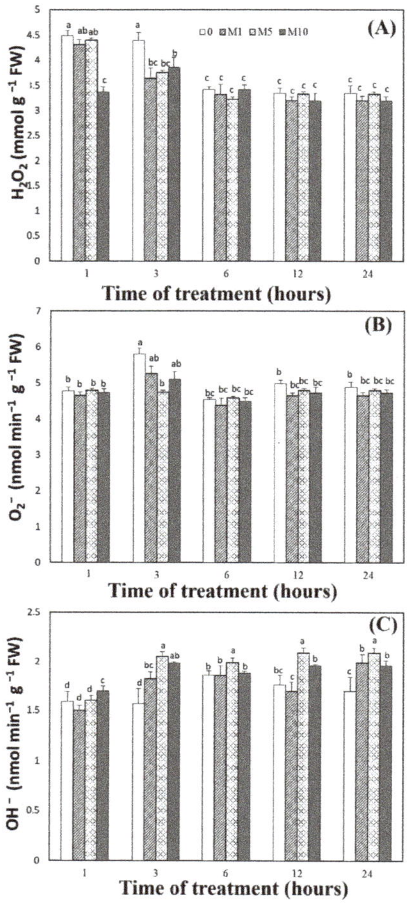


**Figure 3.** Effect of melatonin at (0, 1, 5, and 10 mM) on hydrogen peroxide ((**A**); $H_2O_2$), superoxide radical ((**B**); $O_2^-$) and hydroxyl ((**C**); $-$OH) contents of carnation plants during heat stress condition. Means expressed the average of three replicates ± SE, and means within each graph denoted by the same lowercase letter did not significantly differ according to Duncan test at $p < 0.05$.

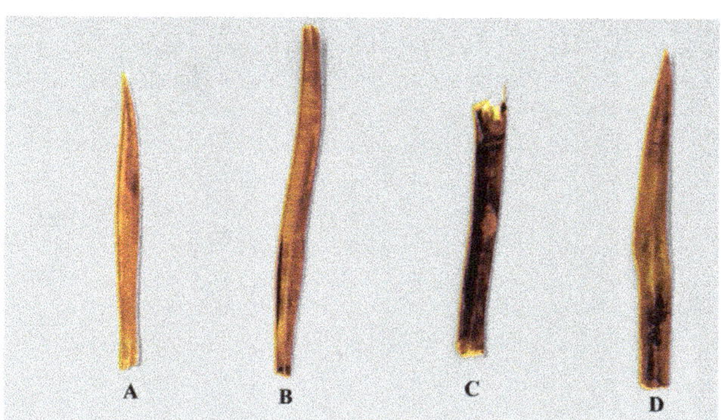


**Figure 4.** Accumulation of $H_2O_2$. Carnation leaves were collected after 12 h after melatonin application at concentration of 0 (control) (**A**), 1 mM (**B**), 5 mM (**C**), and 10 mM (**D**). Carnation leaves treated with DAB solution to measure the hydrogen peroxide, then transferred in darkness one day.

### 3.3. Effects of Melatonin Application on Antioxidant Enzyme Activities and MDA in Carnation Plants under Heat Stress

A relationship between increased antioxidant enzyme activities and reduced ROS levels was clearly evident in our data (Figure 5). The effects of melatonin on carnation plants under high-temperature stress were demonstrated by our finding that POD activity was significantly increased at all testing times with melatonin concentrations of 5 and 10 mM. The lowest levels of POD activity were recorded for control treatments. In addition, POD activity was increased more notably at 12 and 24 h than at earlier times (1–3 and 6 h). SOD enzymes exhibited a similar trend; however, the greatest increase in SOD activity was recorded at 12 h after melatonin application. The lowest value of SOD activity was recorded for untreated carnation plants. Under heat stress, melatonin applications at concentrations of 5 and 10 mM both resulted in significantly increased CAT activity after 1 and 12 h. The application of melatonin at a concentration of 1 mM did not significantly affect CAT in the first hours after treatment (Figure 5C). High-temperature treatment also caused the levels of malondialdehide (MDA) in carnation plants to significantly increase after 1, 12, and 24 h. However, plants which received melatonin application had lower levels of MDA at all tested times compared with untreated plants. Finally, greater increases in MDA activity were recorded at later time points (Figure 5D).

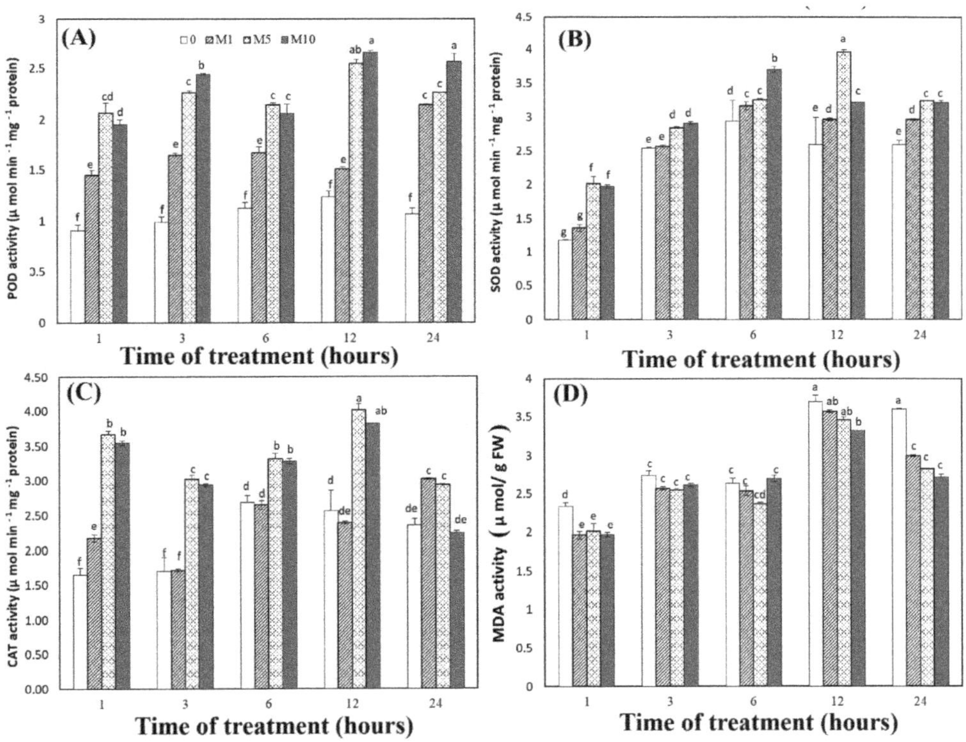


**Figure 5.** Effect of melatonin at (0, 1, 5, and 10 mM) on hydrogen peroxide on major antioxidant enzymes POD (**A**), SOD (**B**), CAT (**C**) and MDA (**D**) under heat stress condition of carnation plants. Means expressed the average of three replicates ± SE, and means within each graph denoted by the same lowercase letter did not significantly differ according to Duncan test at $p < 0.05$.

### 3.4. Effects of Melatonin Application on Gene Expression in Carnation Plants under Heat Stress

In order to study molecular changes in plants treated with melatonin under in vitro heat stress, the levels of expression of genes participating in heat stress were studied. Ten genes were subjected to qRT-PCR analysis. The obtained data shown in Figure 6 show that *GAPDH* was downregulated under heat stress but upregulated when melatonin was applied, especially at concentrations of 1 and 5 Mm. Additionally, the expression of the *GAPDH* gene was higher after 1 and 3 h of heat stress than at any than other time points. The relative expression of *DcPOD1* increased gradually between the first and sixth hours of heat treatments but then declined, as measured at 12 and 24 h (Figure 6B).

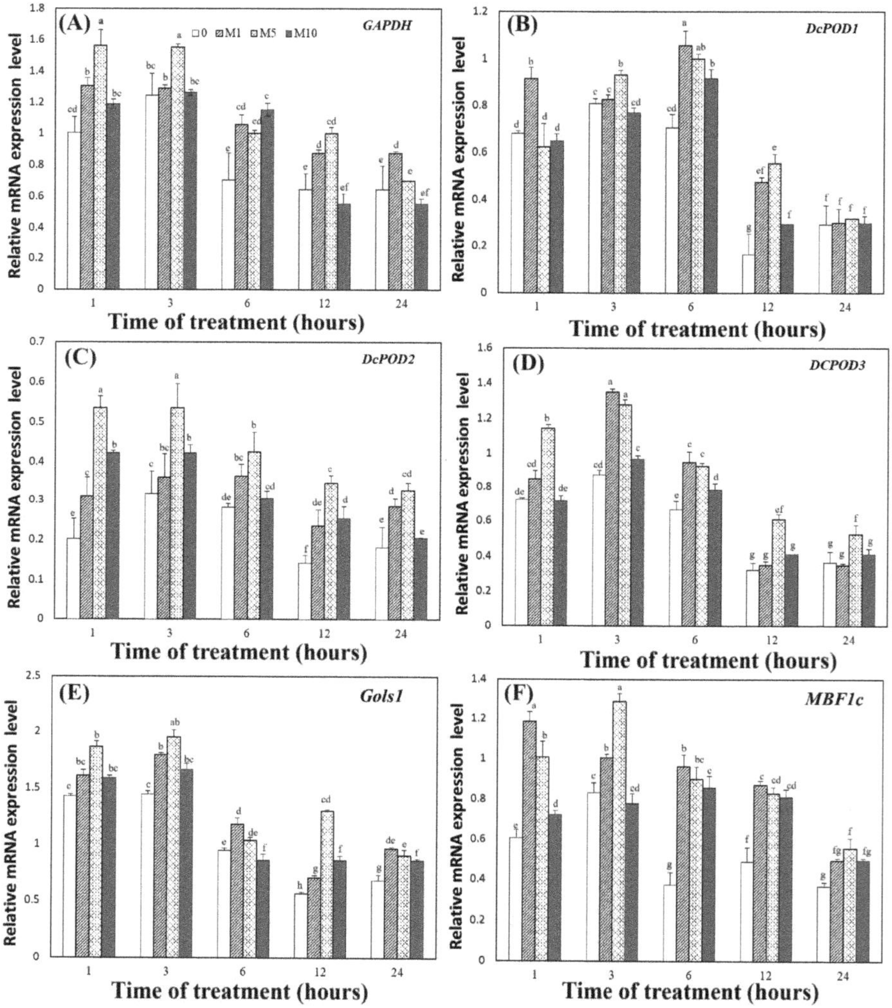


**Figure 6.** Effects melatonin at different concentration on expression pattern of (**A**) *GAPDH*, (**B**) *DcPOD1*, (**C**) *DcPOD2*, (**D**) *DcPOD3*, (**E**) *Gols1* and (**F**) *MBF1c* in carnation plants under heat stress condition by *qRT-PCR*. Microshoots were transferred to MS medium containing melatonin at concentration of 0, 1, 5, and 10 mM), and then heat stress (42 °C) was applied for indicated time. Microshoots were collected immediately for RNA extraction. Notably, 10 genes highly induced by heat stress were used for qPCR analysis and most of them showed significant changes after melatonin applications. Data are the means of three replicates with standard division shown by vertical bars indicates significant differences at $p < 0.05$. Different lowercase letters represent significant differences according to Duncan's tests ($p < 0.05$).

However, treatments with 1 or 5 Mm of melatonin increased the expression of the *DcPOD1* relative gene more than other concentrations, including controls. Similarly, the expressions of *DcPOD2* and *DcPOD3* both decreased in untreated plants at all time points but increased in most plants subjected to melatonin application (Figure 6C,D). In the case of the *DcPOD2* relative gene, at all time points, expression was best in plants treated with a melatonin concentration of 5 Mm. For the *DcPOD3* relative gene, the best results were obtained with concentrations of both 1 and 5 mM. Treatment of carnations with melatonin in vitro increased the relative expression of the *Gols1* gene compared with untreated plants. This increase in expression of the *Gols1* relative gene was greater after 1 and 3 h, compared with other time points (Figure 6E). Similarly, the relative transcript level of *MBF1c* was upregulated in heat-stressed plants treated with melatonin compared with control samples (Figure 6F).

The data presented in Figure 7 show that two relative genes (*HSF30* and *HSP101*) were upregulated in melatonin-treated plants, with the greatest increases recorded for the one- and three-hour time points. Similar data were obtained for the relative genes *HSP70* and *sHSP* (*MT*), as both these genes were upregulated in melatonin-treated samples at the one- and three-hour points (Figure 7C,D). Overall, we recorded increases in the expression of the *GAPDH*, *DcPOD1*, *DcPOD2*, *DcPOD3*, *Gols1*, *MBF1c*, *HSF30*, *HSP101*, *HSP70*, and *sHSP* (MT) genes (Figures 6 and 7), indicating that these genes may be putative targets of melatonin involved in response to heat stress at the transcriptional level.

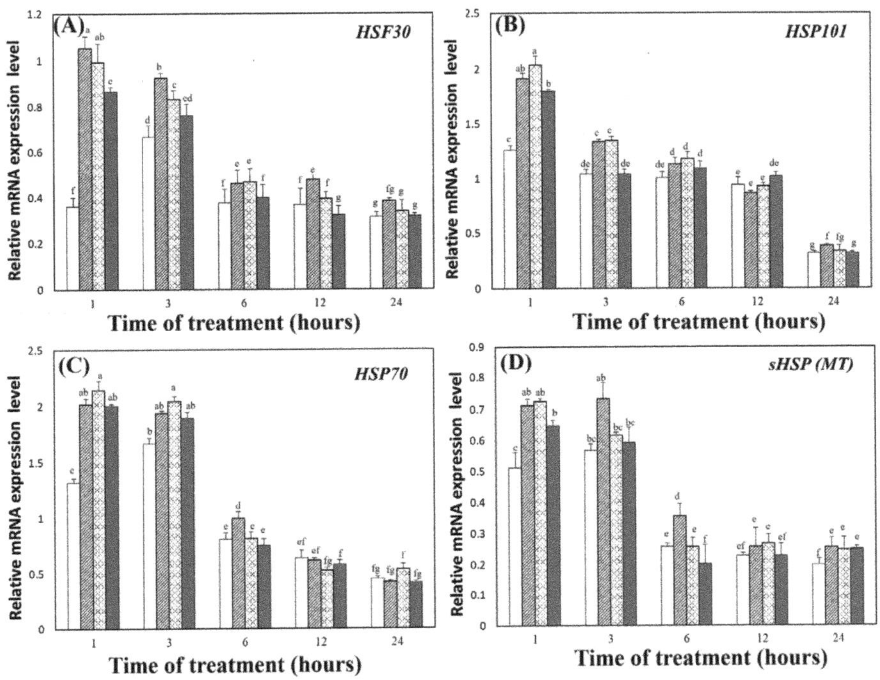


**Figure 7.** Effects of melatonin at different concentration on expression pattern of (**A**) *HSF30*, (**B**) *HSP101*, (**C**) *HSP70*, and (**D**) *sHSP* (*MT*) in carnation plants under heat stress condition by *qRT-PCR*. Microshoots were transferred to MS medium containing melatonin at concentration of 0, 1, 5, and 10 mM), and then heat stress (42 °C) was applied for indicated time. Microshoots were collected immediately for RNA extraction. Notably, 10 genes highly induced by heat stress were used for qPCR analysis and most of them showed significant changes after melatonin applications. Data are the means of three replicates with standard division shown by vertical bars indicates significant differences at $p < 0.05$. Different lowercase letters represent significant differences according to Duncan's tests ($p < 0.05$).

## 4. Discussion

Heat stress inhibits the growth and development of plants. However, melatonin application has been reported to alleviate damage caused by heat stress in plants, such as tall fescue [7], strawberry (*Fragaria* × *ananassa*) [35], and tomato [36]. The present study is the first to evaluate the effects of melatonin application in vitro on carnation plants under heat stress. To assess the effects of melatonin treatments on heat tolerance in carnation plants, measurements of a number of morphological indices, physiological parameters, and molecular changes were obtained and analyzed.

It is well known that leaf senescence reduces green color in plants; low chlorophyll levels in plants may, therefore, be seen as an indicator of leaf senescence [37]. During the photosynthesis process, chlorophyll protects plants from light damage and is involved in the absorption and transmission of light energy [38]. Previous studies have reported that melatonin plays an effective role in increasing chlorophyll levels in several plants, including maize, tomato, tall fescue, *Medicago sativa*, and *Malus hupehensis*, thus protecting plants from leaf damage and the shock caused by high temperatures [7,37–41].

Treatment with melatonin has also been reported to enhance fresh weight, dry weight, and plant height in various plants subjected to heat stress conditions [11,42,43]. In carnations, the data obtained in the present study indicated that melatonin application resulted in increases in chlorophyll content, fresh weight, and plant height, suggesting that substances like melatonin might have a vital role to play in alleviating the effects of heat stress in carnations.

Under heat stress conditions, the accumulation of ROS, such as OH–, $H_2O_2$, and $O_2^-$, and free radicals causes increases in MDA and the leakage of electrolytes, which may be scavenged by plants under stress conditions through the stimulation of the activities of antioxidant enzymes. For example, SOD can inhibit the superoxide radical ($O_2^-$) by converting it to $H_2O_2$, and the catalase enzyme is also able to inhibit it by releasing $H_2O$. It has previously been shown that melatonin treatment alleviates oxidative disfiguration under abiotic stresses by conserving ROS [39,44]. Dewir et al. [2] found that POD enzyme activity was enhanced in carnation plants under heat stress and that this was associated with scavenging ROS in treated plants. In addition, POD enzymes can reduce ROS activity by enhancing secondary metabolites in plants; however, these secondary metabolites protect the plants from heat stress damage by supplying mechanical support to the cultured cells through their participation in the maintenance of cell membrane integrity, further demonstrating that POD plays an important role in lignin synthesis under heat stress conditions [45,46]. Additionally, plants, such as tea plant (*Camellia sinensis*), eggplant (*Solanum melongena*), and carnation, which were treated with melatonin, have all been shown to exhibit low levels of MDA content under heat stress [17,47,48]. In terms of the parameters of minimal fluorescence (F0) and maximal fluorescence (Fm), from which we may obtain maximal variable fluorescence (Fv = Fm − F0) and the photochemical efficiency of PSII (Fv/Fm), we note that melatonin has been shown to play a vital role in repressing heat stress, by reducing leaf curling, blotching, and fraying and increasing the value of Fv/Fm [17,49,50]. A series of metabolic changes are involved in heat stress responses in plants [51]. These include the following: overproduction of ROS and reactive nitrogen species (RNS); lipid peroxidation, production of end products, such as MDA; photoinhibition; protein denaturation and degradation; and accumulation of compatible solutes [52]. Lipid peroxidation is caused by abiotic stress [53]. The end products of lipid peroxidation are reactive aldehydes such as malondialdehide (MDA) and 4-hydroxynonenal (HNE) [54]. In the present study, heat treatments significantly increased MDA content in control carnation plants; contrarily, melatonin application resulted in decreased MDA production. These findings indicate that melatonin alleviates cell membrane damage caused by heat stress. We also found that the content levels of ROS and MDA decreased in carnation plants treated with melatonin (Figures 4 and 5), but levels of the enzymes POD, CAT, and SOD all increased (Figure 5). Taken together, the data reported here confirm the significant effect of melatonin on morphological traits and chlorophyll content (Figure 3); they are in line with the findings of previous studies, which showed that oxidative stress in plants may be alleviated by melatonin treatment [7,8,10,16,55,56].

High temperatures may cause the denaturation of proteins. High-temperature conditions may also cause cell cytotoxicity [57]. Previous studies have reported that, in cells exposed to heat stress, protein homeostasis can be safeguarded and protein aggregation reduced by the interaction of chaperones with stress-denatured proteins [3,17,58]. In addition, heat-induced genes have been found to be the most effective genes in encoding molecular chaperones such as heat shock proteins [19,59]. These genes have been found to be upregulated and induced when cells are exposed to high or low temperatures [3,7,60,61]. In the present study, we evaluated the expression of ten genes and found that heat shock proteins were upregulated under melatonin application. Previous studies have also found that the HSP40 family plays a vital role in enhancing ATPase activity [3]. Exogenous substances can effectively increase the expression of Hsfs and Hsps; this has been found to improve the thermotolerance in creeping bent grass (*Agrostis stolonifera*) [62] and strawberry [35]. The genes that we evaluated in our study have been reported as participating in protein folding and unfolding, and this may enhance the ability of cells to tolerate heat stress [43].

## 5. Conclusions

In the present study, an experiment was performed to determine whether the addition of melatonin to in vitro media would enhance the response of carnation microshoots to heat stress (Figure 8). Based on the measured effects of melatonin on chlorophyll content, fresh weight, and plant height in carnations under heat stress, we may say that melatonin enhances the response to heat stress. We also found that melatonin application improved thermotolerance in carnations by lowering the content levels of ROS and MDA while increasing antioxidant enzyme activities. Through such means, melatonin exerts an inhibitory effect upon heat stress. We also assessed the relationship between melatonin application and the expression of related heat stress genes. We found that the mechanism by which heat stress damage is alleviated is strongly associated with the upregulation of the POD, HSP, and HSF genes. We also found that treatment with melatonin at concentrations of 5 and 10 mM promoted growth and development in carnation plants exposed to 42 °C, in comparison with untreated plants. The supplementation of media with melatonin reduced oxidative damage through scavenging ROS and inhibiting MDA synthesis; it also increased the activities of antioxidant enzymes and upregulated genes involved in heat tolerance. From the results reported here, we may state that melatonin plays a significant role in improving the tolerance of plants to conditions of abiotic stress, especially heat stress.

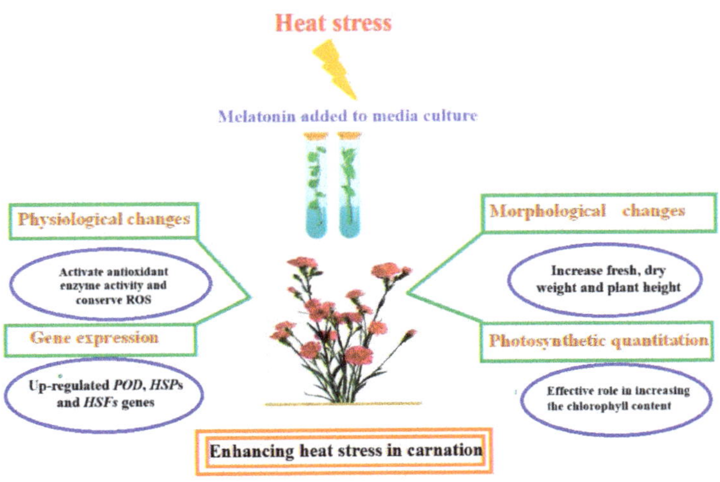


**Figure 8.** The mechanism of melatonin's role in increasing carnation heat tolerance.

**Author Contributions:** Both authors designed the experiment, conducted the experiment, wrote the manuscript. All authors have read and agreed to the published version of the manuscript.

**Funding:** The current work was funded by Mansoura University, Faculty of Agriculture, Egypt.

**Data Availability Statement:** Data are contained within the article.

**Acknowledgments:** The authors extend their gratitude to the anonymous reviewers and editors for their helpful reviews and critical comments. The Corresponding author would like to thank Xiuyun Wang of Zhejiang University's College of Agriculture and Biotechnology for her strong scientific support.

**Conflicts of Interest:** The authors declare no conflicts of interest.

## References

1. Navarro, A.; Elia, A.; Conversa, G.; Campi, P.; Mastrorilli, M. Potted mycorrhizal carnation plants and saline stress: Growth, quality and nutritional plant responses. *Sci. Hortic.* **2012**, *140*, 131–139. [CrossRef]
2. Dewir, Y.H.; Yu, R.; Fan, Y.; Ali, M.B.; Paek, K.Y. Effect of Temperature and gamma-Radiation on Antioxidant Defense System in Suspension Cell Culture of Carnation (*Dianthus caryophyllus*). *Korean J. Hortic. Sci.* **2005**, *46*, 329.
3. Wan, X.L.; Zhou, Q.; Wang, Y.Y.; Wang, W.E.; Bao, M.Z.; Zhang, J.W. Identification of heat-responsive genes in carnation (*Dianthus caryophyllus* L.) by RNA-seq. *Front. Plant Sci.* **2015**, *6*, 519. [CrossRef] [PubMed]
4. In, B.C.; Binder, B.M.; Falbel, T.G.; Patterson, S.E. Analysis of gene expression during the transition to climacteric phase in carnation flowers (*Dianthus caryophyllus* L.). *J. Exp. Bot.* **2013**, *64*, 4923–4937. [CrossRef] [PubMed]
5. Zahra, N.; Hafeez, M.B.; Ghaffar, A.; Kausar, A.; Zeidi, M.A.; Siddique, K.H.M.; Farooq, M. Plant photosynthesis under heat stress: Effects and management. *Environ. Exp. Bot.* **2023**, *206*, 105178. [CrossRef]
6. Wilson, R.A.; Sangha, M.; Banga, S.; Atwal, A.; Gupta, S. Heat stress tolerance in relation to oxidative stress and antioxidants in Brassica juncea. *J. Environ. Biol.* **2014**, *35*, 383.
7. Alam, M.N.; Zhang, L.; Yang, L.; Islam, M.R.; Liu, Y.; Luo, H.; Yang, P.; Wang, Q.; Chan, Z. Transcriptomic profiling of tall fescue in response to heat stress and improved thermotolerance by melatonin and 24-epibrassinolide. *BMC Genom.* **2018**, *19*, 224. [CrossRef]
8. Li, Y.; Zhang, Z.; He, C.; Zhu, K.; Xu, Z.; Ma, T.; Tao, J.; Liu, G. Melatonin protects porcine oocyte in vitro maturation from heat stress. *J. Pineal Res.* **2015**, *59*, 365–375. [CrossRef]
9. Hu, Z.; Fan, J.; Xie, Y.; Amombo, E.; Liu, A.; Gitau, M.M.; Khaldun, A.B.M.; Chen, L.; Fu, J. Comparative photosynthetic and metabolic analyses reveal mechanism of improved cold stress tolerance in bermudagrass by exogenous melatonin. *Plant Physiol. Biochem.* **2016**, *100*, 94–104. [CrossRef]
10. Zhang, N.; Zhang, H.J.; Zhao, B.; Sun, Q.Q.; Cao, Y.Y.; Li, R.; Wu, X.X.; Weeda, S.; Li, L.; Ren, S. The RNA-seq approach to discriminate gene expression profiles in response to melatonin on cucumber lateral root formation. *J. Pineal Res.* **2014**, *56*, 39–50. [CrossRef]
11. Shi, H.; Jiang, C.; Ye, T.; Tan, D.-X.; Reiter, R.J.; Zhang, H.; Liu, R.; Chan, Z. Comparative physiological, metabolomic, and transcriptomic analyses reveal mechanisms of improved abiotic stress resistance in bermudagrass [*Cynodon dactylon* (L). Pers.] by exogenous melatonin. *J. Exp. Bot.* **2014**, *66*, 681–694. [CrossRef]
12. Bajwa, V.S.; Shukla, M.R.; Sherif, S.M.; Murch, S.J.; Saxena, P.K. Role of melatonin in alleviating cold stress in A rabidopsis thaliana. *J. Pineal Res.* **2014**, *56*, 238–245. [CrossRef]
13. Galano, A.; Tan, D.X.; Reiter, R.J. On the free radical scavenging activities of melatonin's metabolites, AFMK and AMK. *J. Pineal Res.* **2013**, *54*, 245–257. [CrossRef] [PubMed]
14. Purushothaman, A.; Sheeja, A.A.; Janardanan, D. Hydroxyl radical scavenging activity of melatonin and its related indolamines. *Free Radic. Res.* **2020**, *54*, 373–383. [CrossRef] [PubMed]
15. Tomás-Zapico, C.; Coto-Montes, A. A proposed mechanism to explain the stimulatory effect of melatonin on antioxidative enzymes. *J. Pineal Res.* **2005**, *39*, 99–104. [CrossRef] [PubMed]
16. Wei, W.; Li, Q.T.; Chu, Y.N.; Reiter, R.J.; Yu, X.M.; Zhu, D.H.; Zhang, W.K.; Ma, B.; Lin, Q.; Zhang, J.S.; et al. Melatonin enhances plant growth and abiotic stress tolerance in soybean plants. *J. Exp. Bot.* **2015**, *66*, 695–707. [CrossRef]
17. Hu, D.; Zhang, X.; Xue, P.; Nie, Y.; Liu, J.; Li, Y.; Wang, C.; Wan, X. Exogenous melatonin ameliorates heat damages by regulating growth, photosynthetic efficiency and leaf ultrastructure of carnation. *Plant Physiol. Biochem.* **2023**, *198*, 107698. [CrossRef]
18. Sarwar, M.; SALEEM, M.F.; Ali, B.; Saleem, M.H.; Rizwan, M.; Usman, K.; KEBLAWY, A.E.; Ali, A.; Afzal, M.; Sheteiwy, M.S. Application of potassium, zinc and boron as potential plant growth modulators in *Gossypium hirsutum* L. under heat stress. *Turk. J. Agric. For.* **2022**, *46*, 567–584. [CrossRef]
19. Sun, Y.; Hu, D.; Xue, P.; Wan, X. Identification of the DcHsp20 gene family in carnation (*Dianthus caryophyllus*) and functional characterization of DcHsp17. 8 in heat tolerance. *Planta* **2022**, *256*, 2. [CrossRef] [PubMed]
20. Wan, X.; Sun, Y.; Feng, Y.; Bao, M.; Zhang, J. Heat stress transcription factor DcHsfA1d isolated from *Dianthus caryophyllus* enhances thermotolerance and salt tolerance of transgenic Arabidopsis. *Biol. Plant* **2022**, *66*, 29. [CrossRef]

21. Xue, P.; Sun, Y.; Hu, D.; Zhang, J.; Wan, X. Genome-wide characterization of DcHsp90 gene family in carnation (*Dianthus caryophyllus* L.) and functional analysis of DcHsp90-6 in heat tolerance. *Protoplasma* **2023**, *260*, 807–819. [CrossRef]
22. Ulhassan, Z.; Huang, Q.; Gill, R.A.; Ali, S.; Mwamba, T.M.; Ali, B.; Hina, F.; Zhou, W. Protective mechanisms of melatonin against selenium toxicity in *Brassica napus*: Insights into physiological traits, thiol biosynthesis and antioxidant machinery. *BMC Plant Biol.* **2019**, *19*, 507. [CrossRef]
23. Gong, Z.; Xiong, L.; Shi, H.; Yang, S.; Herrera-Estrella, L.R.; Xu, G.; Chao, D.-Y.; Li, J.; Wang, P.-Y.; Qin, F. Plant abiotic stress response and nutrient use efficiency. *Sci. China Life Sci.* **2020**, *63*, 635–674. [CrossRef]
24. Murashige, T.; Skoog, F. A revised medium for rapid growth and bio assays with tobacco tissue cultures. *Physiol. Plant.* **1962**, *15*, 473–497. [CrossRef]
25. Frank, W.; Ratnadewi, D.; Reski, R. Physcomitrella patens is highly tolerant against drought, salt and osmotic stress. *Planta* **2005**, *220*, 384–394. [CrossRef] [PubMed]
26. Perveen, S.; Anis, M. Physiological and biochemical parameters influencing ex vitro establishment of the in vitro regenerants of Albizia lebbeck. *Agrofor. Syst.* **2015**, *89*, 721–733. [CrossRef]
27. Halliwell, B.; Gutteridge, J.M.; Aruoma, O.I. The deoxyribose method: A simple "test-tube" assay for determination of rate constants for reactions of hydroxyl radicals. *Anal. Biochem.* **1987**, *165*, 215–219. [CrossRef] [PubMed]
28. Elstner, E.F.; Heupel, A. Inhibition of nitrite formation from hydroxylammoniumchloride: A simple assay for superoxide dismutase. *Anal. Biochem.* **1976**, *70*, 616–620. [CrossRef] [PubMed]
29. Fukao, T.; Yeung, E.; Bailey-Serres, J. The submergence tolerance regulator SUB1A mediates crosstalk between submergence and drought tolerance in rice. *Plant Cell* **2011**, *23*, 412–427. [CrossRef] [PubMed]
30. Elmongy, M.S.; Zhou, H.; Cao, Y.; Liu, B.; Xia, Y. The effect of humic acid on endogenous hormone levels and antioxidant enzyme activity during in vitro rooting of evergreen azalea. *Sci. Hortic.* **2018**, *227*, 234–243. [CrossRef]
31. Góth, L. A simple method for determination of serum catalase activity and revision of reference range. *Clin. Chim. Acta* **1991**, *196*, 143–151. [CrossRef] [PubMed]
32. Sheteiwy, M.; Shen, H.; Xu, J.; Guan, Y.; Song, W.; Hu, J. Seed polyamines metabolism induced by seed priming with spermidine and 5-aminolevulinic acid for chilling tolerance improvement in rice (*Oryza sativa* L.) seedlings. *Environ. Exp. Bot.* **2017**, *137*, 58–72. [CrossRef]
33. Heath, R.L.; Packer, L. Photoperoxidation in isolated chloroplasts: I. Kinetics and stoichiometry of fatty acid peroxidation. *Arch. Biochem. Biophys.* **1968**, *125*, 189–198. [CrossRef] [PubMed]
34. Livak, K.J.; Schmittgen, T.D. Analysis of relative gene expression data using real-time quantitative PCR and the $2-\Delta\Delta CT$ method. *Methods* **2001**, *25*, 402–408. [CrossRef] [PubMed]
35. Manafi, H.; Baninasab, B.; Gholami, M.; Talebi, M.; Khanizadeh, S. Exogenous melatonin alleviates heat-induced oxidative damage in strawberry (*Fragaria* × *ananassa* Duch. cv. Ventana) Plant. *J. Plant Growth Regul.* **2022**, *41*, 52–64. [CrossRef]
36. El-Yazied, A.A.; Ibrahim, M.F.; Ibrahim, M.A.; Nasef, I.N.; Al-Qahtani, S.M.; Al-Harbi, N.A.; Alzuaibr, F.M.; Alaklabi, A.; Dessoky, E.S.; Alabdallah, N.M. Melatonin mitigates drought induced oxidative stress in potato plants through modulation of osmolytes, sugar metabolism, ABA homeostasis and antioxidant enzymes. *Plants* **2022**, *11*, 1151. [CrossRef]
37. Jahan, M.S.; Shu, S.; Wang, Y.; Hasan, M.M.; El-Yazied, A.A.; Alabdallah, N.M.; Hajjar, D.; Altaf, M.A.; Sun, J.; Guo, S. Melatonin pretreatment confers heat tolerance and repression of heat-induced senescence in tomato through the modulation of ABA-and GA-mediated pathways. *Front. Plant Sci.* **2021**, *12*, 650955. [CrossRef]
38. Ahmad, S.; Kamran, M.; Ding, R.; Meng, X.; Wang, H.; Ahmad, I.; Fahad, S.; Han, Q. Exogenous melatonin confers drought stress by promoting plant growth, photosynthetic capacity and antioxidant defense system of maize seedlings. *PeerJ* **2019**, *7*, e7793. [CrossRef]
39. Antoniou, C.; Chatzimichail, G.; Xenofontos, R.; Pavlou, J.J.; Panagiotou, E.; Christou, A.; Fotopoulos, V. Melatonin systemically ameliorates drought stress-induced damage in M edicago sativa plants by modulating nitro-oxidative homeostasis and proline metabolism. *J. Pineal Res.* **2017**, *62*, e12401. [CrossRef]
40. Wassie, M.; Zhang, W.; Zhang, Q.; Ji, K.; Cao, L.; Chen, L. Exogenous salicylic acid ameliorates heat stress-induced damages and improves growth and photosynthetic efficiency in alfalfa (*Medicago sativa* L.). *Ecotoxicol. Environ. Saf.* **2020**, *191*, 110206. [CrossRef]
41. Wang, P.; Sun, X.; Chang, C.; Feng, F.; Liang, D.; Cheng, L.; Ma, F. Delay in leaf senescence of Malus hupehensis by long-term melatonin application is associated with its regulation of metabolic status and protein degradation. *J. Pineal Res.* **2013**, *55*, 424–434. [CrossRef]
42. Kotak, S.; Larkindale, J.; Lee, U.; von Koskull-Doring, P.; Vierling, E.; Scharf, K.D. Complexity of the heat stress response in plants. *Curr. Opin. Plant Biol.* **2007**, *10*, 310–316. [CrossRef]
43. Li, S.; Zhou, X.; Chen, L.; Huang, W.; Yu, D. Functional characterization of Arabidopsis thaliana WRKY39 in heat stress. *Mol. Cells* **2010**, *29*, 475–483. [CrossRef]
44. Meng, J.F.; Xu, T.F.; Wang, Z.Z.; Fang, Y.L.; Xi, Z.M.; Zhang, Z.W. The ameliorative effects of exogenous melatonin on grape cuttings under water-deficient stress: Antioxidant metabolites, leaf anatomy, and chloroplast morphology. *J. Pineal Res.* **2014**, *57*, 200–212. [CrossRef]
45. Chaitanya, K.; Sundar, D.; Masilamani, S.; Ramachandra Reddy, A. Variation in heat stress-induced antioxidant enzyme activities among three mulberry cultivars. *Plant Growth Regul.* **2002**, *36*, 175–180. [CrossRef]

46. Otter, T.; Polle, A. The influence of apoplastic ascorbate on the activities of cell wall-associated peroxidase and NADH oxidase in needles of Norway spruce (*Picea abies* L.). *Plant Cell Physiol.* **1994**, *35*, 1231–1238. [CrossRef]
47. Li, J.; Yang, Y.; Sun, K.; Chen, Y.; Chen, X.; Li, X. Exogenous melatonin enhances cold, salt and drought stress tolerance by improving antioxidant defense in tea plant (*Camellia sinensis* (L.) O. Kuntze). *Molecules* **2019**, *24*, 1826. [CrossRef]
48. Song, L.; Tan, Z.; Zhang, W.; Li, Q.; Jiang, Z.; Shen, S.; Luo, S.; Chen, X. Exogenous melatonin improves the chilling tolerance and preharvest fruit shelf life in eggplant by affecting ROS-and senescence-related processes. *Hortic. Plant J.* **2023**, *9*, 523–540. [CrossRef]
49. Zhang, L.; Hu, T.; Amombo, E.; Wang, G.; Xie, Y.; Fu, J. The alleviation of heat damage to photosystem II and enzymatic antioxidants by exogenous spermidine in tall fescue. *Front. Plant Sci.* **2017**, *8*, 1747. [CrossRef]
50. Fatma, M.; Iqbal, N.; Sehar, Z.; Alyemeni, M.N.; Kaushik, P.; Khan, N.A.; Ahmad, P. Methyl jasmonate protects the PS II system by maintaining the stability of chloroplast D1 protein and accelerating enzymatic antioxidants in heat-stressed wheat plants. *Antioxidants* **2021**, *10*, 1216. [CrossRef]
51. Fan, X.; Zhao, J.; Sun, X.; Zhu, Y.; Li, Q.; Zhang, L.; Zhao, D.; Huang, L.; Zhang, C.; Liu, Q. Exogenous melatonin improves the quality performance of rice under high temperature during grain filling. *Agronomy* **2022**, *12*, 949. [CrossRef]
52. Sanoubar, R.; Cellini, A.; Gianfranco, G.; Spinelli, F. Osmoprotectants and antioxidative enzymes as screening tools for salinity tolerance in radish (*Raphanus sativus*). *Hortic. Plant J.* **2020**, *6*, 14–24. [CrossRef]
53. Talaat, N.B.; Shawky, B.T. Synergistic Effects of Salicylic Acid and Melatonin on Modulating Ion Homeostasis in Salt-Stressed Wheat (*Triticum aestivum* L.) Plants by Enhancing Root $H^+$-Pump Activity. *Plants* **2022**, *11*, 416. [CrossRef] [PubMed]
54. Karumannil, S.; Khan, T.A.; Kappachery, S.; Gururani, M.A. Impact of exogenous melatonin application on photosynthetic machinery under abiotic stress conditions. *Plants* **2023**, *12*, 2948. [CrossRef] [PubMed]
55. Xu, W.; Cai, S.Y.; Zhang, Y.; Wang, Y.; Ahammed, G.J.; Xia, X.J.; Shi, K.; Zhou, Y.H.; Yu, J.Q.; Reiter, R.J. Melatonin enhances thermotolerance by promoting cellular protein protection in tomato plants. *J. Pineal Res.* **2016**, *61*, 457–469. [CrossRef] [PubMed]
56. Yadava, P.; Kaushal, J.; Gautam, A.; Parmar, H.; Singh, I. Physiological and biochemical effects of 24-epibrassinolide on heat-stress adaptation in maize (*Zea mays* L.). *Nat. Sci.* **2016**, *8*, 171–179.
57. Mittler, R.; Vanderauwera, S.; Suzuki, N.; Miller, G.; Tognetti, V.B.; Vandepoele, K.; Gollery, M.; Shulaev, V.; Van Breusegem, F. ROS signaling: The new wave? *Trends Plant Sci.* **2011**, *16*, 300–309. [CrossRef]
58. Qu, A.-L.; Ding, Y.-F.; Jiang, Q.; Zhu, C. Molecular mechanisms of the plant heat stress response. *Biochem. Biophys. Res. Commun.* **2013**, *432*, 203–207. [CrossRef] [PubMed]
59. Scarpeci, T.E.; Zanor, M.I.; Valle, E.M. Investigating the role of plant heat shock proteins during oxidative stress. *Plant Signal. Behav.* **2008**, *3*, 856–857. [CrossRef] [PubMed]
60. Chauhan, H.; Khurana, N.; Nijhavan, A.; Khurana, J.P.; Khurana, P. The wheat chloroplastic small heat shock protein (sHSP26) is involved in seed maturation and germination and imparts tolerance to heat stress. *Plant Cell Environ.* **2012**, *35*, 1912–1931. [CrossRef]
61. Reddy, P.S.; Thirulogachandar, V.; Vaishnavi, C.; Aakrati, A.; Sopory, S.K.; Reddy, M.K. Molecular characterization and expression of a gene encoding cytosolic Hsp90 from Pennisetum glaucum and its role in abiotic stress adaptation. *Gene* **2011**, *474*, 29–38. [CrossRef]
62. Huang, C.; Tian, Y.; Zhang, B.; Hassan, M.J.; Li, Z.; Zhu, Y. Chitosan (CTS) Alleviates heat-induced leaf senescence in creeping Bentgrass by regulating chlorophyll metabolism, antioxidant defense, and the heat shock pathway. *Molecules* **2021**, *26*, 5337. [CrossRef]

horticulturae

MDPI

Article

# Greenhouse Screening for pH Stress in *Rhododendron* Genotypes

**Shusheng Wang [1,2,3], Marie-Christine Van Labeke [2], Emmy Dhooghe [2], Johan Van Huylenbroeck [1] and Leen Leus [1,*]**

1 Plant Sciences Unit, Flanders Research Institute for Agriculture, Fisheries and Food (ILVO), 9090 Melle, Belgium; wangss@lsbg.cn (S.W.); johan.vanhuylenbroeck@ilvo.vlaanderen.be (J.V.H.)
2 Department of Plants and Crops, Faculty of Bioscience Engineering, Ghent University, 9000 Ghent, Belgium; mariechristine.vanlabeke@ugent.be (M.-C.V.L.); emmy.dhooghe@ugent.be (E.D.)
3 Lushan Botanical Garden, Chinese Academy of Sciences, Jiujiang 332900, China
* Correspondence: leen.leus@ilvo.vlaanderen.be

**Abstract:** The genus *Rhododendron* is known for its preference for acidic soils, although some genotypes can tolerate a more neutral or alkaline pH. In this study, a greenhouse experiment was set up for 140 days to examine different parameters to assess pH stress in the progeny of *R. fortunei* and the cross combination *R.* 'Pink Purple Dream' x 'Belami'. Additional cultivars 'Gomer Waterer' and 'Cunningham's White' were included in the greenhouse test. The plants were divided into two groups. One group was planted in a substrate with a neutral pH (treatment, pH 6.3) and the other group of plants was planted in an acidic pH substrate (control, pH 4.5). Tolerance to pH stress was evaluated for the individual genotypes on both substrates 140 days after the start of the experiment. The following parameters were analyzed: shoot length, root development, chlorophyll fluorescence (Fv/Fm), leaf color and weight (fresh and dry). In intolerant genotypes, all parameters except for number of shoots were negatively affected by pH stress; especially, the development of roots was negatively impacted by the neutral pH, resulting in above-ground symptoms of pH stress, including decreased height and lower fresh and dry weight. The results show variation in pH tolerance within the genotypes tested and point to the potential for the selection of *Rhododendron* genotypes with improved tolerance to neutral pH.

**Keywords:** abiotic stress; alkalic soil; calcium; chlorophyll fluorescence; pH stress; *Rhododendron fortunei*

## 1. Introduction

With nearly 1000 species, *Rhododendron* L. is the largest genus in the family of Ericaceae. Despite the popularity of rhododendrons for use as ornamental garden plants, the economic potential of rhododendron is limited as this genus only thrives in acidic soils (pH 4.5–6.0). *Rhododendron* grown in pH-neutral or alkaline soils shows chlorosis symptoms, reduction in growth of shoots and roots, leaf wilting, defoliation and, finally, plant death [1]. In a study that combined databases on soil characteristics with the natural occurrence of *Rhododendron* species in China, 76 rhododendron taxa out of 525 taxa were predicted to have potential lime tolerance [2]. The abundant genetic variation and the variation in pH tolerance allow for breeding by cross-hybridization and selection of rhododendron cultivars that may be better suited for gardens with neutral or high soil pH. The selection of genotypes adapted to higher soil pH requires appropriate bioassays.

Many authors show that, in soils with elevated pH, it is not the calcium level that causes stress in rhododendron but rather the elevated levels of bicarbonate ($HCO_3^-$) that lead to reactions between these soil carbonates, water and carbon dioxide [3–5]. For this reason, rhododendron plants treated with high levels of gypsum ($CaSO_4$) grow better than in substrates with elevated levels of $CaCO_3$ [5,6]. This response might be attributed to bicarbonate ($HCO_3^-$) toxicity [4,5]. In soil with an elevated pH, bicarbonates also create an environment where insoluble forms of iron, manganese, phosphorus and other plant nutrients become unavailable to plant roots [7]. Bicarbonate is a major anion component of calcareous soils; at the concentrations likely to be found in those soils, it can inhibit root growth (cell elongation) in calcifuge plants and disrupt iron uptake, resulting in chlorosis [8].

The breeding and selection of *Rhododendron* genotypes that can grow in more alkaline soils is important to penetrate the market for gardens with neutral or high soil pH. It is known that some *Rhododendron* species, subspecies or botanical varieties have a tolerance to an elevated soil pH as they are found in natural habitats with a soil pH above pH 7. For example, *R. fortunei* (used in the current study) grows in habitats with a higher soil pH [2]. Chaanin [5] studied what pH level should be used to test lime tolerance in rhododendron. A study was performed with 200 lime-treated rhododendron species and hybrids at different pH levels adjusted with $HCO_3^-$. All grew well at a low pH of 4.2, but stunted growth and iron chlorosis were already noticed at a moderate pH of 6.4. At the highest pH of 7.1, all plants died except for a few seedlings of *R. micranthum*, *R. occidentale* and *R. schlippenbachii*.

In a previous study, we tested seedlings of *Rhododendron* genotypes for tolerance to higher soil pH. Using germination experiments in tissue culture, the effects of alkaline pH and extra $Ca^{2+}$ on the germination and growth of seedlings of *R. fortunei*, *R. vernicosum* and *R. chihsinianum* were investigated [9]. Four seedling populations obtained from crossings between commercial rhododendron cultivars were also tested in a similar way [10]. The results indicated no significant effect of alkaline pH on germination, while a dramatic increase in the number of abnormal seedlings and seedling mortality was observed. Wang et al. [9] also described an alkaline pH shock experiment in well plates where seedlings of various rhododendron species were treated with $NaHCO_3$. Already on the second day after treatment, the negative influence of alkaline pH on seedlings could be detected by chlorophyll fluorescence imaging. Those experiments revealed a genotype-dependent response to higher pH, enabling efficient selection of seedlings for tolerance to a higher soil pH.

In the current study, a greenhouse experiment was set up and plants were submitted to two pH levels in a pot experiment with substrate. The aim of the greenhouse experiment was twofold: (1) to validate the differences in pH stress between the different genotypes and (2) to evaluate different parameters to estimate the pH stress in the greenhouse plants.

## 2. Materials and Methods

### 2.1. Plant Material

Two groups of plant material were used for the greenhouse experiment: *Rhododendron* plantlets that had survived the tissue culture selection process in a neutral pH medium and a control group grown in a low pH medium in tissue culture without selection [9]. The genotypes in both groups were seedlings of *R. fortunei* and genotypes of a manual pollination between *R.* 'Pink Purple Dream' × 'Belami' (Table 1). These seedlings were obtained after sowing in tissue culture as described by [9] and clonal propagation in tissue culture, followed by three months of acclimatization in the greenhouse. All *R. fortunei* seedlings were coded as RF and subnumbered per seedling genotype. The *R.* 'Pink Purple Dream' × 'Belami' hybrids were coded as "PB". The subnumbered PB genotypes refer to individual genotypes clonally propagated in tissue culture. Only the PB group without subnumbers refers to a group of individual seedlings not clonally propagated but grown as a group of individual genotypes.

Two additional cultivars were included in the experiment, 'Gomer Waterer' and 'Cunningham's White'. Both are commonly grown and well-known cultivars. 'Cunningham's White' and *R. fortunei* are known to be the parents of the higher-pH tolerant rootstock Inkarho® [5]. Both cultivars were obtained from a commercial grower, Raf Goossens BV (Moerbeke-Waas, Belgium). 'Gomer Waterer' was obtained as a tissue culture propagated plant. 'Cunningham's White' was not propagated in tissue culture and was the only cultivar propagated through cuttings. Before the experiment, all plants were grown in propagation trays with 150 cells of approx. 20 $mm^3$. All plants were between 2 to 6 cm high before they were potted at the start of the experiments. Information on preceding experiments and plant numbers is given in Table 1.

**Table 1.** Genotypes used in the experiments were obtained as tissue culture grown seedlings from *R. fortunei* (RF) and from the cross *R.* 'Pink Purple Dream' × 'Belami' (PB), tissue-cultured plants of 'Gomer Waterer' (GW) and cuttings of 'Cunningham's White' (CW). For genotypes screened during tissue culture in preceding experiments, pH of the medium is mentioned.

| Genotypes | Not Selected (NS)/Selected (S) in Previous Seedling Selection in Tissue Culture | pH Used for Selection | Number of Plants per Treatment |
|---|---|---|---|
| PB-T3-1 | S | 7.7 | 4 |
| PB-T3-4 | S | 7.7 | 4 |
| PB-T4-2 | S | 7.5 | 4 |
| PB | NS | 5.8 | 12 |
| RF-T-2 | S | 8.9 | 3 |
| RF-T-3 | S | 8.9 | 3 |
| RF-T-5 | S | 8.9 | 3 |
| RF-T-6 | S | 8.9 | 3 |
| RF-C-1 | NS | 6.2 | 3 |
| GW | NA | NA | 12 |
| CW | NA | NA | 12 |

NA: not applicable.

### 2.2. Substrate and Experimental Setup

An extra fine-sieved substrate for seedlings and cuttings, Lp307z, was obtained from Greenyard Horticulture (Ghent, Belgium) and comprised sod peat, perlite, white peat and Irish peat and calcium at 0.5 kg $m^{-3}$ (pure calcium) supplemented with 0.1 kg $m^{-3}$ Micromax® premium (ICL, Waardenburg, The Netherlands). The manufacturer's estimate of the pH of the substrate was 4.4 and an EC of 49 µS $cm^{-1}$. In-house analysis resulted in a pH of 4.5 and an EC of 85 µS $cm^{-1}$.

The substrate without addition of extra calcium was used as the control. For the treatment with a neutral soil pH, $CaCO_3$ in a concentration of 2 g powder per L of potting soil was mixed into the substrate for a final pH of 6.3. All 9 RF and PB genotypes as well as 'Cunningham's White' (CW) and 'Gomer Waterer' (GW) (Table 1) were randomly divided into two groups; one group was transplanted in neutral pH and the other in control substrate.

All plants were watered with the ebb and flow system in the greenhouse at a frequency of 3 times per week and were fertigated with NPK(Mg) 20-5-10 (+2) adjusted to an EC of 1 mS $cm^{-1}$. pH levels remained stable throughout the experiment.

### 2.3. Measurements

All measurements were taken 140 days from the start of the treatment. Only root development was scored after 70 and 140 days. Parameters analyzed were fresh and dry weight, chlorophyll a fluorescence, CIELab color analysis and growth of shoots and roots. For the analysis of the fresh weight, the above-ground parts of the plants were cut and analyzed using a balance (Mettler Toledo XS104, Zaventem, Belgium). The individual plants were dried at 70 °C for 3 days before analysis of the dry weight on the same balance.

Chlorophyll fluorescence was analyzed after the plants had undergone at least 1 h of dark adaptation with a pulse amplitude-modulated fluorometer (PAM 2100, Heinz Walz GmbH, Effeltrich, Germany). For every plant, the youngest fully developed leaf was exposed to a low intensity (<0.1 µmol $m^{-1}$ $s^{-1}$) measuring light to estimate the initial (F0) fluorescence value. Consecutively, the leaves were flashed with a saturating light pulse (±8000 µmol $m^{-2}$ $s^{-1}$) for 0.8 s to determine the maximum (Fm) fluorescence. By subtracting F0 from Fm, the variable fluorescence, Fv, was calculated (Fm − F0 = Fv). Then, the parameter Fv/Fm was determined as a measure for the efficiency of excitation energy capture by open photosystem II reaction centers, as Fv/Fm is an indicator for plant stress.

For the analysis of L*a*b* values in the CIELab color space, a portable spectrophotometer (CM-700D, Konica Minolta, Inc., Erfurt, Germany) was used. Color can be quantified using L* as a value for brightness ranging from 0 (black) to +100 (white), a* as a value ranging from −128 (green) to +128 (red) and b* as a value ranging from −128 (blue) to +128 (yellow) [11].

Root development was scored on a scale from 0 to 5. After plants were lifted out of the pot, root development was scored based on the appearance of the root ball. Score 0 was assigned when no roots were visible, score 1 for a small number of roots visible at one side or when a single root appeared on more than one side, score 2 for more than 1 root visible on at least 2 sides, score 3 for some roots visible on all sides, score 4 for roots visible around the entire root ball and score 5 for roots visible from top to bottom and all around the entire root ball (Figure S2). Plant height and length of the youngest shoot flush were measured using a ruler (Figure S3). The number of shoots and flushes was counted per individual plant.

### 2.4. Statistical Analysis

During the experiment, the plant pots were randomly placed in trays on one table in the greenhouse. Prior to measuring and scoring, they were rearranged to a new random placement on the table.

All statistical analyses in this study were performed using R 4.1.3 and R-studio [12]. In the R package "ggboxplot", global $p$-values of variations were calculated by the Kruskal–Wallis test, and pairwise comparisons between each group and the overall group mean (grand mean) were analyzed by a Wilcoxon test. The principal component analysis (PCA) was performed with the package "prcomp", and Pearson's correlation analyses were performed with "ggcorrplot".

## 3. Results

### 3.1. Comparison of pH Stress between Acidic and Neutral Substrate pH

First, an overall analysis of the effect of pH on the plant traits of the 11 genotypes was performed. The overall effect of the higher pH of the neutral substrate showed significant differences for most parameters (Figure 1). The root score (day 70 and 140), plant height, fresh and dry weight, number of flushes, length of the new flush, Fv/Fm and CIELab (a) of plants in the neutral pH substrate significantly decreased when compared with the plants in the acidic (control) substrate ($p < 0.0001$). The CIELab values L* and b* of the plants on neutral pH were significantly higher than in the control ($p < 0.0001$). No significant difference was found in the number of shoots.

### 3.2. Comparison of pH Stress between Genotypes

To evaluate differences in the performance of different genotypes, the individual genotypes were compared with the grand mean of all genotypes for all parameters. The grand mean is defined as the sum of all data for all genotypes divided by the total number of samples.

Differences in root development were found between genotypes. After 70 days in the control substrate, significantly better root development was found in PB-T3-4 ($p < 0.01$), while CW showed a significantly lower development of roots ($p < 0.01$) when compared with the grand mean of all genotypes (Figure 2A,B). The root scores of the different genotypes in the neutral substrate showed no significant differences on day 70 (Figure 2B). On day 140, however, the root score of PB-T3-4 was significantly ($p < 0.001$) higher than the grand mean both in the control and in the neutral pH. Only PB-T2-4 showed a significantly lower value ($p < 0.001$) compared with the grand mean in the neutral pH (Figure 2D).

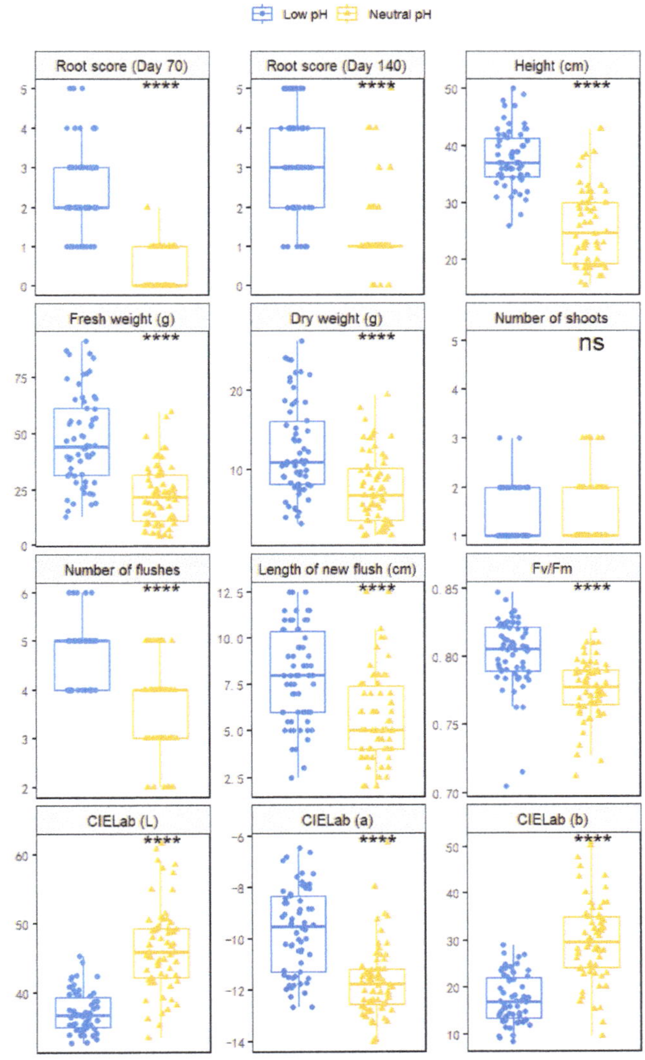


**Figure 1.** Overall effects of low pH (control) and neutral pH ($CaCO_3$ supplemented) substrate on root score, plant height, fresh weight, dry weight, number of shoots, number of flushes, length of the new flush, Fv/Fm and CIELab (L, a, b) of all genotypes. Pairwise comparisons against the control were calculated by a Wilcoxon test and indicated by **** ($p < 0.0001$) and ns (not significant), respectively.

In the randomly selected progeny plants of R. 'Pink Purple Dream' × 'Belami' (PB), the fresh weight was significantly ($p < 0.05$) higher than the grand mean (Figure 2A) in the control treatment and showed the largest variation when compared with the other genotypes. In the neutral pH treatment, fresh weight was approximately equal to the grand mean. The dry weight of PB genotypes (Figure 2C,D) showed similar results. Both genotypes GW ('Gomer Waterer') and PB-T3-4 performed significantly better than the grand mean in the control and neutral pH (both $p < 0.01$) treatment for fresh weight and dry weight. The genotype CW ('Cunningham's White') performed significantly worse ($p < 0.0001$) when compared with the grand mean.

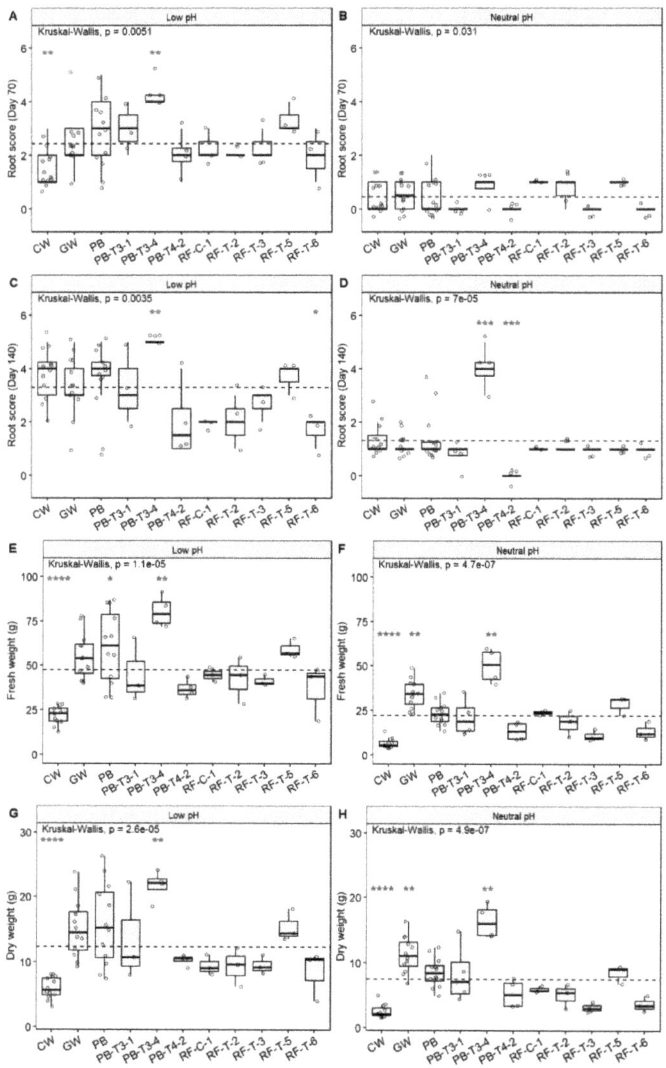


**Figure 2.** Performance of different genotypes evaluated by (**A**,**B**) root score on day 70 and (**C**,**D**) root score on day 140, (**E**,**F**) fresh and (**G**,**H**) dry weight on day 140 under low and neutral ($CaCO_3$) pH treatments. Global $p$-values of variations among genotypes were calculated by the Kruskal–Wallis test. Pairwise comparisons against all (grand mean, illustrated by the dotted line) were calculated by the Wilcoxon test indicated by **** ($p < 0.0001$), *** ($p < 0.001$), ** ($p < 0.01$) and * ($p < 0.05$) and no marks (not significant), respectively.

Above-ground growth was measured after 140 days. Also for growth parameters differences in performance were found among genotypes. Genotype PB-T3-4 grew tallest in both the control and neutral pH substrates and was significantly ($p < 0.01$) taller than the grand mean (Figure 3A,B). Genotype PB-T4-2 grew significantly less tall in the control and neutral pH substrate compared with the grand mean ($p < 0.05$). Only in the neutral pH substrate were genotypes CW and GW significantly less tall and taller than the grand mean, respectively ($p < 0.05$) (Figure 3A,B). GW developed significantly ($p < 0.05$) fewer shoots than the grand mean in the low and neutral pH. PB developed significantly more shoots in the neutral pH ($p < 0.05$) (Figure 3C,D). The number of shoot flushes was significantly higher in RF-T-2 in the control ($p < 0.05$) but not in the neutral pH. In the neutral pH, significantly fewer flushes developed in CW ($p < 0.001$) and significantly more flushes developed in RF-C-1 ($p < 0.05$) (Figure 3E,F). The length of the new flush of genotype PB-T3-4 was only significantly higher than the grand mean in the neutral substrate but not in the control ($p < 0.01$). In two other genotypes, RF-C-1 ($p < 0.01$) and RF-T-3 ($p < 0.05$), the flushes were significantly less long than the grand mean compared in the neutral pH, and, for RF-C-1 and RF-T-2, the flush length was also significantly less long in low pH ($p < 0.05$) (Figure 3G,H).

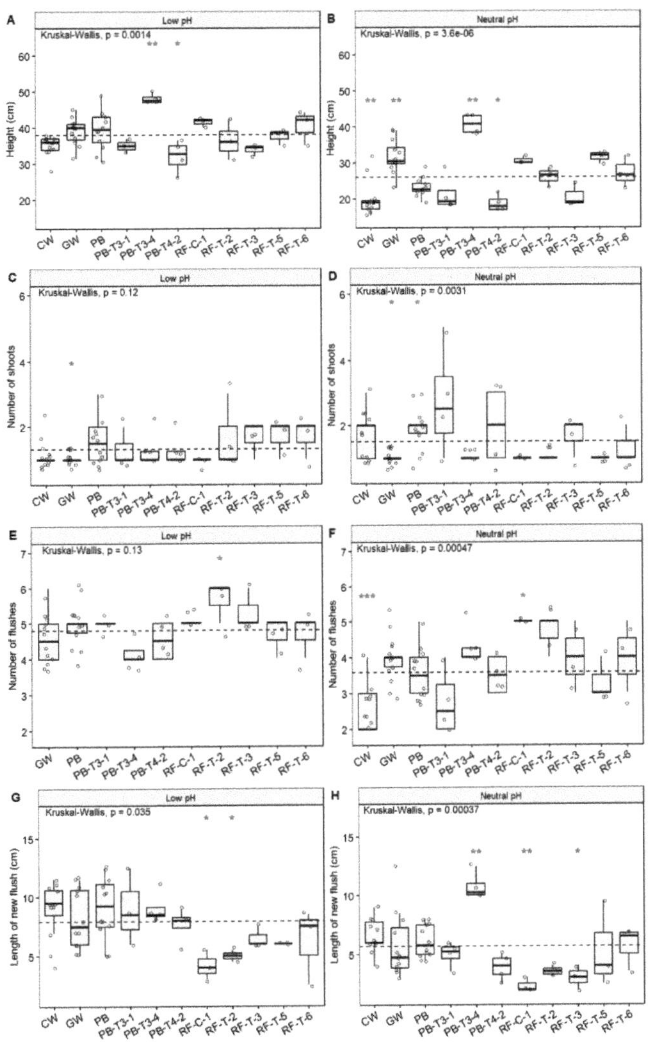


**Figure 3.** Performance of different genotypes evaluated by (**A**,**B**) height, (**C**,**D**) number of shoots, (**E**,**F**) number of flushes (CW not included) and (**G**,**H**) length of the new flush on day 140 under control and neutral ($CaCO_3$) pH treatments. Global *p*-values of variations among genotypes were calculated by the Kruskal–Wallis test. Pairwise comparisons against all (grand mean, illustrated by the dotted line) were calculated by Wilcoxon's test, and indicated by *** ($p < 0.001$), ** ($p < 0.01$) and * ($p < 0.05$) and no marks (not significant), respectively.

Chlorophyll fluorescence was measured as Fv/Fm in this experiment. In the control (low pH), the Fv/Fm values were all around 0.8. Values were significantly higher than the grand mean in two genotypes, CW and RF-T-3 (both $p < 0.05$), and lower in PB ($p < 0.01$) than the grand mean in the control (Figure 4C). In the neutral pH, the grand mean value dropped to around 0.75 and was significantly lower for CW ($p < 0.01$) than the grand mean and higher in GW ($p < 0.01$) and PB-T3-4 ($p < 0.05$) (Figure 4A,B). The value for PB-T3-4 was the highest among all genotypes in the neutral pH; this was higher than the Fv/Fm value this genotype had in the low pH substrate.

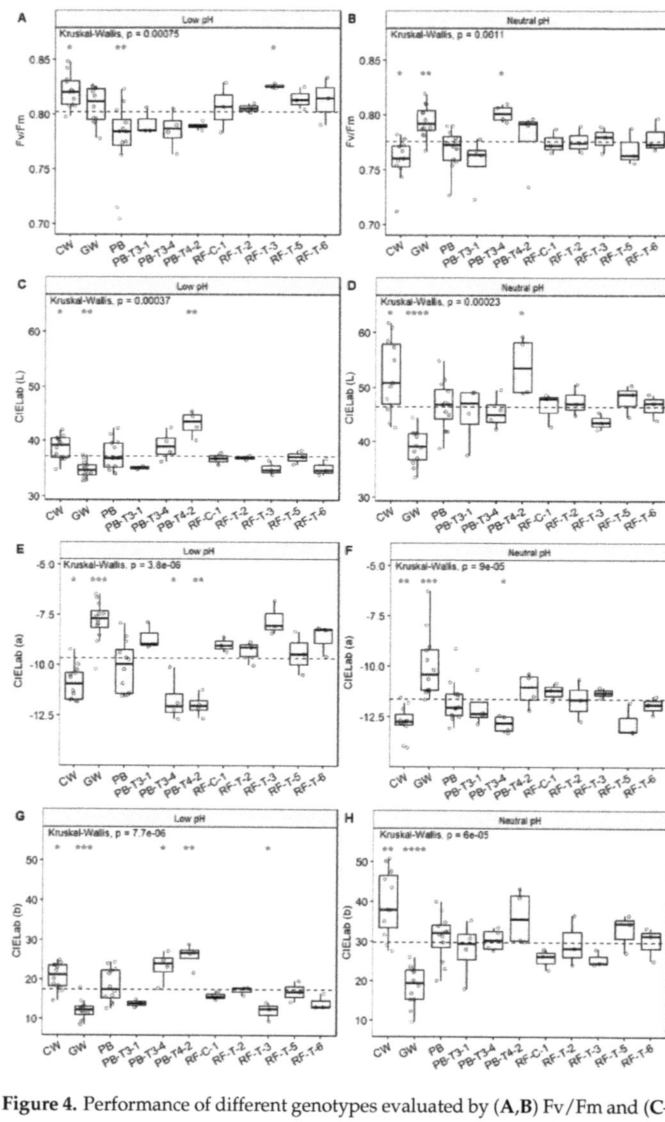


**Figure 4.** Performance of different genotypes evaluated by (**A**,**B**) Fv/Fm and (**C**–**H**) CIELab (L*, a*, b*) on day 140 under control and neutral ($CaCO_3$) pH treatments. Global *p*-values of variations among genotypes were calculated by the Kruskal–Wallis test. Pairwise comparisons against all (grand mean, illustrated by the dotted line) were calculated by the Wilcoxon test, and indicated by **** ($p < 0.0001$), *** ($p < 0.001$), ** ($p < 0.01$), * ($p < 0.05$) and no marks (not significant), respectively. Performance of different genotypes evaluated by height, number of shoots and number of flushes.

L*a*b* values in the CIELab color space measured by the portable spectrophotometer CM-700D were important non-destructive parameters for plant phenotyping. A higher L* value indicated a whiter color, a lower a* value indicated more green and a higher b* value indicated more yellow in the leaves. Differences in CIELab values were found among different genotypes. L* and b* of CW were significantly higher than the grand mean in both the control and neutral substrate, while L* and b* of GW showed the opposite. The darker colored leaves of GW can be seen in Figure S4. In both the control and neutral pH, significant differences from the grand mean were found for L* in CW, GW and PB-T4-2 (Figure 4C,D). These variations only showed a genetic difference in leaf color without the influence of pH. For a* CW, GW and PB-T3-4 differed from the grand mean in both pH levels, while only PB-T4-2 showed a significantly lower a* value when compared with the grand mean in the low pH ($p < 0.05$). In the neutral pH, the value was higher than but not significantly different from the base mean (Figure 4E,F). For b*, significantly different values in the low pH were found for CW, GW, PB-T3-4, PB-T4-2 and RF-T-3. In the neutral pH, there was no longer a significant difference for PB-T3-4, PB-T4-2 and RF-T-3 when compared with the grand mean (Figure 4G,H).

A PCA analysis of root score day 70 and 140, fresh and dry weight, height, number of shoots, length of new flush, Fv/Fm and CIELab of selected (T) and non-selected (C) genotypes in neutral and control substrates indicated that the two PCs (PC1 and PC2) accounted for 61.7% and 17.0% variance, respectively (Figure 5).

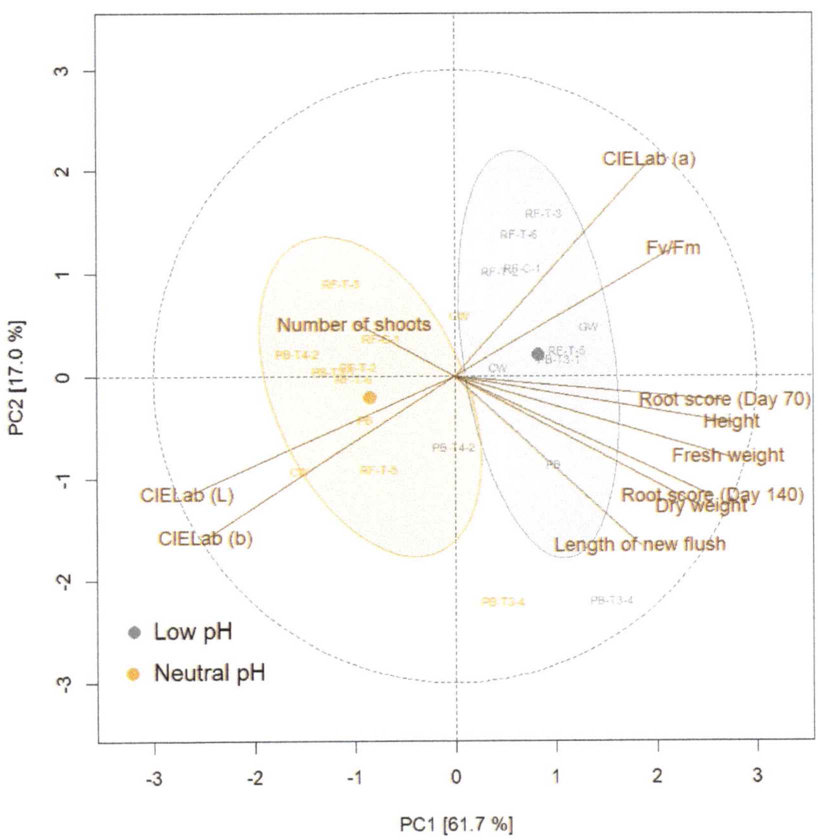


**Figure 5.** PCA biplots of PC1 and PC2 of root and shoot growth parameters, Fv/Fm and CIELab of the different genotypes in neutral pH and acidic control substrates.

The growth parameters of root score, height, length of new flush, fresh and dry weight had the same direction and were close to each other, so they were highly positively correlated, while the number of shoots was in the opposite direction and was thus negatively correlated. The chlorophyll fluorescence parameter Fv/Fm had the same direction and was highly positively correlated with CIELab (a*), while it was negatively correlated with CIELab (L* and b*). The genotypes in the control (indicated in blue and green) and neutral (in grey and yellow) substrate had a different group/ellipse structure that could be distinguished by PC1, indicating that the genotypes growing in the neutral substrate had different characteristics to the genotypes in the control substrate. Differences in performance between genotypes could be shown. The genotype PB-T3-4 was best performing and far from the other genotypes both in the control and neutral substrates (Figures 5 and S4). For PB-T3-4, the root score after growing the seedlings for 70 days in the neutral pH was significantly ($p < 0.05$) worse than the control (Figure S1). However, after 140 days, the root score of PB-T3-4 in the neutral pH was not significantly different from the control. In contrast, for other genotypes such as CW, GW, PB, PB-T3-1 and PB-T4-2, the root score after 140 days in the neutral pH was still significantly lower than the control (Figure S1). In addition, after 140 days in PB-T3-4, significant ($p < 0.05$) differences for height, fresh weight and b* were observed, but no significant differences were noted for dry weight, number of shoots, number of flushes, length of new flush, Fv/Fm, L* and a* (Figure S1).

Further, the correlation analysis indicated that most parameters were significantly correlated, except for the number of shoots, which only negatively correlated with height (Figure 6). CIELab (L* and b*) had the highest correlation (0.98), followed by fresh and dry weight (0.96). The chlorophyll fluorescence parameter Fv/Fm had a significant negative correlation with b* (r = −0.76) and L* (r = −0.75) and a significant positive correlation with height (r = 0.63), root score (r = 0.52 both on day 70 and 140) and fresh weight (r = 0.46). CIELab (L* and b*) had a significant negative correlation with most root and shoot growth parameters except for the number of shoots and the length of the new flush. CIELab (a*) showed only a significant positive correlation with the root score (day 70) and height. The root score (day 70 and 140) had a significant positive correlation with all shoot parameters except for the number of shoots; for example, a good correlation was found with the length of the new shoot (r = 0.81).

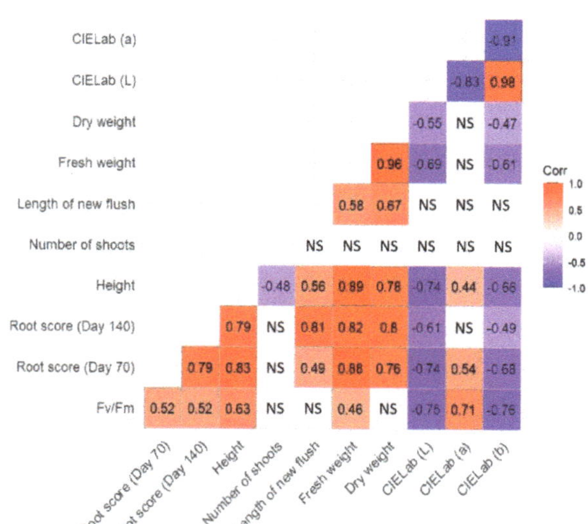


**Figure 6.** Correlation plot of the root and shoot growth parameters, Fv/Fm and CIELab of all tested genotypes. The values in the squares with different colors show the correlation coefficients (Pearson's r) between two parameters, and "NS" indicates no significant correlation ($p > 0.05$).

## 4. Discussion

In calcifuge plants, the higher pH of neutral or alkaline soils can damage plant root structure, reduce root activity and affect plant nutrient utilization, which leads to yellowing of leaves and, in severe cases, death [13]. The damage caused in plants by alkaline stress is a complex physiological and biochemical process that affects the entire growth and development of the plant from seed germination onward [14]. *Rhododendron* is a textbook example of a calcifuge plant.

Previously, we selected germinating seedlings in media with a higher pH [9,10]. As *R. fortunei* again showed tolerance for a higher pH, we included seedlings of this species (RF) in the current study [9]. Seedlings from crosses between commercial rhododendrons 'Pink Purple Dream' × 'Belami' (PB) were also screened in tissue culture and showed genotypic variation in lime tolerance [10]. The results of the aforementioned studies show that a high pH stress has no significant effect on the germination of rhododendron seeds but does inhibit the growth of seedlings. Moreover, the differences in alkalinity tolerance during the germination period are dependent upon the genotype [9]. Genotypes used in the current study were the progeny of RF and PB selected on tissue culture media with a high pH and seedlings that had not undergone selection. All seedlings were propagated in tissue culture. Additionally, plantlets of two cultivars ('Cunningham's White' (CW) and 'Gomer Waterer' (GW)) were used to test pH stress in a greenhouse pot experiment. 'Gomer Waterer' and 'Cunningham's White' were used as tissue cultured plantlets and cuttings, respectively. In total, seven of the tested genotypes (RF and PB) were preselected on an alkaline medium; four genotypes, one PB, one RF, CW and GW were not preselected in the seedling stage. The number of not-selected plants was too small to make a good comparison between tissue culture selection in the seedling stage and greenhouse performance. When the individual genotypes selected on a high pH tissue culture were compared with not-selected genotypes, in general no better performance was found in the selected genotypes, the RF progeny, or the hybrids of PB. This was in contrast with what was expected. Nevertheless, the best performing genotype, PB-T3-4, was obtained after selection in the seedling stage in a tissue culture medium with an elevated pH level.

In the acidic control substrate, a large variation in the parameters tested was found between the genotypes. These were considered as genotypical differences, while, in the neutral pH, differences reflected the response between genotype (G) and environment (E) (or a so called genotype–environment interaction (GxE)). The effect of a pH of 6.3 on plant growth could be shown visually by changes in biomass indicators. In the present experiment, fresh and dry weights of all plants were significantly reduced after 140 days in the neutral pH substrate. Other discriminative growth parameters involved were root score, plant height, number of flushes and the length of the new flush. The number of shoots was not dependent upon the substrate pH, probably because not many plants developed lateral shoots. The plants were not pruned during the experiment and no other action was taken to induce the formation of lateral shoots. The PCA plot shows very well how the different parameters, except for the number of shoots, were commonly influenced by the neutral pH. The correlation matrix shows good correlations, especially between plant height, fresh and dry weight and root scores after 70 or 140 days. The PCA shows how the genotypes clustered together per treatment. The results therefore confirm that alkaline stress inhibits the normal growth of rhododendrons, which is in agreement with previous studies on rhododendron [6,15–17] and is observed in other pH susceptible plants [18–20]. The growth and development of the roots and the inter-root environment directly affected the growth and development of the whole plant. In our experiment, it was shown that in rhododendron roots subjected to long-term (140 d) alkaline soil stress, plant height, number of shoots, number of flushes and the length of the new flush were significantly inhibited in all genotypes of *R.* 'Pink Purple Dream' × 'Belami' and *R. fortunei*. Especially, the lack of development of roots was remarkable; in the stressed plants, roots did not develop in the neutral pH substrate and they did not grow beyond the original transplanted root plug volume. Chaanin and Preil [5] noticed a complete inhibition of root formation of rhododendron in tissue culture supplemented with $NaHCO_3$. Also, the results of Turner et al. [21] suggest that root morphology and function might be the limiting factor under alkaline conditions. They tested rhododendrons grown 49 days in a pH nutrient solution with pH 6.5 and observed smaller root systems than plants grown in an optimal pH. In nutrient solutions with pH 6.5, the roots developed as clusters of short highly branched roots, which was not the case for plants grown in nutrient solutions with a pH of 5.5. Demasi et al. [22], who also used a hydroponic system, observed no changes in root system development in higher pH after 21 days. In an experiment with potted plants and a longer experimental timeline, Demasi et al. [23] found that growth and biomass of different evergreen azalea species and cultivars decreased at a high pH in varying degrees, showing that certain genotypes are more tolerant of a high pH than others. Roots were not evaluated in that study.

Interestingly, after 70 and 140 days in the control, the root score of genotype PB-T3-4 showed a significant growth advantage above the mean baseline. This significantly better root growth was also found in the acidic substrate after 140 days, while this was not reflected after 70 days of pH stress. The root system is the main organ for nutrient uptake in plants and the first to sense changes in inter-root environmental conditions, requiring a certain adaptation period to adapt to changes in the environment [24]. In addition, the analysis of above-ground phenotype and growth indicators like plant height and the length of the new flush confirmed this superior pH tolerance in PB-T3-4.

Chlorophyll fluorescence was used as a parameter to assess plant stress. Fv/Fm reflects the maximum light energy conversion efficiency of PSII, which is usually significantly affected by biotic or abiotic stress and can be used to determine the damage to plant photosynthetic organs. This makes it an effective indicator of the photosynthetic physiology of plants under stress [25]. For the detection of abiotic stresses, chlorophyll fluorescence measurements have been applied and examples can be found for stress related to heat [26], cold [27,28], drought [29,30], salt [31] and nutrient deficiency [32] and (high) light stress [33,34]. Also, for biotic stresses, chlorophyl fluorescence has been used to examine stress caused by nematodes [34], pathogens [35] and insects [36]. Chlorophyll fluorescence imaging is also an efficient way to perform high-throughput screening of rhododendron seedlings for their tolerance to alkaline growing media [9]. Most genotypes of the species *R. fortunei* and the cross *R.* 'Pink Purple Dream' × 'Belami' and also CW showed a significant down-regulation of Fv/Fm after 140 days in a neutral pH when compared with plants in a low pH. In the genotype PB-T3-4, the values for Fv/Fm were slightly higher in a neutral pH when compared with a low pH. Also, GW had a good Fv/Fm value of around 0.8 in the neutral substrate, although this was significantly lower than the control.

The CIELab (L*) and CIELab (b*) indicators of each genotype were significantly up-regulated after alkaline stress as photosynthetic efficiency was reduced, which indicated that alkaline stress caused chlorosis of the leaves. Our results indicated that, other than Fv/Fm, CIELab (especially L* and b*) was also a good predictor for the biomass of PB and RF. However, unlike Fv/Fm, the correlations between CIELab and biomass were genotype-specific. Recorded chlorosis under lime stress increased significantly after 4 to 7 weeks in evergreen potted azalea depending on the genotype. In our greenhouse test, GW responded differently from the other genotypes as its L* value remained lower, meaning the leaves were still dark-colored. A typical symptom of pH stress is interveinal chlorosis caused by iron deficiency [37]. The ability to take up iron depends on plant enzymes like ferric chelate reductase (FCR) that reduce $Fe^{3+}$ to $Fe^{2+}$. Genotypic differences in levels of ferric chelate reductase activity under iron deficiency were found by Demasi et al. [1]. Lower foliar $Fe^{2+}$ concentrations were measured by Demasi et al. [23], but the authors concluded that the pH of the substrate hampered azalea ornamental performance more than Fe deficiency. No relationship between iron levels and chlorophyll content was observed. In a study on the content of elements in the leaves of rhododendron plants in their natural habitat, it was found that deficiencies of manganese, and to a lesser extent iron, related to the growth of rhododendron on alkaline soils [38,39]. Also, Chaanin and Preil [4] found a reduction in iron and manganese levels in young leaves of rhododendron artificially subjected to high levels of $CaCO_3$ in the substrate.

Although 'Cunningham's White' is known to be one of the parents of the high pH-tolerant rootstock Inkarho® [40], its performance in our greenhouse test was poorer when compared with 'Gomer Waterer' and many PB and RF genotypes. Preil and Ebbinghaus [15] mention a reduced shoot and root development for 'Cunningham's White' grown on a substrate with $CaCO_3$, but chlorosis was less profound in comparison with other genotypes. They found that shoot and root development did not always respond in the same way to pH stress. The lime tolerance screens from Preil and Ebbinghaus [15] have led to the development of the Inkarho® (Interessengemeinschaft Kalktolerante Rhododendron) rootstock tolerant of soil pH 6.5–7.0, derived from a cross between *R. fortunei* and 'Cunningham's White' [5]. Other authors describe 'Cunningham's White' as a nineteenth-century cultivar with moderate lime tolerance used extensively as rootstock [6,40]. Another breeding product is Bloombux®, released in 2014 as a lime-tolerant hybrid from *R. micranthum* and *R. hirsutum* advertised as tolerant of soils up to pH 7.5 [40].

In the present study, the *Rhododendron* genotype PB-T3-4, well performing in both the control and neutral pH, has the potential to grow in soils with an elevated pH. The identification of this genotype is important for the development of new cultivars that can thrive in alkaline soils, which may be useful for landscape gardening and for the development of new ornamental plants. Further studies are needed to investigate the underlying mechanisms that enable *Rhododendron* genotypes to grow in alkaline soils and to identify other candidate genotypes and select for other ornamental characteristics to obtain a marketable cultivar with enhanced adaptability to alkaline soils.

## 5. Conclusions

Various evaluation methods were used to assess pH stress in rhododendron. After 140 days of growth in substrate with a pH of 6.3, root development was especially negatively affected. This observation had a good correlation with observed plant height, flush length and fresh and dry plant weight. The genotype PB-T3-4, progeny of 'Pink Purple Dream' × 'Belami', resulted from this selection process and outperformed the other genotypes tested, including seedlings of *R. fortunei* and the commercial rhododendrons 'Cunningham's White' and 'Gomer Waterer'.

**Supplementary Materials:** The following supporting information can be downloaded at: https://www.mdpi.com/article/10.3390/horticulturae9121302/s1, Figure S1: Effects of neutral ($CaCO_3$) treatment on root score (Day 70 and 140), height, fresh weight, dry weight, number of shoots, number of flushes, length of new flush, Fv/Fm and CIELab (L*, a*, b*) of different genotypes. Pairwise comparisons against the control treatment were calculated by Wilcoxon test, and indicated by **** ($p < 0.0001$), *** ($p < 0.001$), ** ($p < 0.01$), * ($p < 0.05$) and "ns" (not significant) respectively; Figure S2: Root scores 0 to 5, with score 0: no roots are visible; score 1: a little bit of roots visible at one side or just a single root on a few sides of the root ball; score 2: roots at two sides appear; score 3: roots are visible around the pot; score 4: roots are present around the root ball; score 5: roots from top to bottom are all around the potting soil; Figure S3: Measurements of above ground growth (height) and number of flushes; Figure S4: Pictures taken at the end of the experiment 140 days after potting in acidic or neutral pH.

**Author Contributions:** Conceptualization, S.W., L.L., J.V.H. and M.-C.V.L.; methodology, S.W. and L.L.; formal analysis, S.W.; writing—original draft preparation, S.W. and L.L.; writing—review and editing, S.W., L.L., E.D., J.V.H. and M.-C.V.L.; supervision, L.L., E.D. and M.-C.V.L. All authors have read and agreed to the published version of the manuscript.

**Funding:** This work was supported by the China Scholarship Council (201608360101).

**Data Availability Statement:** The data presented in this study are available on request from the corresponding author.

**Acknowledgments:** The authors would like to thank the technical staff of ILVO for the excellent plant care in the greenhouse and Miriam Levenson for the English language corrections.

**Conflicts of Interest:** The authors declare no conflict of interest.

## References


1. Demasi, S.; Handa, T.; Scariot, V. Ferric chelate reductase activity under iron deficiency stress in Azalea. *Int. J. Hortic. Florcult.* **2015**, *3*, 157–160.
2. Wang, S.; Leus, L.; Van Labeke, M.-C.; Van Huylenbroeck, J. Prediction of lime tolerance in *Rhododendron* based on herbarium specimen and geochemical data. *Front. Plant Sci.* **2018**, *9*, 1538. [CrossRef] [PubMed]
3. Boxma, R. Bicarbonate as the most important soil factor in lime-induced chlorosis in The Netherlands. *Plant Soil* **1972**, *37*, 233–243. [CrossRef]
4. Chaanin, A.; Preil, W. Influence of bicarbonate on iron deficiency chlorosis in *Rhododendron*. *Acta Hortic.* **1994**, *364*, 71–77. [CrossRef]
5. Chaanin, A. Lime tolerance in rhododendron. *Comb. Proc. IPPS* **1998**, *48*, 180–182.
6. Giel, P.; Bojarczuk, K. Effects of high concentrations of calcium salts in the substrate and its pH on the growth of selected. *Rhododendron cultivars*. *Acta Soc. Bot. Pol.* **2011**, *80*, 105–114. [CrossRef]
7. Ström, L.; Owen, A.G.; Godbold, D.L.; Jones, D.L. Organic acid behaviour in a calcareous soil—Implications for rhizosphere nutrient cycling. *Soil Biol. Biochem.* **2005**, *37*, 2046–2054. [CrossRef]
8. Lee, J.A.; Woolhouse, H.W. A comparative study of bicarbonate inhibition of root growth in calcicole and calcifuge grasses. *New Phytol.* **1969**, *68*, 1–11. [CrossRef]
9. Wang, S.; Leus, L.; Lootens, P.; Van Huylenbroeck, J.; Van Labeke, M.-C. Germination kinetics and chlorophyll fluorescence imaging allow for early detection of alkalinity stress in *Rhododendron* species. *Horticulturae* **2022**, *8*, 823. [CrossRef]
10. Wang, S.; Zheng, T.; Leus, L.; Van Labeke, M.-C.; Van Huylenbroeck, J. Screening and evaluation of *Rhododendron* progenies for alkaline pH tolerance. *Acta Hort.* **2021**, *1331*, 95–100. [CrossRef]
11. ACTTR. Available online: https://www.acttr.com/en/en-report/en-report-technology/383-en-tech-color-space-conversion.html (accessed on 4 October 2023).
12. The R Project for Statistical Computing. Available online: https://www.R-project.org (accessed on 4 October 2023).
13. Li, C.; Fang, B.; Yang, C.; Shi, D.; Wang, D. Effects of various salt–alkaline mixed stresses on the state of mineral elements in nutrient solutions and the growth of alkali resistant halophytechloris virgata. *J. Plant Nutr.* **2009**, *32*, 1137–1147. [CrossRef]
14. Wei, L.; Zhang, R.; Zhang, M.; Xia, G.; Liu, S. Functional analysis of long non-coding RNAs involved in alkaline stress responses in wheat. *J. Exp. Bot.* **2022**, *73*, 5698–5714. [CrossRef] [PubMed]
15. Preil, W.; Ebbinghaus, R. Breeding of lime tolerant *Rhododendron* rootstocks. *Acta Hort.* **1994**, *364*, 61–70. [CrossRef]
16. Susko, A.Q. Phenotypic and Genetic Variation for Rhizosphere Acidification, a Candidate Trait for pH Adaptability, in Deciduous Azalea (*Rhododendron* sect. *Pentanthera*). Master's Thesis, University of Minnesota Twin Cities, Minneapolis, MN, USA, 2016.
17. Susko, A.; Rinehart, T.A.; Bradeen, J.; Hokanson, S.C. An evaluation of two seedling phenotyping protocols to assess pH adaptability in deciduous azalea (*Rhododendron* sect. *Pentanthera, G. Don*). *HortScience* **2018**, *53*, 268–274. [CrossRef]
18. Rosen, C.J.; Allan, D.L.; Luby, J.J. Nitrogen form and solution pH influence growth and nutrition of two vaccinium clones. *J. Am. Soc. Hortic. Sci.* **1990**, *115*, 83–89. [CrossRef]
19. Marrs, R.H.; Bannister, P. Response of several members of the Ericaceae to soils of contrasting pH and base-status. *J. Ecol.* **1978**, *66*, 829–834. [CrossRef]
20. Finn, C.E.; Rosen, C.J.; Luby, J.J. Nitrogen form and solution pH effects on root anatomy of cranberry. *HortScience* **1990**, *25*, 1419–1421. [CrossRef]
21. Turner, A.J.; Arzola, C.I.; Nunez, G.H. High pH stress affects root morphology and nutritional status of hydroponically grown rhododendron (*Rhododendron* spp.). *Plants* **2020**, *9*, 1019. [CrossRef]
22. Demasi, S.; Caser, M.; Kobayashi, N.; Kurashige, Y.; Scariot, V. Hydroponic screening for iron deficiency tolerance in evergreen azaleas. *Not. Bot. Horti Agrobot. Cluj-Napoca* **2015**, *43*, 210–213. [CrossRef]
23. Demasi, S.; Caser, M.; Handa, T.; Koboyashi, N. Adaptation to iron deficiency and high pH in evergreen azaleas (*Rhododendron* spp.): Potential resources for breeding. *Euphytica* **2017**, *213*, 148. [CrossRef]
24. Li, X.; Dong, J.L.; Chu, W.Y.; Chen, Y.; Duan, Z. The relationship between root exudation properties and root morphological traits of cucumber grown under different nitrogen supplies and atmospheric $CO_2$ concentrations. *Plant Soil* **2018**, *425*, 415–432. [CrossRef]
25. Ji, Y.; Zhang, X.Q.; Peng, Y.; Liang, X.Y.; Huang, L.K.; Ma, X.; Ma, Y.M. Effects of drought stress on lipid peroxidation, osmotic adjustment and activities of protective enzymes in the roots and leaves of orchardgrass. *Acta Pratacult. Sin.* **2014**, *23*, 144–151. [CrossRef]
26. Robson, J.K.; Ferguson, J.N.; McAusland, L.; Atkinson, J.A.; Tranchant-Dubreuil, C.; Cubry, P.; Sabot, F.; Wells, D.M.; Price, A.H.; Wilson, Z.A.; et al. Chlorophyll fluorescence-based high-throughput phenotyping facilitates the genetic dissection of photosynthetic heat tolerance in African (*Oryza glaberrima*) and Asian (*Oryza sativa*) rice. *J. Exp. Bot.* **2023**, *74*, 5181–5197. [CrossRef] [PubMed]
27. Lootens, P.; Devacht, S.; Baert, J.; Van Waes, J.; Van Bockstaele, E.; Roldan-Ruiz, I. Evaluation of cold stress of young industrial chicory (*Cichorium intybus* L.) by chlorophyll a fluorescence imaging. II. Dark relaxation kinetics. *Photosynthetica* **2011**, *49*, 185–194. [CrossRef]

28. Mishra, A.; Heyer, A.G.; Mishra, K.B. Chlorophyll fluorescence emission can screen cold tolerance of cold acclimated arabidopsis thaliana accessions. *Plant Meth.* **2014**, *10*, 38. [CrossRef] [PubMed]
29. Mohammadi, K.; Jiang, Y.L.; Wang, G.L. Flash drought early warning based on the trajectory of solar-induced chlorophyll fluorescence. *Proc. Natl. Acad. Sci. USA* **2022**, *119*, 32. [CrossRef]
30. Arief, M.A.A.; Kim, H.; Kurniawan, H.; Nugroho, A.P.; Kim, T.; Cho, B.K. Chlorophyll fluorescence imaging for early detection of drought and heat stress in strawberry plants. *Plants* **2023**, *12*, 1387. [CrossRef]
31. Shomali, A.; Aliniaeifard, S.; Bakhtiarizadeh, M.R.; Lotfi, M.; Mohammadian, M.; Vafaei Sadi, M.S.; Rastogi, A. Artificial neural network (ANN)-based algorithms for high light stress phenotyping of tomato genotypes using chlorophyll fluorescence features. *Plant Phys. Biochem.* **2023**, *201*, 107893. [CrossRef]
32. Latifinia, E.; Eisvand, H.R. Soybean physiological properties and grain quality responses to nutrients, and predicting nutrient deficiency using chlorophyll fluorescence. *J. Soil Sci. Plant Nutr.* **2022**, *22*, 2346. [CrossRef]
33. Zheng, L.; Steppe, K.; Van Labeke, M.C. Spectral quality of monochromatic LED affects photosynthetic acclimation to high-intensity sunlight of *Chrysanthemum* and *Spathiphyllum*. *Physiol. Plant.* **2020**, *169*, 10–26. [CrossRef]
34. Senesi, G.S.; De Pascale, O.; Marangoni, B.S.; Caires, A.R.L.; Nicolodelli, G.; Pantaleo, V.; Leonetti, P. Chlorophyll fluorescence imaging (CFI) and laser-induced breakdown spectroscopy (LIBS) applied to investigate tomato plants infected by the root knot nematode (RKN) *Meloidogyne incognita* and tobacco plants infected by cymbidium ringspot virus. *Photonics* **2022**, *9*, 627. [CrossRef]
35. Meng, L.J.; Mestdagh, H.; Ameye, M.; Audenaert, K.; Hofte, M.; Van Labeke, M.C. Phenotypic variation of *Botrytis cinerea* isolates is influenced by spectral light quality. *Front. Plant Sci.* **2020**, *11*, 1233. [CrossRef] [PubMed]
36. Valadares, N.R.; Soares, M.A.; Ferreira, E.A.; de Sa, V.G.M.; Azevedo, A.M.; Leite, G.L.D.; Zanuncio, J.C. Physiological responses in genetically modified cotton and its isohybrid attacked by *Aphis gossypii* Glover (Hemiptera: Aphididae). *Arthropod-Plant Interact.* **2023**, *17*, 167–172. [CrossRef]
37. Guerinot, M.L.; Yi, Y. Iron: Nutritious, noxious, and not readily available. *Plant Physiol.* **1994**, *104*, 815–820. [CrossRef]
38. Kaisheva, M.E. The Effect of Metals and Soil pH on the Growth of *Rhododendron* and Other Alpine Plants in Limestone Soil. Ph.D. Dissertation, University of Edinburgh, Edinburgh, UK, 2006. Available online: https://www.era.lib.ed.ac.uk/handle/1842/2606 (accessed on 1 October 2023).
39. Krebs, S.L. *Rhododendron*. In *Ornamental Crops*; Van Huylenbroeck, J., Ed.; Springer: Cham, Switzerland, 2018; pp. 673–718. [CrossRef]
40. Dunemann, F.; Kahnau, R.; Stange, I. Analysis of complex leaf and flower characters in *Rhododendron* using a molecular linkage map. *Theor. Appl. Genet.* **1999**, *98*, 1146–1155. [CrossRef]

horticulturae

MDPI

*Review*

# Visual Analysis of Research Progress on the Impact of Cadmium Stress on Horticultural Plants over 25 Years

**Zhouli Liu [1,2,3,*], Benyang Hu [1,2,3], Yi Zhao [4,*], Shuyan Zhang [1,2,3], Xiangbo Duan [1,2,3], Hengyu Liu [1,2,3] and Luyang Meng [1,2,3]**

1 College of Life Science and Engineering, Shenyang University, Shenyang 110044, China
2 Northeast Geological S&T Innovation Center of China Geological Survey, Shenyang 110000, China
3 Key Laboratory of Black Soil Evolution and Ecological Effect, Ministry of Natural Resources, Shenyang 110000, China
4 School of Chemistry and Environmental Engineering, Liaoning University of Technology, Jinzhou 121001, China
* Correspondence: zlliu@syu.edu.cn (Z.L.); zhao_yi_66@163.com (Y.Z.)

**Abstract:** In recent years, there has been a significant growth in scholarly attention to the effects of Cd stress on horticultural plants, as reflected by the abundance of research articles on this issue in academic publications. Therefore, it is necessary to conduct a review of current research and provide a comprehensive perspective to quickly grasp the latest developments and future trends in the research field of "horticultural plants-Cd responses". By utilizing a visualizing bibliometric analysis software CiteSpace, this study integrated and analyzed a total of 4318 relevant research records—2311 from the Web of Science (WOS) database and 2007 from the China National Knowledge Infrastructure (CNKI) database—related to "horticultural plants-Cd responses", covering the period from 1999 to 2024. A visual analysis was conducted in the form of knowledge mappings, including the current research status of "horticultural plants-Cd responses", as well as the differences in publications' temporal distribution, spatial distribution (cooperation networks) and intellectual base between China and foreign countries, precisely uncovering the core aspects of research topics related to the field. The results indicated the following: (1) Scientific research on "horticultural plants-Cd responses" has experienced a significant increase in publication volume and has entered a phase of rapid development. Globally, there has been an annual average increase of 217 articles in the WOS since 2019, while in China, the annual average increase has been 134 articles in the CNKI since 2015. (2) China is the most productive country in terms of publication volume (1165 articles, 52.79%), engaging in active partnerships with other countries worldwide. Chinese scholars (Lin L. and Liao M.) are leading researchers in both domestic and international research fields of "horticultural plants-Cd responses". The network of collaborations among authors and institutions in the WOS database seemed denser compared to that in the CNKI database. (3) International research hotspots have focused on accumulation, tolerance and oxidative stress, while domestically, the focus has been on antioxidant enzymes, growth and seed germination. Phytoremediation, subcellular distribution and the transcriptome are the world's emerging topics, while in China, growth and physiological characteristics are still emerging topics. (4) In comparison, China exhibited a lagging development trend, which is reflected in the fact that it began to focus on gene expression and transcriptome research only after the global frontier shifted towards biochar and cadmium co-stress and yield response. Based on these, this study provides a systematic theoretical basis for subsequent research on "horticultural plants-Cd responses", aiding scholars in their efforts to understand the dynamic frontiers and address the challenges in this field.

**Keywords:** cadmium; horticultural plants; CiteSpace; research hotspots; development trend

 https://doi.org/10.3390/horticulturae11010028

## 1. Introduction

The contamination of heavy metals (HMs) has become one of the most significant environmental issues worldwide because of several anthropogenic activities (such as industrials melting and electroplating, agricultural fertilization and improper disposal of used batteries) [1–4]. The HM contamination can result in altering the natural chemical properties of soil, affecting crop yields and food safety and disrupting the balance of ecosystems [5–8]. Cadmium (Cd) is regarded as a highly toxic HM element, which has those features of difficulty of degradation, easy migration and high carcinogenicity [9–11]. Cd can not only inhibit the growth of plants (such as reducing seed germination rates and affecting root system formation) and disrupt the physiological activities of plants (such as reducing antioxidant enzyme activities and inhibiting photosynthesis) but also severely affect animal and human health through respiratory intake by atmospheric deposition and esophageal ingestion by the food chain [12–16]. Cd stress is bound to cause intricate environmental challenges and heighten ecological threats [17–19]. Therefore, research on the effects of Cd contamination has become a global hot topic.

Horticultural plants cultivated for human consumption or ornamental purposes, generally include plants such as vegetables, fruit trees, flowers and medicinal plants [20–25]. The resources of horticultural plants around the world are extremely abundant and incredibly diverse. Horticultural plants can be found everywhere in our daily lives, and they are also one of the earliest varieties of plants to be cultivated by human beings [26–28]. The relationship between horticultural plants and human beings is extremely close [29,30]. Cd contamination resulting from anthropogenic activities will inevitably have an effect on horticultural plants, and greater focus is needed on attaching importance to this problem [31–33].

In recent years, there has been a significant growth in scholarly attention to the effects of Cd stress on horticultural plants, as reflected by the abundance of research articles on this issue in academic publications. As scholarly interest grows in the effects of Cd stress on horticultural plants, numerous studies have been undertaken by scientists to explore this research field such as those on the response process of seed germination, plant growth, photosynthesis, antioxidant systems and gene expression [34–40]. These studies are often presented as independent research to indicate the changes in certain indicators or functions of plants under environmental stress, having less information on providing a comprehensive perspective to quickly grasp the latest developments in the research field. Therefore, it is necessary to conduct a review of current research, which can not only integrate and summarize a wide range of information from diverse studies but also point out unresolved issues and future directions in the research field, as a guiding significance for researchers in selecting research topics and designing experiments.

Despite many years of research exploring the effects of Cd stress on horticultural plants, to date, there is still a lack of systematic analysis and comprehensive understanding of how horticultural plants respond to Cd stress on a global scale, including within China. CiteSpace, as a visualizing bibliometric analysis software, has garnered widespread attention in recent years [41–45]. It provides data support for quantitative analysis, aiding in the evaluation of future research directions by examining the expansion of articles in specific areas of research [46,47]. The software of CiteSpace can transform complex bibliometric analyses into intuitive visual outcomes, helping researchers understand and explore data from the scientific literature more effectively [48–50]. This technology offers researchers a novel approach to grasp the most recent development trends and future research directions of China and the world in various research fields [51]. It is a precious instrument for obtaining insights and staying updated with the newest breakthroughs [52–54]. Therefore, to obtain more comprehensive research findings, this study integrates and analyzes data from the Web of Science (WOS) and the China National Knowledge Infrastructure (CNKI) databases, covering the period from 1999 to 2024, using CiteSpace 6.3.R1 software. The objectives of this study are to conduct a visual analysis in the form of knowledge mappings, including the current research status of "horticultural plants-Cd responses", as well as the differences in publications' temporal distribution, spatial distribution (cooperation networks) and intellectual bases between China and foreign countries. It will be helpful for scholars to gain a quantitative and intuitive grasp of the current state of research and to track the research hotspots and dynamic frontiers in the field of "horticultural plants-Cd responses". This study also provides a systematic theoretical basis for subsequent research on "horticultural plants-Cd responses", aiding scholars in their efforts to understand and address the challenges in this field.

## 2. Materials and Methods

### 2.1. Data Sources

All data in this study were derived from the databases of Web of Science (WOS) and China National Knowledge Infrastructure (CNKI), which cover a wide range of high-quality academic research studies from around the world, ensuring the comprehensiveness and authority of the data sources. The search period extended from January 1999 to September 2024. On 6 September 2024, research studies related to "horticultural plants and cadmium" were retrieved from the databases of WOS and CNKI. In the database of WOS, the retrieval formula was set as follows: (TS = "horticultural plant" OR "horticultural flower" OR "vegetable" OR "fruit tree" OR "fruit" OR "horticultural flower" OR "ornamental plant" OR "ornamental flower" OR "garden flower" OR "flower plant" OR "flowering plant" OR "landscape greening plant" OR "landscape plant" OR "greening plant" + TS = "cadmium" OR "Cd"). In the database of CNKI, the retrieval formula was set as follows: (subject = "horticultural plant" OR "horticultural flower" OR "vegetable" OR "fruit tree" OR "horticultural flower" OR "ornamental plant" OR "ornamental flower" OR "garden flower" OR "flower plant" OR "flowering plant" OR "landscape greening plant" OR "landscape plant" OR "greening plant" + TS = "cadmium" OR "Cd"). Through careful reading and screening, 1598 unrelated or duplicate research records were excluded, leaving a total of 4318 relevant research records–2311 from the WOS database and 2007 from the CNKI database–related to "horticultural plants-Cd responses". These retained records served as the foundational data for analysis.

### 2.2. Analysis Method

CiteSpace is a sophisticated tool for visualizing and analyzing scholarly literature, which generates graphical representations derived from quantitative analytical methods [55,56]. It was initially developed by Dr. Chen and his research group in 2004 and has been progressively refined and disseminated ever since [57,58]. Currently, CiteSpace is widely applied in research fields such as environmental pollution and plant science [59–62]. In the present study, CiteSpace 6.3.R1 (https://citespace.podia.com/) was used to conduct a visualizing bibliometric analysis based on those retained research records from the databases of WOS and CNKI. In the analysis of literature growth trends and predominant publication outlets, we utilized data that had been processed through CiteSpace. Subsequently, we refined and arranged these data to create visual graphs for bibliometric examination. CiteSpace was also employed to graphically represent the distribution of authors, countries, research institutions and keywords since 1999. CiteSpace emphasizes the use of connecting lines to illustrate the prominence of each subject area and dissects the clustering connections among various nodes [63–66]. As a result, the software is capable of precisely uncovering the core aspects of research topics related to "horticultural plants-Cd responses". Additionally, an analysis of the general publication trends was carried out using Microsoft Office Excel 2020 and Origin 2022 to create the corresponding charts.

## 3. Results and Discussion

### *3.1. Temporal Distribution of Publications on "Horticultural Plants-Cd Responses"*

Research literature analysis is the process of examining the volume, trajectory and composition of scholarly publications within a specific academic domain. This type of analysis enables researchers to understand the annual publication volume, cumulative publication volume and annual average increase and to forecast future trends in scholarly output within their field of study [67]. The study analyzed a total of 4318 relevant research records—2311 from the Web of Science (WOS) database and 2007 from the China National Knowledge Infrastructure (CNKI) database—related to "horticultural plants-Cd responses". A statistic analysis of research studies on "horticultural plants-Cd responses" by year is shown in Table 1. In the WOS database, there are a greater number of studies related to "horticultural plants-Cd responses" compared to the CNKI database; however, overall, there is an upward trend, indicating that scholars from both domestic and international backgrounds have a significant interest in this field of research.

**Table 1.** Statistical chart of research studies on "horticultural plants-Cd responses" by year.

| Years | 1999 | 2000 | 2001 | 2002 | 2003 | 2004 | 2005 | 2006 | 2007 |
|---|---|---|---|---|---|---|---|---|---|
| WOS | 7 | 4 | 7 | 17 | 15 | 21 | 27 | 27 | 31 |
| CNKI | 1 | 1 | 6 | 2 | 9 | 9 | 17 | 25 | 56 |
| Years | 2008 | 2009 | 2010 | 2011 | 2012 | 2013 | 2014 | 2015 | 2016 |
| WOS | 52 | 64 | 61 | 77 | 55 | 74 | 78 | 107 | 108 |
| CNKI | 71 | 91 | 82 | 71 | 77 | 76 | 82 | 121 | 121 |
| Years | 2017 | 2018 | 2019 | 2020 | 2021 | 2022 | 2023 | 2024 | |
| WOS | 112 | 125 | 159 | 203 | 181 | 258 | 258 | 183 | |
| CNKI | 114 | 132 | 168 | 147 | 141 | 162 | 133 | 92 | |

The academic interest in the effects of Cd on plants can be traced back to the early 1990s. One of the earliest studies on the effects of cadmium on plants is the 1990 study by Assche F.V. and Clijsters H., titled "Effects of metals on enzyme activity in plants" published in the journal "*Plant, Cell & Environment*" [68]. This study explored the impact of metals on the activity of plant enzymes, laying the foundation for subsequent research on the effects of Cd on plant growth. In addition, the literature reviewed by Wang W. in 1991, titled "Literature review on higher plants for toxicity testing", published in the journal "*Water, Air, & Soil Pollution*", also involved the study of Cd in plant toxicity testing [69]. Nonetheless, research on "horticultural plants-Cd responses" started relatively late. As shown in Table 1, seven articles were published in the WOS database, while only one article was published in the CNKI database.

As shown in Figure 1, the global research on "horticultural plants-Cd responses" in the WOS database exhibits an overall publication trend that divides the study into three distinct research stages: StageI, the initial exploration stage (1999–2007); StageII, the steady growth stage (2008–2018); and StageIII, the rapid development stage (2019–2024). In comparison, from the overall publication trend in the CNKI database, the study related to "horticultural plants-Cd responses" is also divided into the three distinct research stages: StageI, the initial exploration stage (1999–2006); StageII, the steady growth stage (2007–2014); and StageIII, the rapid development stage (2015–2024). Over time, the focus on "horticultural plants-Cd responses" has varied, with the annual publication count in a particular field serving as an indicator of the research's developmental progress [70]. Scientific research on "horticultural plants-Cd responses" has experienced a significant increase in publication volume and has entered a phase of rapid development. Globally, there has been an annual average increase of 216.60 articles in the WOS since 2019, while in China, the annual average increase has been 134.44 articles in the CNKI since 2015.

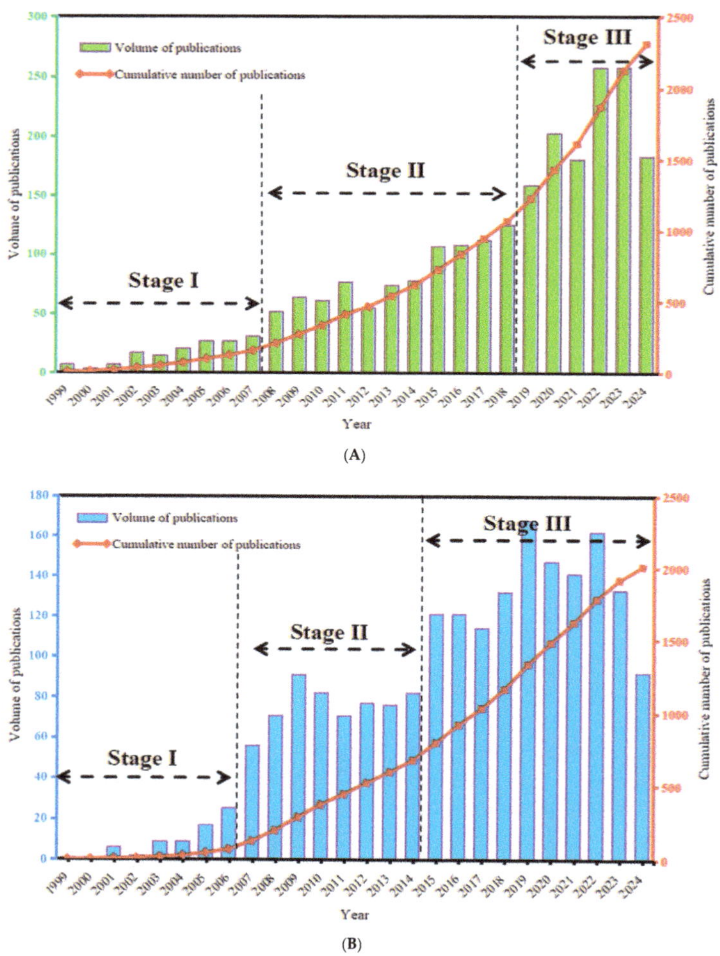


**Figure 1.** Annual publication volume and cumulative publication volume at different stages from the databases of WOS (**A**) and CNKI (**B**) (1999–2024).

As shown in Figure 2 and Table 1, the annual publication volume on "horticultural plants-Cd responses" exhibited a trend of fluctuating growth in both domestic and international research contexts. In terms of annual publication volume, from 1999 to 2006, the international research studies on "horticultural plants-Cd responses" were published more than that in China; however, from 2007 to 2019, China's publication volume of research studies began to surpass the international publication volume. From 2020 to the present, the international publication volume on "horticultural plants-Cd responses" has been significantly higher than that of China. This reflects the trend observed in various research domains, where the global scientific community has consistently maintained a higher output of publications compared to China during this period. It is important to note that publication volumes can be influenced by numerous factors, including research funding, academic collaborations and the focus of research interests at a global level. As shown in Figure 2, compared to China, the international cumulative publication volume has increased by 15.15%, indicating a year-by-year rise in the number of publications and a continuous increase in global scientific attention to "horticultural plants-Cd responses".

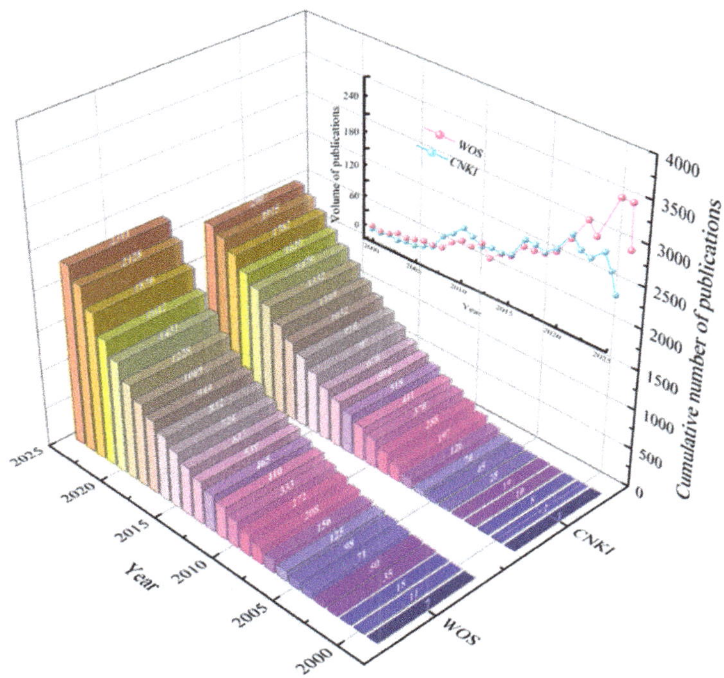


**Figure 2.** Trend comparison of annual publication volume and cumulative publication volume from the databases of WOS and CNKI (1999–2024). Different color and number indicated the varying annual publication volume.

### *3.2. Spatial Distribution of Publications on "Horticultural Plants-Cd Responses"*

#### 3.2.1. Countries' Cooperation Networks

The progression of research on "horticultural plants-Cd responses" varies across nations. Examining the geographic spread of scholarly efforts and collaborative initiatives provides insight into the research trends within this specific domain [67,71,72]. As shown in Figure 3, the number of nodes in the country cooperation networks from the database of WOS is N = 93, E = 512 and D = 0.1197. "N = 93" stands for the number of nodes, which represent 93 distinct countries in the country cooperation networks. "E = 512" stands for the number of edges, which represent 512 connections or relationships between nodes in the country cooperation networks. "D = 0.1197" stands for "density," which refers to the ratio of actual edges (connections) to the maximum possible connections in the country cooperation networks [73]. Therefore, the co-occurrence knowledge graph in the realm of "horticultural plants-Cd responses" revealed a network comprising 93 distinct countries interconnected by 512 collaborative ties, with a network density of 11.97%, which demonstrated a certain level of collaborative activity between distinct countries, being neither too extensive to manage nor too limited to lack diversity. The cooperation among these countries is relatively frequent but not overly dense, with numerous potential opportunities for collaboration yet to be tapped into, allowing for further strengthening of partnerships among multiple nations. The top 10 productive countries from the database of WOS are shown in Table 2. In terms of publication volume, the five most productive countries are the People's Republic of China (1165 articles), India (200 articles), Pakistan (176 articles), the USA (144 articles), and Saudi Arabia (120 articles). Their respective proportions are 52.79%, 9.06%, 7.97%, 6.52%, and 5.44%, indicating that China is the most productive country, actively engaging in partnerships with other countries worldwide.

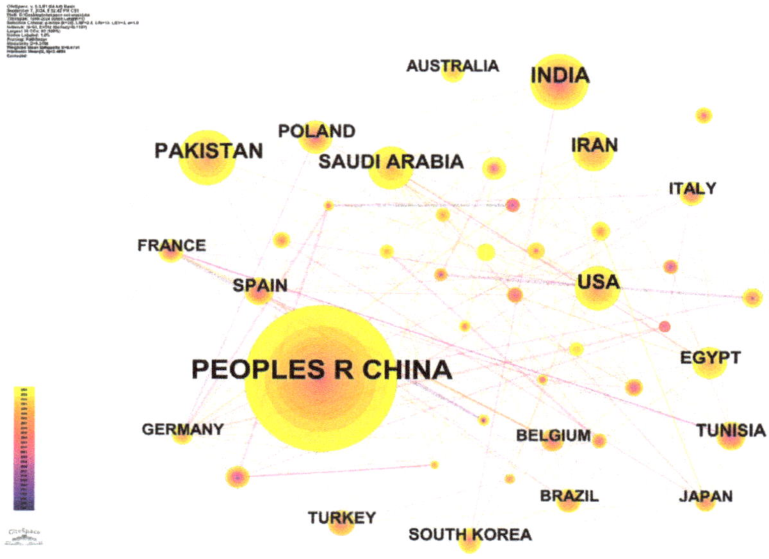


**Figure 3.** Mapping of country collaboration networks from the database of WOS.

**Table 2.** Top 10 productive countries from the database of WOS.

| Rank | Countries | Publication Number | Percentage (%) | First Published Year |
|---|---|---|---|---|
| 1 | PEOPLES R CHINA | 1165 | 52.79 | 2002 |
| 2 | INDIA | 200 | 9.06 | 1999 |
| 3 | PAKISTAN | 176 | 7.97 | 2008 |
| 4 | USA | 144 | 6.52 | 2001 |
| 5 | SAUDI ARABIA | 120 | 5.44 | 2011 |
| 6 | IRAN | 105 | 4.76 | 2002 |
| 7 | POLAND | 88 | 3.99 | 2001 |
| 8 | EGYPT | 73 | 3.31 | 2009 |
| 9 | SPAIN | 72 | 3.26 | 1999 |
| 10 | TUNISIA | 64 | 2.90 | 2003 |

### 3.2.2. Authors' Cooperation Networks

As shown in Figure 4, there are N = 274 nodes in the authors' cooperation network, E = 342 connections, and a network density (D) of 0.0091 within the WOS database. The number of nodes (N = 274) represents a broad base of 274 researchers contributing to the study field of "horticultural plants-Cd responses", and the number of edges (E = 342) shows the 342 connections between these authors through a research collaboration or co-authorship on a published paper. A density of 0.0091, or 0.91%, is quite low, suggesting that while there is collaboration, it is not widespread across the entire network. Most authors likely collaborate with only a small fraction of the total number of authors in the network. The authors' collaboration network within WOS shows a significant number of authors and collaborations, but with a low overall density, indicating a decentralized structure with opportunities for further collaboration and network growth.

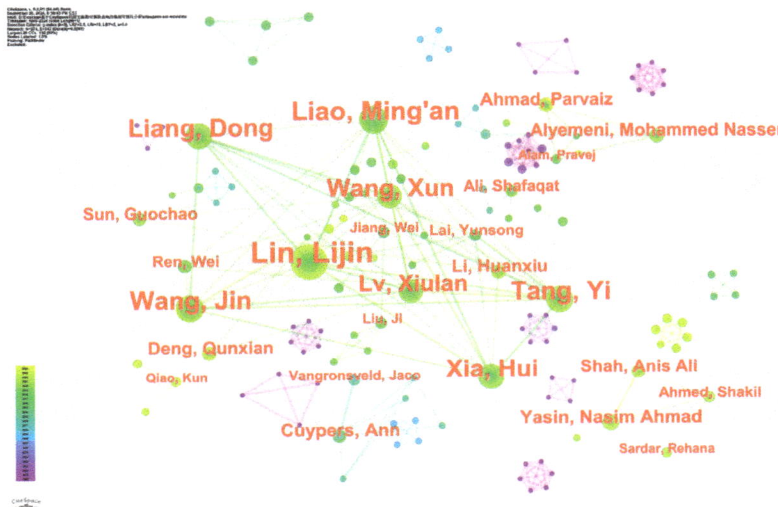


(**A**)

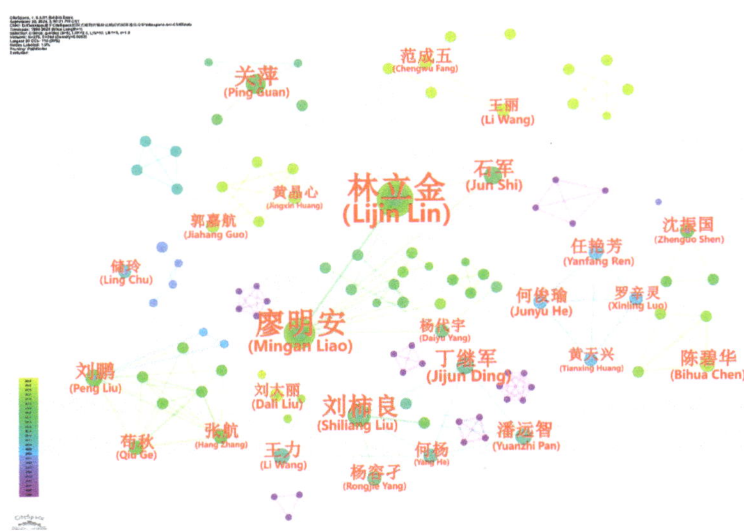


(**B**)

**Figure 4.** Mapping of author collaboration networks from the databases of WOS (**A**) and CNKI (**B**).

Comparatively, there are N = 278 nodes in the authors' cooperation network, E = 244 connections, and a network density (D) of 0.0063 within the CNKI database. The number of nodes (N = 278) represents 278 distinct authors or research entities involved in the network, suggesting a broad engagement in collaborative research within the field of "horticultural plants-Cd responses". The number of edges (E = 244) shows a moderate level of collaboration through co-authorship on a publication, indicating that there is an active exchange of ideas and research efforts. A density of 0.0063, or 0.63%, is considered sparse, which means that while there are collaborations, they represent only a small fraction of all possible connections among the authors. The authors' collaboration network within CNKI shows a moderate level of collaboration with low-density rate, indicating the network is not fully connected but has considerable potential for expansion and strengthening of collaborative ties. Further analysis could reveal key authors or keyword clustering that could be targeted for interventions to enhance collaboration. In summary, the network of collaborations among authors in the WOS database seemed denser compared to that in the CNKI database.

In the authors' cooperation network, the circles (nodes) represent individuals such as authors or researchers in the network, and the number and thickness of the lines (edges) reflect the collaborative relationships and tensity between these individuals (authors or researchers) [74]. As shown in Figure 4, whether it is the WOS database or the CNKI database, it is evident that among the research teams in the field of "horticultural plants-Cd responses", the team centered around Lin L. is the largest. Additionally, the teams centered around Liao M. are also gradually growing. The top 10 productive authors from the databases of WOS and CNKI (1999–2024) are summarized in Table 3. In the WOS database, Lin L. has a total of 75 research articles published, followed by Liao M. with 43, Wang J. with 38, Tang Y. with 37, and Xia H. with 34. In the CNKI database, Lin L. has a total of 35 research articles published, followed by Liao M. with 25, Liu S. with 14, Guan P. with 10, and Shi J. with 8. It is observed that Chinese scholars (Lin L. and Liao M.) are leading researchers in both domestic and international research fields for "horticultural plants-Cd responses". In recent years, the research teams led by Lin L. and Liao M. from Sichuan Agricultural University have maintained active research interests in the field of "horticultural plants-Cd responses". They have been focusing particularly on the effects of intercropping, straw application, and exogenous substances on horticultural plants under Cd stress [75–78].

**Table 3.** Top 10 productive authors from the databases of WOS and CNKI (1999–2024).

| WOS | | | | CNKI | | | |
|---|---|---|---|---|---|---|---|
| Rank | Author | Quantity | Year | Rank | Author | Quantity | Year |
| 1 | Lin L | 75 | 2015 | 1 | Lin L | 35 | 2013 |
| 2 | Liao M | 43 | 2014 | 2 | Liao M | 25 | 2013 |
| 3 | Wang J | 38 | 2016 | 3 | Liu S | 14 | 2013 |
| 4 | Tang Y | 37 | 2016 | 4 | Guan P | 10 | 2014 |
| 5 | Xia H | 34 | 2016 | 5 | Shi J | 8 | 2015 |
| 6 | Liang D | 34 | 2016 | 6 | Ding J | 8 | 2013 |
| 7 | Wang X | 33 | 2017 | 7 | Liu P | 7 | 2008 |
| 8 | Lv X | 25 | 2016 | 8 | Wang L | 6 | 2013 |
| 9 | Cuypers A | 12 | 2010 | 9 | Pan Z | 6 | 2013 |
| 10 | Yasin NA | 12 | 2020 | 10 | Chen B | 6 | 2020 |

3.2.3. Institution Cooperation Networks

Analyzing the institutions and collaborations behind publications related to "horticultural plants-Cd responses" can provide an accurate understanding of the distribution of research efforts in this field [79,80]. As shown in Figure 5, there are N = 291 nodes in the institutions' cooperation network, E = 718 connections, and a network density (D) of 0.017 within the WOS database. Nevertheless, there are N = 254 nodes in the institutions' cooperation network, E = 87 connections, and a network density (D) of 0.0027 within the

CNKI database. Comparatively, the collaboration network in the WOS database is larger than that in the CNKI database, which may reflect differences in the research fields and regions covered by the different databases. The number of connections in WOS is much higher than in CNKI, indicating that collaborations recorded in WOS are more frequent. Both databases have low network densities, but the network density of WOS is higher than that of CNKI, meaning that the collaborative ties in WOS are relatively more intensive compared to the maximum possible number of connections. The low density of both networks indicates a substantial opportunity to increase the number of collaborative relationships, especially in the CNKI database, where the potential for collaboration may be even greater. These data can help us understand the current state and potential for collaboration between different research fields and regions, providing a reference for promoting future collaborations.

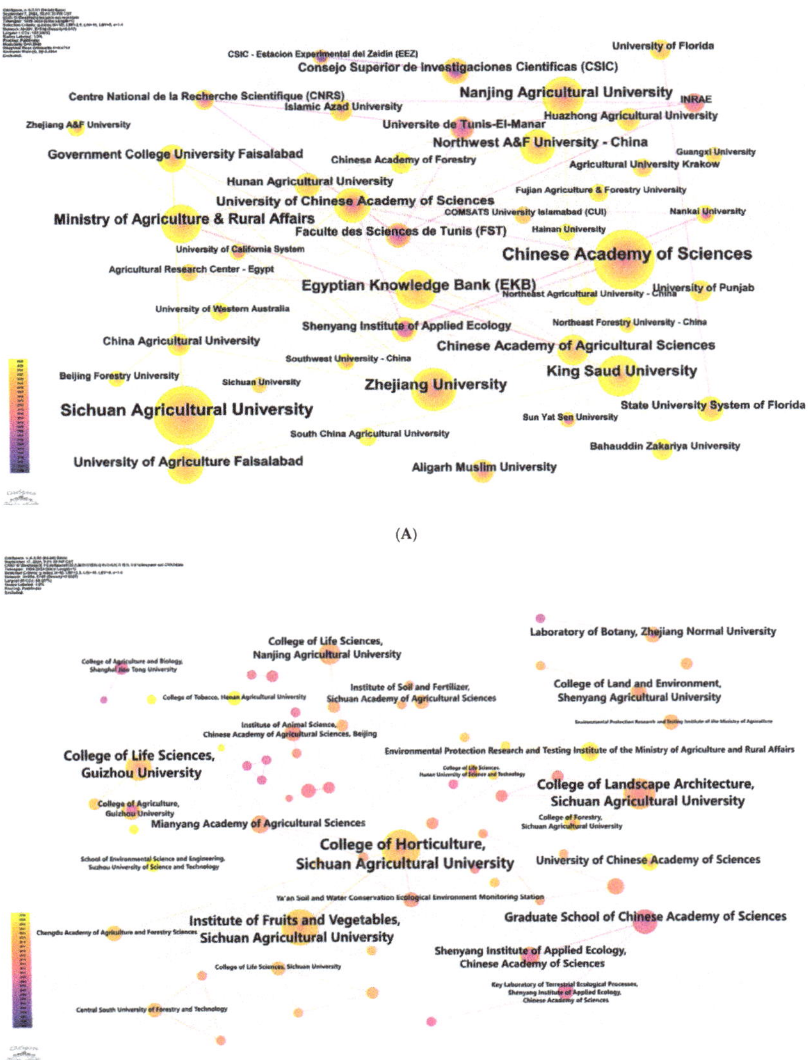


**Figure 5.** Mapping of institution collaboration networks from the databases of WOS (**A**) and CNKI (**B**).

The top 10 productive institutions from the databases of WOS and CNKI (1999–2024) are summarized in Table 4. In terms of publication volume from the database of WOS, the five most productive institutions are Chinese Academy of Sciences (159 articles), Sichuan Agricultural University (139 articles), Zhejiang University (81 articles), King Saud University (72 articles), and Egyptian Knowledge Bank (71 articles). Chinese Academy of Sciences, as a leading research institution, has a substantial output (159 articles), reflecting its broad research capabilities and extensive resources. Sichuan Agricultural University (139 articles), with a high number of publications, is particularly active in the field of "horticultural plants-Cd responses", indicating a focused research effort and potentially strong faculty in this area. Zhejiang University's significant publication volume (81 articles) suggests a strong commitment to research on horticultural plants' responses to Cd, indicating well-funded programs and collaborative efforts. The productivity of King Saud University (72 articles) indicates a substantial contribution to this field, suggesting a research focus that aligns with environmental and agricultural priorities. The significant number of publications from Egyptian Knowledge Bank (71 articles) reflects a dedication to advancing knowledge in the agricultural sciences, which is crucial for a country where agriculture is a vital sector.

**Table 4.** Top 10 productive institutions from the databases of WOS and CNKI (1999–2024).

| Database | Rank | Research Institute | Count | Year |
|---|---|---|---|---|
| WOS | 1 | Chinese Academy of Sciences | 159 | 2004 |
| | 2 | Sichuan Agricultural University | 139 | 2014 |
| | 3 | Zhejiang University | 81 | 2002 |
| | 4 | King Saud University | 72 | 2011 |
| | 5 | Egyptian Knowledge Bank (EKB) | 71 | 2011 |
| | 6 | Nanjing Agricultural University | 70 | 2008 |
| | 7 | Ministry of Agriculture & Rural Affairs | 65 | 2009 |
| | 8 | Chinese Academy of Agricultural Sciences | 51 | 2009 |
| | 9 | University of Chinese Academy of Sciences | 46 | 2005 |
| | 10 | University of Agriculture Faisalabad | 46 | 2015 |
| CNKI | 1 | College of Horticulture, Sichuan Agricultural University | 37 | 2013 |
| | 2 | Institute of Fruits and Vegetables, Sichuan Agricultural University | 31 | 2015 |
| | 3 | College of Landscape Architecture, Sichuan Agricultural University | 23 | 2013 |
| | 4 | College of Life Sciences, Guizhou University | 16 | 2014 |
| | 5 | Graduate School of Chinese Academy of Sciences | 14 | 2009 |
| | 6 | College of Life Science and Technology, Xinxiang University | 13 | 2016 |
| | 7 | College of Resources and Environment, Yunnan Agricultural University | 12 | 2012 |
| | 8 | College of Horticulture and Landscape Architecture, Henan University of Science and Technology | 11 | 2020 |
| | 9 | Shenyang Institute of Applied Ecology, Chinese Academy of Sciences | 10 | 2007 |
| | 10 | College of Life Sciences, Nanjing Agricultural University | 9 | 2003 |

Comparatively, in terms of publication volume from the database of CNKI, the five most productive institutions are College of Horticulture, Sichuan Agricultural University (37 articles), Institute of Fruits and Vegetables, Sichuan Agricultural University (31 articles), College of Landscape Architecture, Sichuan Agricultural University (23 articles), College of Life Sciences, Guizhou University (16 articles), and Graduate School of Chinese Academy of Sciences (14 articles). College of Horticulture, Sichuan Agricultural University (37 articles) focusing on horticulture suggests specialized research in plant cultivation and responses to environmental stressors like Cd. The focus of Institute of Fruits and Vegetables, Sichuan Agricultural University (31 articles) signifies research aimed at understanding and mitigating the impact of Cd on these essential food crops. It is observed that Sichuan Agricultural University is highly productive across both databases, indicating a strong focus on horticultural research related to Cd. The prominence of Sichuan Agricultural University in both databases suggests that this region of China is a hub for research on "horticultural plants-Cd responses". Zhejiang University and the Chinese Academy of Sciences also show consistent productivity. The higher publication volumes in WOS suggest a wider scope of research or more extensive international collaborations. The CNKI data show a more concentrated effort within specific colleges and universities in China.

### *3.3. Visual Analysis of Intellectual Base on "Horticultural Plants-Cd Responses"*

The intellectual base, which constitutes a comprehensive collection of prior references in a given field, primarily focuses on analyzing the most frequently cited topics, their interconnectedness, and the relationships among these central topics [81]. The top 10 highest cited articles from the database of WOS are shown in Table 5; these have been influential in shaping the discourse and have significantly contributed to the intellectual base, which encompasses the foundational theories, concepts, and empirical evidence that form the core of the scholarly discipline or field of "horticultural plants-Cd responses".

**Table 5.** Top 10 highest cited articles from the database of WOS.

| Title | Author | Year | Journal | Article Types | Institutions | Journal Impact Factor (5 Years) | Citation | Reference |
|---|---|---|---|---|---|---|---|---|
| The effects of cadmium toxicity | Genchi, G. et al. | 2020 | International Journal of Environmental Research and Public Health | Review | Università della Calabria | 4.4 | 1124 | [82] |
| Unravelling cadmium toxicity and tolerance in plants: Insight into regulatory mechanisms | Gallego, SM. et al. | 2012 | Environmental and Experimental Botany | Review | Universidad de Buenos Aires | 5.2 | 891 | [83] |
| Cadmium in soils and groundwater: A review | Kubier, A. et al. | 2019 | Applied Geochemistry | Review | University of Bremen | 3.4 | 634 | [84] |
| Cadmium toxicity in plants: Impacts and remediation strategies | Haider, FU. et al. | 2021 | Ecotoxicology and Environmental Safety | Review | Gansu Agricultural University | 6.3 | 618 | [85] |

**Table 5.** *Cont.*

| Title | Author | Year | Journal | Article Types | Institutions | Journal Impact Factor (5 Years) | Citation | Reference |
|---|---|---|---|---|---|---|---|---|
| Cadmium in plants: uptake, toxicity, and its interactions with selenium fertilizers | Ismael, MA. et al. | 2019 | Metallomics | Critical Review | Huazhong Agricultural University | 3.7 | 386 | [86] |
| Cadmium bioavailability, uptake, toxicity and detoxification in soil-plant system | Shahid, M. et al. | 2017 | Reviews of Environmental Contamination and Toxicology | Review | COMSATS Institute of Information Technology | 7.1 | 369 | [87] |
| A critical review on effects, tolerance mechanisms and management of cadmium in vegetables | Rizwan, M. et al. | 2017 | Chemosphere | Critical Review | Government College University | 7.7 | 341 | [88] |
| Cadmium stress in plants: A critical review of the effects, mechanisms, and tolerance strategies | El Rasafi, T. et al. | 2020 | Critical Reviews in Environmental Science and Technology | Critical Review | University Mohammed 6 Polytechnic | 14.5 | 243 | [89] |
| Morphological and physiological responses of plants to cadmium yoxicity: A review | He, SY. et al. | 2017 | Pedosphere | Review | Zhejiang Gongshang University | 5.3 | 238 | [90] |
| Toxicity of cadmium and its competition with mineral nutrients for uptake by plants: A review | Qin, SY. et al. | 2020 | Pedosphere | Review | Henan Agricultural University | 5.3 | 237 | [91] |

Among these 10 articles, the most highly cited was "The Effects of Cadmium Toxicity", authored by Genchi G. et al., with 1124 citations. It was published in the "International Journal of Environmental Research and Public Health" (IF = 4.4) in 2020 [82]. The article pointed out the main sources, routes of transmission, and half-life of Cd. The study also demonstrated that Cd can induce a variety of epigenetic changes in mammalian cells both in vivo and in vitro, leading to pathological risks and the development of various types of cancers. The research also proposed that plants such as sunflowers (*Helianthus annuus* L.) and Indian mustard (*Brassica juncea*) possess the ability to remove cadmium from contaminated soil and water. The second most highly cited was "Unravelling cadmium toxicity and tolerance in plants: Insight into regulatory mechanisms", authored by Gallego S.M. et al., with 891 citations. It was published in "*Environmental and Experimental Botany*" (IF = 5.2) in 2012 [83]. This article provided a comprehensive review of the mechanisms by which plants absorb, transport, and accumulate Cd, as well as an overview of the detoxification processes that plants employ to combat Cd. Additionally, it discussed the oxidative stress caused by Cd and the antioxidant defenses plants activate in response. Furthermore, the article elucidated the impact of reactive oxygen and nitrogen species on Cd-induced toxicity in plants. The third most highly cited was "Cadmium in soils and groundwater: A review", authored by Kubier A. et al., with 634 citations. It was published in "*Ap-*

*plied Geochemistry*" (IF = 3.4) in 2019 [84]. This article examined the concentration of Cd in rocks, sediments, and soils and discussed anthropogenic sources of Cd. It also explored the hydrochemical behavior of Cd, including its solubility, complexation, and sorption. Additionally, the article presented case studies of elevated Cd concentrations in soil and groundwater. The fourth most highly cited was "Cadmium toxicity in plants: Impacts and remediation strategies", authored by Haider F.U. et al., with 618 citations. It was published in "*Ecotoxicology and Environmental Safety*" (IF = 6.3) in 2021 [85]. This article reviewed the sources of Cd contamination in the environment, the soil factors that affect Cd uptake, and the dynamics of Cd in the soil rhizosphere. It also discussed the mechanisms of Cd uptake, translocation, and toxicity in plants. The article highlighted how Cd toxicity in horticultural plants such as tobacco, rapeseed, pea and *Thlaspi caerulescens* L. can reduce the uptake and translocation of nutrients and water, increase oxidative damage, disrupt plant metabolism, and inhibit plant growth and development. Furthermore, the article explored plant defense mechanisms against Cd toxicity and potential remediation strategies. These strategies include the use of biochar, mineral nutrients, compost, organic manure, growth regulators, and hormones, as well as the application of phytoremediation, bioremediation, and chemical methods. This article, published in 2021, has garnered 618 citations, indicating widespread attention and recognition within the international academic community.

In general, these 10 highest cited papers theoretically analyzed the main sources of Cd, critical thresholds, pollution pathways, and toxicity characteristics. They systematically dissected the stress responses and detoxification mechanisms of plants to Cd, and in practice, they proposed targeted suggestions for the management methods and remediation technologies for Cd. Additionally, as shown in Table 5, these 10 highly cited papers are all review articles. The reason might be multifaceted. On one hand, the topics covered by these reviews could be at the forefront of research, attracting attention from a broad spectrum of scholars and consequently leading to frequent citations. On the other hand, cadmium pollution and its impact on plants and the environment are of interest across various disciplines, including environmental science, toxicology, plant biology, and public health. These reviews might serve as interdisciplinary bridges, appealing to a wide academic audience.

### 3.4. *Visual Analysis of Research Hotspots on"Horticultural Plants-Cd Responses"*

#### 3.4.1. Keyword Co-Occurrence Network Analysis

Keywords serve to encapsulate the principal content of a research article, extracting essential details like objectives, methodologies, and perspectives [92]. The frequency analysis of these keywords is vital for pinpointing trending issues and evolutionary trends within a specific domain [67,93,94]. As shown in Figure 6, in keyword co-occurrence network analysis within the WOS database, there are 260 nodes (N = 260), 1254 edges (E = 1254), and a network density of 0.0372. In contrast, within the CNKI database, there are 285 nodes (N = 285), 751 connections (E = 751), and a network density of 0.0186. Keywords with higher centrality are highlighted in pink circles on the map. For instance, in the WOS database, keywords such as "plants", "phytoremediation", and "glutathione" are emphasized. Similarly, in the CNKI database, keywords like "cadmium stress", "antioxidant enzymes", "cadmium contamination", "growth", "phytoremediation", "rapeseed", "heavy metal", "photosynthesis" and "Cd stress" stand out. This visual representation helps researchers quickly identify key themes and important concepts in their research field. The size and position of the pink circles can intuitively show the importance of the keywords, that is, their frequency of occurrence in the literature and the degree of association with other keywords. In the WOS co-occurrence network, the plant-related terms were *Phaseolus vulgaris* [95] and *Brassica juncea* [96]. Additionally, "bean Phaseolus vulgaris" is also mentioned. In the CNKI co-occurrence network, the plant-related terms included 油菜 (rapeseed), 小白菜 (pakchoi), 紫花苜蓿 (*Medicago sativa* L.), 龙葵 (*Solanum nigrum* L.), 金银花 (*Lonicera Japonica* Thunb.), among others. The plants mentioned are likely to be the focus of significant research within the respective databases, indicating areas of interest for scientists and scholars. Despite having fewer nodes (N = 260) compared with the CNKI, WOS has more connections within the field of "horticultural plants-Cd responses", indicating a higher degree of interlinkage between keywords. This could mean that the keywords in WOS are more frequently co-occurring in the research articles, suggesting a denser network of research topics in the field of "horticultural plants-Cd responses". This could imply a more interdisciplinary research environment or more consistent use of certain combinations of keywords. The density of the network in WOS (Density = 0.0372) is almost double that of CNKI (Density = 0.0186), indicating a relatively more interconnected network of keywords in WOS. A higher density suggests a more closely knit group of topics that frequently appear together in the context of the research articles indexed. The sparser network in CNKI might suggest opportunities for researchers to explore new intersections between existing areas of research. Conversely, the denser network in WOS could indicate a more established body of research where connections between topics are well explored. The comparison of keyword co-occurrence networks between WOS and CNKI databases highlighted differences in the structure and interconnectivity of research topics. WOS showed a denser network with more connections, suggesting a more interconnected research landscape. CNKI, with a higher number of nodes and a sparser network, might indicate a broader range of distinct research areas with potential for further integration. Understanding these differences can help researchers identify trends, opportunities for collaboration, and areas ripe for interdisciplinary exploration.

Analyzing a knowledge graph allows for the identification of research topics and trends within the field of "horticultural plants-Cd responses" by examining the significance and connections of keywords. It is recognized that in Citespace, centrality is a measure of how much a node affects the shortest paths between other nodes; a higher centrality value suggests a greater level of influence [23]. A node is considered to be a key node when its centrality value exceeds 0.1. The top 10 high-frequency keywords from the databases of WOS and CNKI are shown in Table 6. International research findings from the databases of WOS showed that the top 10 high-frequency keywords are accumulation (752), tolerance (587), heavy metals (562), oxidative stress (505), toxicity (496), growth (493), responses (270), Cd (269), cadmium stress (232), and phytoremediation (222); moreover, the centrality value of phytoremediation is above 0.1, indicating that globally, research hotspots have focused on topics such as accumulation, tolerance, and oxidative stress, as well as the formation of core keywords like phytoremediation. Domestic research findings from the databases of CNKI showed that the top 10 high-frequency keywords are cadmium stress (625), antioxidant enzymes (115), cadmium contamination (104), growth (91), seed germination (90), phytoremediation (85), rapeseed (83), physiological characteristics (79), heavy metal (78) and photosynthesis (63); moreover, the centrality values of cadmium stress, antioxidant enzymes, cadmium contamination, growth, phytoremediation, and rapeseed all exceed 0.1, indicating that in China, research hotspots have focused on topics such as antioxidant enzymes, growth, and seed germination, as well as the formation of multiple core keywords including antioxidant enzymes, growth and seed germination.

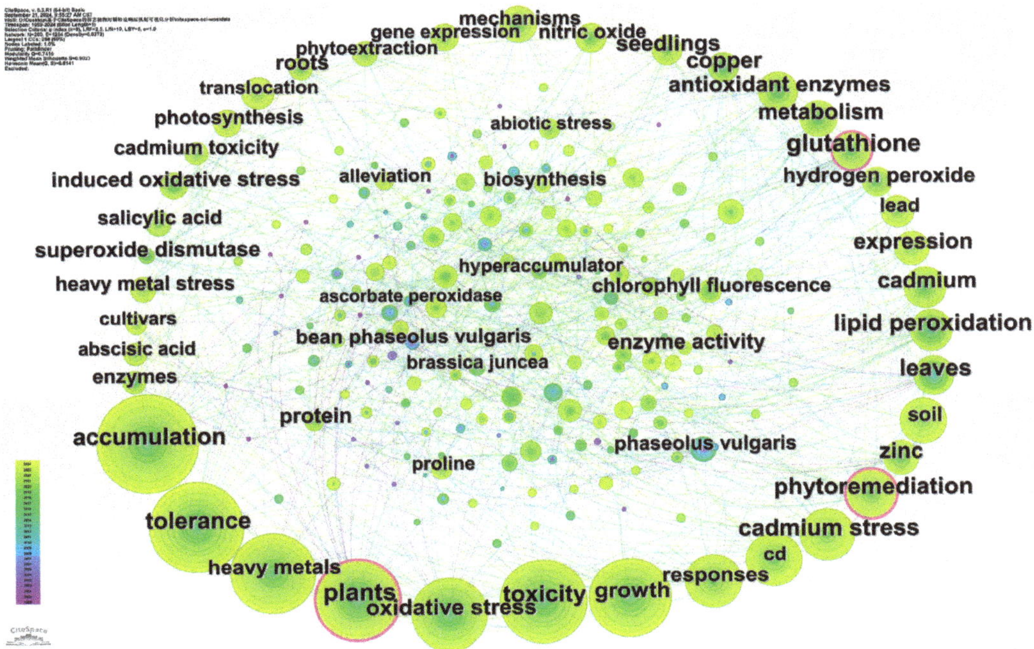


(A)

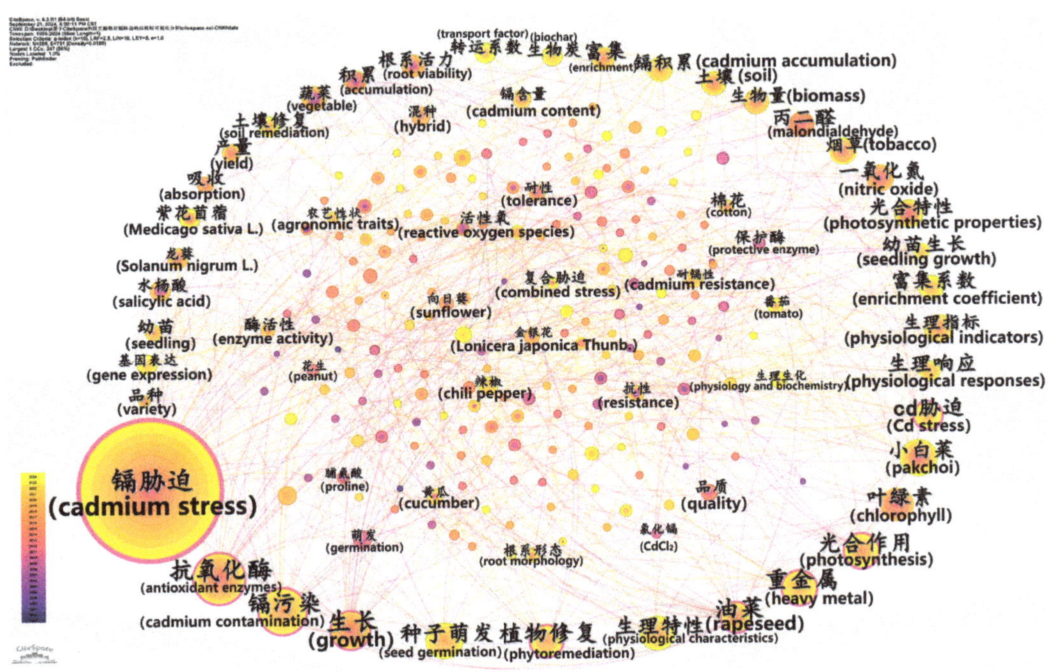


(B)

**Figure 6.** Mapping of keyword co-occurrence networks from the databases of WOS (A) and CNKI (B).

**Table 6.** Top 10 high-frequency keywords from the databases of WOS and CNKI.

| Database | Rank | Keyword | Count | Centrality | Year |
|---|---|---|---|---|---|
| WOS | 1 | accumulation | 752 | 0.08 | 1999 |
| | 2 | tolerance | 587 | 0.08 | 2002 |
| | 3 | heavy metals | 562 | 0.05 | 1999 |
| | 4 | oxidative stress | 505 | 0.04 | 2001 |
| | 5 | toxicity | 496 | 0.05 | 2006 |
| | 6 | growth | 493 | 0.05 | 2002 |
| | 7 | responses | 270 | 0.05 | 2005 |
| | 8 | Cd | 269 | 0.03 | 2002 |
| | 9 | cadmium stress | 232 | 0.08 | 2005 |
| | 10 | phytoremediation | 222 | 0.11 | 2004 |
| CNKI | 1 | cadmium stress | 625 | 0.43 | 2001 |
| | 2 | antioxidant enzymes | 115 | 0.16 | 2004 |
| | 3 | cadmium contamination | 104 | 0.13 | 2000 |
| | 4 | growth | 91 | 0.16 | 2007 |
| | 5 | seed germination | 90 | 0.06 | 2007 |
| | 6 | phytoremediation | 85 | 0.14 | 2001 |
| | 7 | rapeseed | 83 | 0.18 | 2001 |
| | 8 | physiological characteristics | 79 | 0.08 | 2006 |
| | 9 | heavy metal | 78 | 0.08 | 2005 |
| | 10 | photosynthesis | 63 | 0.09 | 2007 |

#### 3.4.2. Keyword Clustering Analysis

The CiteSpace software evaluates the effectiveness of a knowledge map by considering two key metrics: the Q value and the S value [97]. The Q value, also known as modularity, ranges from 0 to 1, with a score above 0.3 denoting that the divided community structure is significant. The S value, also known as weighted mean silhouette, which stands for the silhouette score, measures the average clarity of the clusters [98]. An S value above 0.5 suggests that the clustering outcomes are reasonable, while an S value above 0.7 indicates that the clustering outcomes are reasonable and persuasive [67,93]. As shown in Figure 7, in keyword clustering analysis within the WOS database, Q = 0.3598, and S = 0.6751. The Q value (0.3598) is slightly above the threshold of 0.3, indicating that the community structure identified in the clustering analysis is statistically significant, meaning that the nodes (keywords) within these communities are more closely connected than they would be if distributed randomly. The S value (0.6751) exceeds the threshold of 0.5 but is below 0.7. This suggests that the clustering results are reasonable and the clustering effect is good, but there may still be room for improvement, and the optimal clustering effect has not yet been achieved. The international clusters from the WOS database are #0 cadmium stress, #1 phytoremediation, #2 heavy metals, #3 subcellular distribution, #4 phaseolus vulgaris, #5 transcriptome, #6 chlorophyll fluorescence.

Comparatively, in keyword clustering analysis within the CNKI database, we have Q = 0.4425, and S = 0.7693. The Q value (0.4425) being greater than the threshold of 0.3 indicates that the community structure identified in the clustering analysis is significant, meaning that the nodes (keywords) within these communities are more closely connected than they would be if distributed randomly, demonstrating a good clustering effect. An S value (0.7693) exceeding the threshold of 0.7 suggests that the clustering results are persuasive, meaning that the clustering effect is not only reasonable but also has good discrimination and internal consistency. The domestic clusters from the CNKI database are #0 油菜(rapeseed), #1 叶绿素 (chlorophyll), #2 抗氧化酶 (antioxidant enzymes), #3 生理指标 (physiological indicators), #4 生长 (growth), #5 镉含量 (cadmium content), #6 镉污染 (cadmium contamination), and #7 镉胁迫 (cadmium stress). By observing the connections between different clusters, potential interdisciplinary cross-points can be discovered. The results of clustering analysis revealed that clusters like phytoremediation, subcellular distribution, transcriptome, and chlorophyll fluorescence are emerging topics globally. Meanwhile, in China, research on growth and physiological characteristics, including chlorophyll and antioxidant enzymes, continues to be an area of emerging interest.

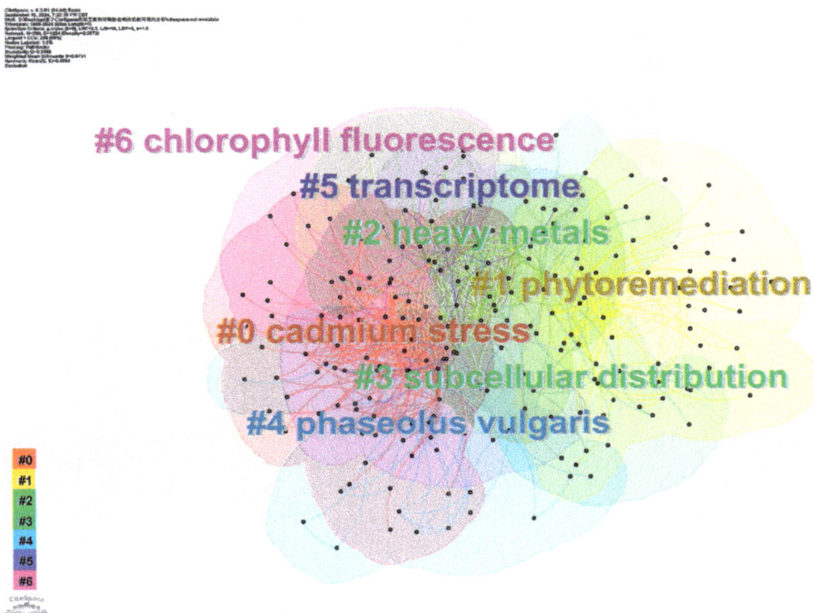


(A)

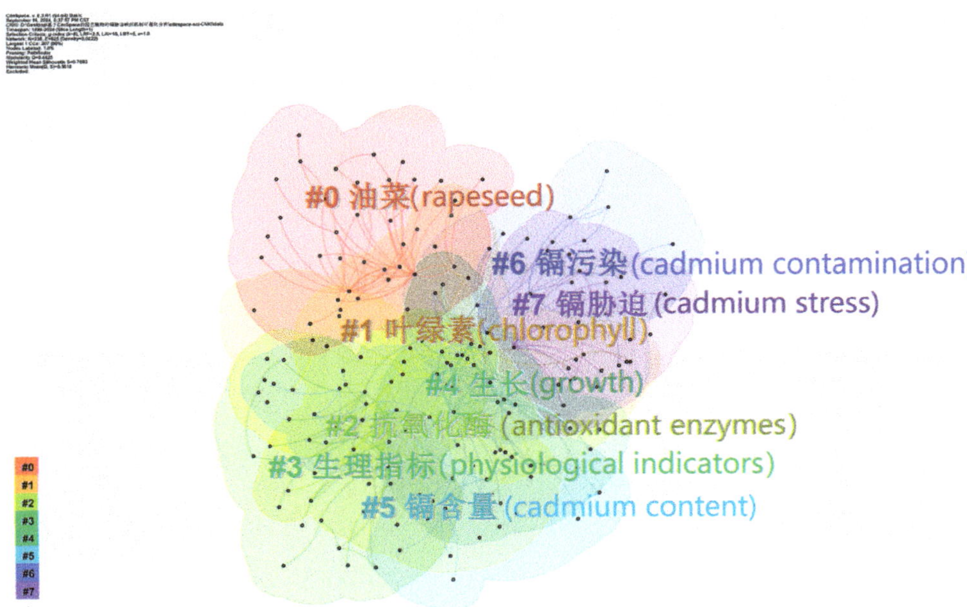


(B)

**Figure 7.** Mapping of keyword clustering from the databases of WOS (**A**) and CNKI (**B**).

Through keyword clustering, it is possible to identify research trends and hotspots within a specific field, as well as help researchers discover areas or topics that have not been fully explored [99,100]. As shown in Table 7, the symbol "#" denotes a cluster, while "note" indicates the count of citation references within each cluster "#". The mean (year) of publications within a cluster "#" acts as a simple but informative metric, highlighting whether the articles are relatively recent or dated [101]. A summary of the international keyword clustering from the databases of WOS showed the following: #0 (cadmium stress), with 65 nodes; #1 (phytoremediation), with 62 nodes; #2 (heavy metals), with 35 nodes; #3 (subcellular distribution), with 32 nodes; #4 (phaseolus vulgaris), with 28 nodes; #5 (transcriptome), with 27 nodes; #6 (chlorophyll fluorescence), with 9 nodes. In contrast, domestic keyword clustering summary from the databases of CNKI showed the following: #0 油菜 (rapeseed), with 33 nodes; #1 叶绿素 (chlorophyll), with 31 nodes; #2 抗氧化酶 (antioxidant enzymes), with 31 nodes; #3 生理指标 (physiological indicators), with 24 nodes; #4 生长 (growth), with 23 nodes; #5 镉含量 (cadmium content), with 22 nodes; #6 镉污染 (cadmium contamination), with 22 nodes; and #7 镉胁迫 (cadmium stress), with 21 nodes. The largest cluster from the databases of WOS was "cadmium stress", labeled 2011, consisting of 65 keywords, including primary terms such as cadmium stress, nitric oxide, antioxidant enzymes, gene expression, and phytoremediation. In contrast, the largest cluster from the databases of CNKI was "rapeseed", labeled 2010, consisting of 33 keywords, including rapeseed, phytoremediation, cadmium, cadmium stress, and enrichment. Keywords were extracted from each cluster based on a weighting algorithm. The five most significant keywords in each cluster are displayed from left to right, arranged according to their level of importance.

**Table 7.** The comparison of keyword clustering from the databases of WOS and CNKI (1999–2024).

| Database | Label | Node | S Value | Mean (Year) | Keywords |
|---|---|---|---|---|---|
| WOS | #0 | 65 | 0.622 | 2011 | cadmium stress (48.72, $1.0 \times 10^{-4}$; nitric oxide (37.96, $1.0 \times 10^{-4}$); antioxidant enzymes (33.25, $1.0 \times 10^{-4}$); gene expression (24.24, $1.0 \times 10^{-4}$); phytoremediation (20.33, $1.0 \times 10^{-4}$) |
| | #1 | 62 | 0.658 | 2012 | phytoremediation (65.32, $1.0 \times 10^{-4}$); cadmium stress (41.58, $1.0 \times 10^{-4}$); oxidative stress (35.32, $1.0 \times 10^{-4}$); hyperaccumulator (33.78, $1.0 \times 10^{-4}$); ornamental plant (32.78, $1.0 \times 10^{-4}$) |
| | #2 | 35 | 0.798 | 2005 | heavy metals (20.15, $1.0 \times 10^{-4}$); nutrients (14.44, 0.001); yield (12.66, 0.001); root morphology (12.03, 0.001); nutrient uptake (12.03, 0.001) |
| | #3 | 32 | 0.689 | 2017 | subcellular distribution (24.82, $1.0 \times 10^{-4}$); antioxidant activity (17.15, $1.0 \times 10^{-4}$); piriformospora indica (9.58, 0.005); pectin (9.24, 0.005); heavy metals accumulation (9.24, 0.005) |
| | #4 | 28 | 0.654 | 2005 | phaseolus vulgaris (10.17, 0.005); nitrogen metabolism (9.49, 0.005); senescence (9.18, 0.005); glutamine synthetase (9.18, 0.005); ammonium (9.18, 0.005) |
| | #5 | 27 | 0.625 | 2015 | transcriptome (31.09, $1.0 \times 10^{-4}$); cd stress (21.63, $1.0 \times 10^{-4}$); qrt-pcr (19.03, $1.0 \times 10^{-4}$); genes (14.27, 0.001); expression pattern (14.27, 0.001) |
| | #6 | 9 | 0.865 | 2010 | chlorophyll fluorescence (31.64, $1.0 \times 10^{-4}$); ascorbate-glutathione cycle (20.61, $1.0 \times 10^{-4}$); photosynthesis (14.08, 0.001); stomatal conductance (12.75, 0.001); spectral reflectance (12.75, 0.001) |

**Table 7.** *Cont.*

| Database | Label | Node | S Value | Mean (Year) | Keywords |
|---|---|---|---|---|---|
| CNKI | #0 | 33 | 0.809 | 2010 | rapeseed (81.78, 1.0 × $10^{-4}$); phytoremediation (76.45, 1.0 × $10^{-4}$); cadmium (34.73, 1.0 × $10^{-4}$); cadmium stress (29.26, 1.0 × $10^{-4}$); enrichment (28.24, 1.0 × $10^{-4}$) |
| | #1 | 31 | 0.798 | 2013 | chlorophyll (51.5, 1.0 × $10^{-4}$); stress (42.96, 1.0 × $10^{-4}$); malondialdehyde (34.65, 1.0 × $10^{-4}$); heavy metal (33.31, 1.0 × $10^{-4}$); seed germination (31.06, 1.0 × $10^{-4}$) |
| | #2 | 31 | 0.794 | 2010 | antioxidant enzymes (39.92, 1.0 × $10^{-4}$); physiological characteristics (39.56, 1.0 × $10^{-4}$); Cd stress (38.7, 1.0 × $10^{-4}$); reactive oxygen species (28.2, 1.0 × $10^{-4}$); seedling growth (24.37, 1.0 × $10^{-4}$) |
| | #3 | 24 | 0.684 | 2016 | physiological indicators (58.05, 1.0 × $10^{-4}$); pakchoi (50.45, 1.0 × $10^{-4}$); nutrient elements (16.25, 1.0 × $10^{-4}$); seedling (13.61, 0.001); cotton (12.54, 0.001) |
| | #4 | 23 | 0.836 | 2011 | growth (66.1, 1.0 × $10^{-4}$); photosynthetic properties (29.97, 1.0 × $10^{-4}$); photosynthesis (28.68, 1.0 × $10^{-4}$); enrichment coefficient (24.04, 1.0 × $10^{-4}$); physiological responses (18.38, 1.0 × $10^{-4}$) |
| | #5 | 22 | 0.753 | 2015 | cadmium content (23.78, 1.0 × $10^{-4}$); biochar (22.5, 1.0 × $10^{-4}$); yield (20.38, 1.0 × $10^{-4}$); soil (18.49, 1.0 × $10^{-4}$); resistance (17.07, 1.0 × $10^{-4}$) |
| | #6 | 22 | 0.83 | 2013 | cadmium contamination (78.99, 1.0 × $10^{-4}$); biomass (25.74, 1.0 × $10^{-4}$); physiology and biochemistry (18.42, 1.0 × $10^{-4}$); transport factor (17.39, 1.0 × $10^{-4}$); cadmium (15.88, 1.0 × $10^{-4}$) |
| | #7 | 21 | 0.605 | 2014 | cadmium stress (151, 1.0 × $10^{-4}$); cadmium (54.98, 1.0 × $10^{-4}$); cadmium contamination (24.91, 1.0 × $10^{-4}$); heavy metal (17.4, 1.0 × $10^{-4}$); Cd stress (12.6, 0.001) |

### 3.5. *Visual Analysis of Dynamic Frontier*

#### 3.5.1. Evolution Trends

Through CiteSpace analysis, a timeline mapping of keywords regarding the field of "horticultural plants-Cd responses" was generated, revealing the temporal scope and evolution trends of emerging research themes. The timeline mapping of keywords from the databases of WOS and CNKI progresses from left to right, with the circular nodes representing keywords (Figure 8). The size of each node corresponds to the occurrence rate of its respective keyword throughout the timeline, and a higher occurrence rate implies an area of intense research interest. The year a node appears on the timeline marks the initial emergence of the keyword, with a more extended presence indicating a more enduring research focus [102]. The international timeline settings were defined in the following way: time span = 1999–2024, slice length = 1, and g-index (k = 8). There were 260 nodes and 1254 connection lines in the network mapping. The node with the biggest size, labeled 1999, was "accumulation", subsequently joined by increasingly prominent high-frequency keywords such as "tolerance" (labeled 2002), "heavy metals" (labeled 1999), "oxidative stress" (labeled 2001), "toxicity" (labeled 2006), and "growth" (labeled 2002). The current research emphasis is centered on rhizosphere, subcellular distribution, biochar, and yield.

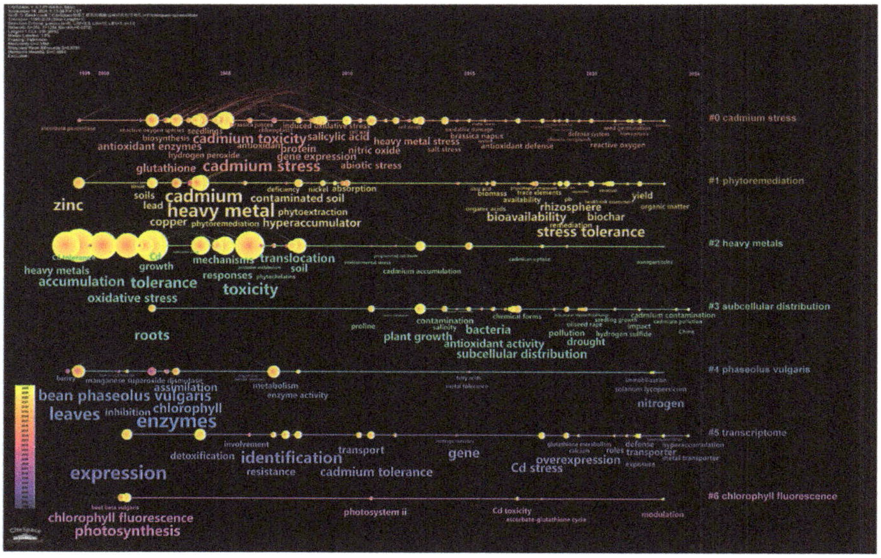


**(A)**

**(B)**

**Figure 8.** Timeline mapping of keywords from the databases of WOS (**A**) and CNKI (**B**).

Comparatively, in timeline mapping of keywords from the CNKI database, there were 238 nodes and 652 connection lines in the network mapping. The node with the biggest size, labeled 2001, was "cadmium stress", subsequently joined by increasingly prominent high-frequency keywords such as "antioxidant enzymes" (labeled 2004), "cadmium contamination" (labeled 2000), "growth" (labeled 2007), "seed germination" (labeled 2007), and "phytoremediation" (labeled 2001). The current research emphasis is centered on transport factor, cadmium absorption, physiological responses, gene expression, and transcriptome. The results above showed that keywords in CNKI tended to describe specific biochemical processes and molecular mechanisms, while those in WOS focused more on plant physiology and the responses of the entire organism. Both databases demonstrated a sustained interest in phytoremediation and heavy metal pollution research, but CNKI might concentrate more on molecular and genetic mechanisms, while WOS might place more emphasis on physiological and applied studies. This evolutionary process, in terms of timeline mapping of keywords, intuitively showcases the gradual developmental trajectory of international and domestic research in "horticultural plants-Cd responses".

3.5.2. Research Frontiers

CiteSpace's burst detection feature was utilized to quantify the prevalence and chronological emergence of keywords within scholarly texts, subsequently establishing the dynamic frontiers and prospective orientation of the field's research on "horticultural plants-Cd responses" [103,104]. Keywords that emerge and continue to appear for at least two years often represent a trend, which can assist in predicting future research hotspots and trajectories [67]. Table 8 lists the top 20 keywords with the strongest citation bursts from the 1999–2024 co-occurrence networks in both the WOS and CNKI databases, which provides a clear view of the timing and magnitude of keyword outbursts. The term "Year" indicates the initial appearance of a term in our dataset, "Strength" measures the magnitude of its mutation, "Begin" marks the start of its mutation period, and "End" denotes its termination. International keywords with strong citation bursts, arranged in descending order of strength, from the databases of WOS were "leaves" (28.2), "copper" (15.34), "*thlaspi caerulescens*" (12.27), "zinc" (11.13), "lipid peroxidation" (9.99), "superoxide dismutase" (8.98), "bean phaseolus vulgaris" (8.69), "metabolism" (8.22), "hydrogen peroxide" (8.17), "rhizosphere" (8.07), etc., indicating that the listed keywords are at the forefront of research, garnering increased attention from the global scientific community. Emerging topics emphasized the importance of the temporal scope [26]. In light of this, the temporal scope of keyword bursts on the research of "horticultural plants-Cd responses" in WOS showed a sustained span: "leaves" (13 years), "zinc" (13 years), "copper" (12 years),"bean phaseolus vulgaris" (11 years), "chlorophyll" (10 years), "*thlaspi caerulescens*" (10 years), "lipid peroxidation" (10 years), "nickel" (8 years), etc. This analysis indicates that in the research on "horticultural plants-Cd responses", certain themes and indicators have received long-term and continuous attention within the international research community. The enduring research interest in these keywords may be related to their importance in plant physiology, nutrient cycling, and environmental pollution remediation. Researchers are likely exploring the tolerance, accumulation mechanisms, and potential impacts on human health of horticultural plants under Cd stress through these keywords [14,105–108].

In comparison, domestic keywords with strong citation bursts, arranged in descending order of strength, from the databases of CNKI were as follows: "rapeseed" (7.23), "chlorophyll" (7.08), "malondialdehyde" (5.7), "gene expression" (5.52), "protective enzyme" (5.36), "transport factor" (4.94), "germination" (4.89), "cadmium absorption" (4.82), "growth" (4.69), "vegetable" (4.66), etc. The analysis suggests that these keywords are correlated with the research interests of domestic scientists. The temporal scope of keyword bursts on the research of "horticultural plants-Cd responses" in CNKI showed a sustained span: "chlorophyll" (11 years), "rapeseed" (10 years), "protective enzyme" (10 years), "accumulation" (8 years), "malondialdehyde" (8 years), etc. The data suggest that in the research on "horticultural plants-Cd responses" within the CNKI database, certain keywords have maintained a consistent focus over the years. These continuous research interests could be attributed to the desire to understand the physiological and biochemical changes in plants under Cd stress, develop Cd-tolerant or low-Cd-accumulating crop varieties, and ensure food safety by reducing the uptake of toxic elements in edible parts of plants.

**Table 8.** Mapping of outburst keywords from the databases of WOS (**A**) and CNKI (**B**).

| Top 20 Keywords with the Strongest Citation Bursts (A) | | | | | |
|---|---|---|---|---|---|
| **Keywords** | **Year** | **Strength** | **Begin** | **End** | **1999–2024** |
| leaves | 1999 | 28.2 | 1999 | 2012 | |
| zinc | 1999 | 11.13 | 1999 | 2012 | |
| bean phaseolus vulgaris | 1999 | 8.69 | 1999 | 2010 | |
| copper | 2003 | 15.34 | 2003 | 2015 | |
| chlorophyll | 2003 | 7.56 | 2003 | 2013 | |
| thlaspi caerulescens | 2004 | 12.27 | 2004 | 2014 | |
| lipid peroxidation | 2004 | 9.99 | 2004 | 2014 | |
| superoxide dismutase | 2004 | 8.98 | 2007 | 2011 | |
| metabolism | 2007 | 8.22 | 2007 | 2011 | |
| hydrogen peroxide | 2004 | 8.17 | 2007 | 2012 | |
| nickel | 2008 | 6.74 | 2008 | 2016 | |
| hyperaccumulator | 2010 | 7.49 | 2010 | 2015 | |
| metal accumulation | 2010 | 7.88 | 2016 | 2019 | |
| induced oxidative stress | 2009 | 7.19 | 2016 | 2018 | |
| brassica napus | 2009 | 6.84 | 2017 | 2020 | |
| rhizosphere | 2018 | 8.07 | 2018 | 2020 | |
| subcellular distribution | 2017 | 7.07 | 2019 | 2021 | |
| oilseed rape | 2020 | 6.78 | 2020 | 2021 | |
| biochar | 2021 | 7.62 | 2021 | 2024 | |
| yield | 2022 | 7.89 | 2022 | 2024 | |

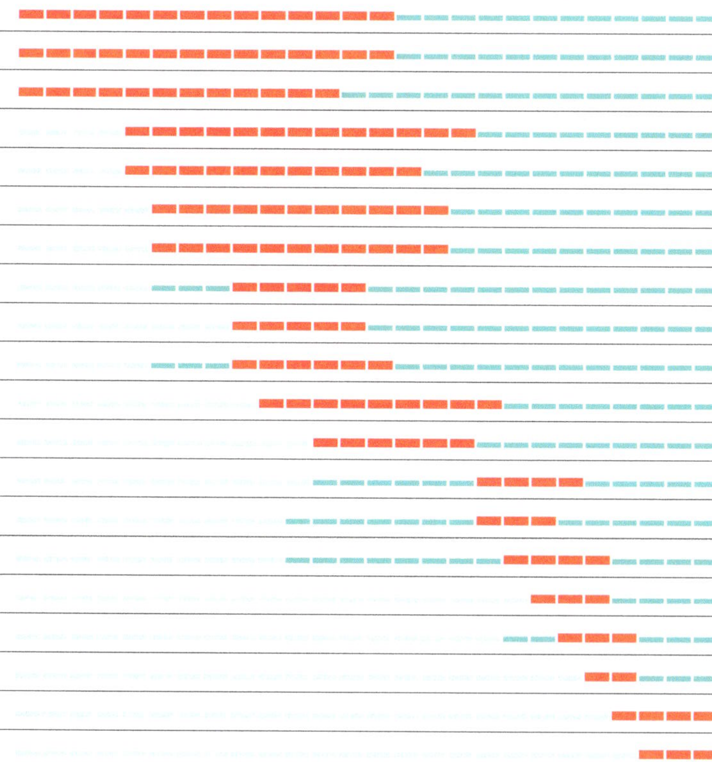

**Table 8.** *Cont.*

| Top 20 Keywords with the Strongest Citation Bursts (B) | | | | | |
|---|---|---|---|---|---|
| Keywords | Year | Strength | Begin | End | 1999–2024 |
| rapeseed | 2001 | 7.23 | 2001 | 2011 | |
| protective enzyme | 2001 | 5.36 | 2001 | 2011 | |
| vegetable | 2005 | 4.66 | 2005 | 2009 | |
| chlorophyll | 2006 | 7.08 | 2006 | 2017 | |
| accumulation | 2007 | 3.87 | 2007 | 2015 | |
| cabbage | 2007 | 3.64 | 2007 | 2011 | |
| malondialdehyde | 2007 | 5.7 | 2008 | 2016 | |
| growth | 2007 | 4.69 | 2008 | 2012 | |
| germination | 2009 | 4.89 | 2009 | 2012 | |
| quality | 2010 | 3.65 | 2010 | 2015 | |
| phytoremediation | 2001 | 3.82 | 2015 | 2018 | |
| *Solanum nigrum* L. | 2015 | 3.68 | 2015 | 2017 | |
| cadmium content | 2016 | 3.49 | 2016 | 2018 | |
| cadmium accumulation | 2011 | 3.66 | 2018 | 2022 | |
| transport factor | 2019 | 4.94 | 2019 | 2024 | |
| cadmium absorption | 2019 | 4.82 | 2019 | 2020 | |
| chili pepper | 2020 | 4.48 | 2020 | 2021 | |
| physiological responses | 2009 | 4.18 | 2020 | 2021 | |
| gene expression | 2021 | 5.52 | 2021 | 2024 | |
| transcriptome | 2021 | 4.6 | 2021 | 2024 | |

The burst strength indicators for the terms "biochar" and "yield" in the WOS database, which were 7.62 and 7.89, respectively, suggest that these topics have gained notable research attention since 2021 and 2022, respectively. Their sustained interest up to the present (2024) indicate that they are emerging as new frontiers in international scientific inquiry. In the CNKI database, "gene expression" and "transcriptome" show burst strengths of 5.52 and 4.6, respectively. These topics, whose notable research interest began in 2021, persist to date (2024), indicating that they have emerged as new domestic research frontiers in China. The comparison above clearly demonstrates the research interest and trends of different topics in the two databases, while also revealing the differences in research focus between international and domestic scientific communities. On the whole, China has exhibited a lagging development trend, which is reflected in the fact that it began to focus on gene expression and transcriptome research only after the global frontier shifted towards biochar and cadmium co-stress and yield response.

## 4. Conclusions and Outlook

As a highly carcinogenic environmental contaminant, the persistence of Cd in the environmental medium and its toxic effects on horticultural plants have garnered widespread attention from researchers around the world in recent years, as reflected by the abundance of research articles on this issue in academic publications. By utilizing a visualizing bibliometric analysis software, CiteSpace, this study integrated and analyzed a total of 4318 relevant research records—2311 from the WOS database and 2007 from the CNKI database—related to "horticultural plants-Cd responses", covering the period from 1999 to 2024. From the temporal distribution of publications, domestically and internationally, research on "horticultural plants-Cd responses" has experienced three distinct phases: StageI, the initial exploration stage; StageII, the steady growth stage; and StageIII, the rapid development stage. The annual publication volume on "horticultural plants-Cd responses" has exhibited a trend of fluctuating growth in both domestic and international research contexts, while, compared to China, the international cumulative publication volume has increased by 15.15%, indicating a year-by-year rise in the number of publications and a continuous increase in global scientific attention to "horticultural plants-Cd responses". From countries', authors' and institutions' cooperation networks, in terms of publication volume, the five most productive countries are the People's Republic of China (1165 articles), India (200 articles), Pakistan (176 articles), the USA (144 articles), and Saudi Arabia (120 articles). Their respective proportions of the total global output are 52.79%, 9.06%, 7.97%, 6.52%, and 5.44%, indicating that China is the most productive country and is actively engaging in partnerships with other countries worldwide. Chinese scholars (Lin L. and Liao M.) are leading researchers in both domestic and international research fields for "horticultural plants-Cd responses". The five most productive institutions, with their respective article counts, are Chinese Academy of Sciences (159 articles), Sichuan Agricultural University (139 articles), Zhejiang University (81 articles), King Saud University (72 articles), and Egyptian Knowledge Bank (71 articles). This distribution reflects the broad research capabilities and extensive resources of these institutions, with the Chinese Academy of Sciences leading in research output. The analyses above underscore the importance of research output from countries, authors, and institutions as an indicator of the field's development and the focus areas within the field of "horticultural plants-Cd responses". They also highlight the value of monitoring the temporal distribution trends of publications to understand research priorities and achievements at the national, author, and institutional levels. The highest cited articles have been influential in shaping the discourse and have significantly contributed to the intellectual base, which encompasses the foundational theories, concepts, and empirical evidence that form the core of the scholarly discipline or field of "horticultural plants-Cd responses". From keyword co-occurrence network and clustering analysis, international research hotspots have focused on accumulation, tolerance, and oxidative stress, while in China, the focus has been on antioxidant enzymes, growth, and seed germination. Phytoremediation, subcellular distribution, and the transcriptome are the world's emerging topics, while in China, growth and physiological characteristics are still emerging topics. The analysis of timeline mapping and outburst keywords revealed the temporal scope and evolution trends and research frontiers of emerging research themes. On comparison, it is seen that China exhibited a lagging development trend, which is reflected in the fact that it began to focus on gene expression and transcriptome research only after the global frontier shifted towards biochar and cadmium co-stress and yield response.

To enhance future research on "horticultural plants-Cd responses" in China, the following suggestions could be considered:

(1) Emphasize interdisciplinary research: Collaboration should be encouraged between different disciplines such as biology, soil science, environmental science, and public health to develop a comprehensive understanding of Cd stress and its impacts on agriculture and the environment.
(2) Focus on biochar and co-stress studies: Since the global research frontier has shifted towards biochar and co-stress studies, Chinese researchers should focus more on how biochar can mitigate Cd stress and how plants respond to multiple stresses such as Cd and drought or exogenous substances.
(3) Promote phytoremediation research: Research should be supported on phytoremediation techniques using horticultural plants that can hyperaccumulate Cd, such as *Thlaspi caerulescens*, *Lonicera japonica* Thunb., to remediate contaminated soils.
(4) Invest in plant breeding and genetic engineering: Cd-resistant horticultural plant varieties should be developed through traditional breeding and genetic engineering techniques to reduce the impact of Cd on agriculture and food safety.
(5) Leverage International Collaboration: International research institutions should collaborate to share knowledge, resources, and expertise and to stay abreast of the latest advancements in cadmium research.
(6) Foster innovation in nanotechnology: The use of nanotechnology should be explored in addressing Cd stress, e.g., nanomaterials for soil remediation or nanosensors for monitoring Cd levels in horticultural plants and soil.

By implementing these strategies, China can improve its research capabilities in the area of "horticultural plants-Cd responses" and contribute to global efforts in managing this significant environmental and health challenge.

**Author Contributions:** Conceptualization, Z.L., B.H. and Y.Z.; data curation, Z.L. and Y.Z.; formal analysis, Z.L. and S.Z.; funding acquisition, Z.L.; methodology, Z.L. and X.D.; software, Z.L. and L.M.; writing—original draft, Z.L. and B.H.; writing—review and editing, Z.L. and H.L. All authors have read and agreed to the published version of the manuscript.

**Funding:** This study is financially supported by Liaoning Revitalization Talents Program (XLYC2203070), Liaoning Province Science and Technology Plan Joint Project Natural Science Foundation-General Program (2024-MSLH-506), the funding project of Northeast Geological S&T Innovation Center of China Geological Survey (QCJJ2022-44), the National Natural Science Foundation of China (32301437), the Young Scientists Fund of the National Natural Science Foundation of China (32201730, 42307433), the Project of Education Department of Liaoning Province (LJKZ0616, JYTMS20231166), the Innovation and Entrepreneurship Training Program for College Students (202411035196), the Graduate Education and Teaching Reform Project of Shenyang University.

**Data Availability Statement:** Data are contained within the article.

**Conflicts of Interest:** The authors declare no conflicts of interest.

## References

1. Mukherjee, A.G.; Wanjari, U.R.; Renu, K.; Vellingiri, B.; Gopalakrishnan, A.V. Heavy metal and metalloid-induced reproductive toxicity. *Environ. Toxicol. Pharmacol.* **2022**, *92*, 103859. [CrossRef] [PubMed]
2. Tomczyk, P.; Wdowczyk, A.; Wiatkowska, B.; Szymańska-Pulikowska, A. Assessment of heavy metal contamination of agricultural soils in Poland using contamination indicators. *Ecol. Indic.* **2023**, *156*, 111161. [CrossRef]
3. Bai, S.Y.; Han, X.; Feng, D. Shoot-root signal circuit: Phytoremediation of heavy metal contaminated soil. *Front. Plant Sci.* **2023**, *14*, 1139744. [CrossRef] [PubMed]
4. Huang, S.Q.; Lu, Z.Q.; Zhao, X.X.; Tan, W.B.; Wang, H.; Liu, D.L.; Xing, W. Molecular basis of energy crops functioning in bioremediation of heavy metal pollution. *Agriculture* **2024**, *14*, 914. [CrossRef]
5. Zhu, W.G.; Zhu, D.Y.; He, J.M.; Lian, X.X.; Chang, Z.B.; Guo, R.C.; Li, X.H.; Wang, Y.L. Phytoremediation of soil co-contaminated with heavy metals (HMs) and tetracyclines: Effect of the co-contamination and HM bioavailability analysis. *J. Soils Sediments* **2022**, *22*, 2036–2047. [CrossRef]
6. Borah, G.; Deka, H. Crude oil associated heavy metals (HMs) contamination in agricultural land: Understanding risk factors and changes in soil biological properties. *Chemosphere* **2023**, *310*, 136890. [CrossRef]
7. Yılmaz, C.H. Heavy metals and their sources, potential pollution situations and health risks for residents in Adıyaman province agricultural lands, Türkiye. *Environ. Geochem. Health* **2023**, *45*, 3521–3539. [CrossRef]
8. Du, Y.B.; Tian, Z.J.; Zhao, Y.F.; Wang, X.R.; Ma, Z.Z.; Yu, C.H. Exploring the accumulation capacity of dominant plants based on soil heavy metals forms and assessing heavy metals contamination characteristics near gold tailings ponds. *J. Environ. Manag.* **2024**, *351*, 119838. [CrossRef]
9. Ma, J.Y.; Dong, X.M.; Yu, L.J.; Zhang, Y.H. Response and function of *Solanum lycopersicum* L. SlSGR2 gene under cadmium stress. *Horticulturae* **2022**, *8*, 1002. [CrossRef]
10. Arruebarrena, M.A.; Hawe, C.T.; Lee, Y.M.; Branco, R.C. Mechanisms of cadmium neurotoxicity. *Int. J. Mol. Sci.* **2023**, *24*, 16558. [CrossRef]
11. Long, H.Y.; Feng, G.F.; Fang, J. In-situ remediation of cadmium contamination in paddy fields: From rhizosphere soil to rice kernel. *Environ. Geochem. Health* **2024**, *46*, 404. [CrossRef] [PubMed]
12. Iqbal, B.; Javed, Q.; Khan, I.; Tariq, M.; Ahmad, N.; Elansary, H.O.; Jalal, A.; Li, G.L.; Du, D.L. Influence of soil microplastic contamination and cadmium toxicity on the growth, physiology, and root growth traits of *Triticum aestivum* L. *S. Afr. J. Bot.* **2023**, *160*, 369–375. [CrossRef]
13. Liu, Z.L.; Lu, Q.X.; Zhao, Y.; Wei, J.B.; Liu, M.; Duan, X.B.; Lin, M.S. Ameliorating effects of graphene oxide on cadmium accumulation and eco-physiological characteristics in a greening hyperaccumulator (*Lonicera japonica* Thunb.). *Plants* **2024**, *13*, 19. [CrossRef] [PubMed]
14. Elbagory, M.; Farrag, D.K.; Hashim, A.M.; Omara, A.E.D. The combined effect of Pseudomonas stutzeri and biochar on the growth dynamics and tolerance of lettuce plants (*Lactuca sativa*) to cadmium stress. *Horticulturae* **2021**, *7*, 430. [CrossRef]
15. Liu, Z.L.; Chen, M.D.; Lin, M.S.; Chen, Q.L.; Lu, Q.X.; Yao, J.; He, X.Y. Cadmium uptake and growth responses of seven urban flowering plants: Hyperaccumulator or bioindicator? *Sustainability* **2022**, *14*, 619. [CrossRef]
16. Vemić, A.; Popović, V.; Miletić, Z.; Radulović, Z.; Rakonjac, L.; Lučić, A. Effect of cadmium (Cd) and lead (Pb) soil contamination on the development of Hymenoscyphus fraxineus on *Fraxinus excelsior* and *F. angustifolia* seedlings. *Iforest* **2023**, *16*, 307. [CrossRef]
17. Wang, W.; Man, Z.; Li, X.L.; Zhao, Y.Y.; Chen, R.Q.; Pan, T.T.; Wang, L.P.; Dai, X.R.; Xiao, H.; Liu, F. Multi-phenotype response and cadmium detection of rice stem under toxic cadmium exposure. *Sci. Total Environ.* **2024**, *917*, 170585. [CrossRef]
18. Chen, Y.; Wu, X.L.; Lin, Z.X.; Teng, D.Z.; Zhao, Y.M.; Chen, S.N.; Hu, X.F. Screening of cadmium resistant bacteria and their growth promotion of *Sorghum bicolor* (L.) Moench under cadmium stress. *Ecotoxicol. Environ. Saf.* **2024**, *272*, 116012. [CrossRef]
19. Sivakoff, F.S.; McLaughlin, R.; Gardiner, M.M. Cadmium soil contamination alters plant-pollinator interactions. *Environ. Pollut.* **2024**, *356*, 124316. [CrossRef]
20. Zhao, W.X.; Zhao, H.Y.; Wang, H.S.; He, Y. Research progress on the relationship between leaf senescence and quality, yield and stress resistance in horticultural plants. *Front. Plant Sci.* **2022**, *13*, 1044500. [CrossRef]
21. Noor, I.; Sohail, H.; Sun, J.; Nawaz, M.A.; Li, G.; Hasanuzzaman, M.; Liu, J. Heavy metal and metalloid toxicity in horticultural plants: Tolerance mechanism and remediation strategies. *Chemosphere* **2022**, *303*, 135196. [CrossRef] [PubMed]
22. Gao, F.; Zhang, X.D.; Zhang, J.; Li, J.; Niu, T.H.; Tang, C.N.; Wang, C.; Xie, J.M. Zinc oxide nanoparticles improve lettuce (*Lactuca sativa* L.) plant tolerance to cadmium by stimulating antioxidant defense, enhancing lignin content and reducing the metal accumulation and translocation. *Front. Plant Sci.* **2022**, *13*, 1015745. [CrossRef] [PubMed]
23. Pan, Y.J.; Xu, X.N.; Lang, Q.Q.; Liao, S.Q.; Li, Y.M. Three different fertilizers enhance spinach growth and reduce spinach Cd concentration in Cd contaminated alkaline soil. *Horticultura* **2023**, *9*, 445. [CrossRef]

24. Li, Y.T.; Gao, Y.; Chen, W.; Zhang, W.G.; Lu, X. Shifts in bacterial diversity, interactions and microbial elemental cycling genes under cadmium contamination in paddy soil: Implications for altered ecological function. *J. Hazard. Mater.* **2024**, *461*, 132544. [CrossRef]
25. Zhao, X.Y.; Peng, J.; Zhang, L.; Yang, X.; Qiu, Y.J.; Cai, C.C.; Hu, J.T.; Huang, T.; Liang, Y.; Wang, Z. Optimizing the quality of horticultural crop: Insights into pre-harvest practices in controlled environment agriculture. *Front. Plant Sci.* **2024**, *15*, 1427471. [CrossRef]
26. Liu, Z.L.; He, X.Y.; Chen, W.; Zhao, M.Z. Ecotoxicological responses of three ornamental herb species to cadmium. *Environ. Toxicol. Chem.* **2013**, *32*, 1746–1751. [CrossRef]
27. Aslam, M.M.; Okal, E.J.; Waseem, M. Cadmium toxicity impacts plant growth and plant remediation strategies. *Plant Growth Regul.* **2023**, *99*, 397–412. [CrossRef]
28. Liu, C.C.; Wen, L.; Cui, Y.J.; Ahammed, G.J.; Cheng, Y. Metal transport proteins and transcription factor networks in plant responses to cadmium stress. *Plant Cell Rep.* **2024**, *43*, 218. [CrossRef]
29. Ziaf, K.; Talha, H.M.; Ghani, M.A.; Ahmad, I.; Anwar, R.; Ali, B.; Majeed, Y.; Shakeel, A.; Iqbal, M.; Zaid, A. Differential accumulation pattern of cadmium in plant parts of pea varieties in response to varying cadmium levels. *S. Afr. J. Bot.* **2023**, *161*, 599–606. [CrossRef]
30. Moravčíková, D.; Žiarovská, J. The effect of cadmium on plants in terms of the response of gene expression level and activity. *Plants* **2023**, *12*, 1848. [CrossRef]
31. Liu, Z.L.; Chen, W.; He, X.Y.; Jia, L.; Huang, Y.Q.; Zhang, Y.; Yu, S. Cadmium-induced physiological response in *Lonicera japonica* Thunb. *Clean-Soil Air Water* **2013**, *41*, 478–484. [CrossRef]
32. Hassan, H.; Elaksher, S.H.; Shabala, S.; Ouyang, B. Cadmium uptake and detoxification in tomato plants: Revealing promising targets for genetic improvement. *Plant Physiol. Biochem.* **2024**, *214*, 108968. [CrossRef] [PubMed]
33. Liu, J.; Kang, L.Y.; Du, L.F.; Liao, S.Q.; Dong, W.; Ma, M.T.; Zou, G.Y.; Li, S.J. Distribution, accumulation and translocation of the heavy metal Cd in various varieties of edible rapeseed under Cd stress. *Sustainability* **2024**, *16*, 2876. [CrossRef]
34. Ndlovu, S.; Pullabhotla, R.V.; Ntuli, N.R. Response of *Corchorus olitorius* leafy vegetable to cadmium in the soil. *Plants* **2020**, *9*, 1200. [CrossRef]
35. Szerement, J.; Szatanik-Kloc, A. Cell-wall pectins in the roots of Apiaceae plants: Adaptations to Cd stress. *Acta Physiol. Plant.* **2022**, *44*, 53. [CrossRef]
36. Jia, K.Y.; Zhan, Z.P.; Wang, B.Q.; Wang, W.H.; Wei, W.J.; Li, D.W.; Huang, W.; Xu, Z.M. Exogenous selenium enhances cadmium stress tolerance by improving physiological characteristics of cabbage (*Brassica oleracea* L. var. *capitata)* seedlings. *Horticulturae* **2023**, *9*, 1016. [CrossRef]
37. Liu, Z.L.; He, X.Y.; Chen, W. Effects of cadmium hyperaccumulation on the concentrations of four trace elements in *Lonicera japonica* Thunb. *Ecotoxicology* **2011**, *20*, 698–705. [CrossRef]
38. Chen, L.; Hu, W.F.; Long, C.; Wang, D. Exogenous plant growth regulator alleviate the adverse effects of U and Cd stress in sunflower (*Helianthus annuus* L.) and improve the efficacy of U and Cd remediation. *Chemosphere* **2021**, *262*, 127809. [CrossRef]
39. Huang, Y.F.; Chen, J.H.; Sun, Y.M.; Wang, H.X.; Zhan, J.Y.; Huang, Y.N.; Zou, J.W.; Wang, L.; Su, N.N.; Cui, J. Mechanisms of calcium sulfate in alleviating cadmium toxicity and accumulation in pak choi seedlings. *Sci. Total Environ.* **2022**, *805*, 150115. [CrossRef]
40. Wang, H.F.; Xu, Y.X.; Yang, M.Q.; Jin, X.; Li, X.F.; Dai, Z.; Zhang, D.L.; Zhou, K.X.; Lin, L.J.; Wang, J. Exogenous strigolactone improves the cadmium accumulation in *Solanum nigrum* var. *Humile*. *Environ. Prog. Sustain. Energy* **2024**, *43*, 14379. [CrossRef]
41. Xu, Z.P.; Shao, T.J.; Dong, Z.B.; Li, S.L. Research progress of heavy metals in desert—Visual analysis based on CiteSpace. *Environ. Sci. Pollut. R.* **2022**, *29*, 43648–43661. [CrossRef] [PubMed]
42. Geng, Y.Q.; Zhang, N.G.; Zhu, R.J. Research progress analysis of sustainable smart grid based on CiteSpace. *Energy Strategy Rev.* **2023**, *48*, 101111. [CrossRef]
43. Chen, J.H.; Liu, X.R.; Zhou, S.R.; Kang, J. Knowledge mapping analysis of resilient shipping network using CiteSpace. *Ocean Coast. Manag.* **2023**, *244*, 106775. [CrossRef]
44. Li, H.K.; Li, B. The state of metaverse research: A bibliometric visual analysis based on CiteSpace. *J. Big Data* **2024**, *11*, 14. [CrossRef]
45. Geng, Y.Q.; Jiang, X.Y.; Bai, W.Q.; Yan, Y.; Gao, J. Research progress of tourism marketing over 30 years: Bibliometrics based on CiteSpace. *Ecol. Indic.* **2024**, *162*, 112059. [CrossRef]
46. Xiao, Y.T.; Sun, C.X.; Wang, D.Z.; Li, H.Q.; Guo, W. Analysis of hotspots in subsurface drip irrigation research using CiteSpace. *Agriculture* **2023**, *13*, 1463. [CrossRef]
47. Ye, H.R.; Du, Y.; Jin, Y.T.; Liu, F.Y.; He, S.S.; Guo, Y.H. Articles on hemorrhagic shock published between 2000 and 2021: A CiteSpace-Based bibliometric analysis. *Heliyon* **2023**, *9*, e18840. [CrossRef]
48. Wang, F.R.; Tan, B.; Chen, Y.; Fang, X.Y.; Jia, G.W.; Wang, H.Y.; Chen, G.; Shao, Z.Z. A visual knowledge map analysis of mine fire research based on CiteSpace. *Environ. Sci. Pollut. R.* **2022**, *29*, 77609–77624. [CrossRef]

49. Huang, J.H.; Xiang, S.; Chen, S.Q.; Wu, W.; Huang, T.Y.; Pang, Y. Perfluoroalkyl substance pollution: Detecting and visualizing emerging trends based on CiteSpace. *Environ. Sci. Pollut. R.* **2022**, *29*, 82786–82798. [CrossRef]
50. Chen, J.H.; Zhang, X.T.; Xu, L.; Xu, J.H. Trends of digitalization, intelligence and greening of global shipping industry based on CiteSpace Knowledge Graph. *Ocean Coast. Manag.* **2024**, *255*, 107206. [CrossRef]
51. Cui, L.H.; Tang, W.J.; Deng, X.S.; Jiang, B. Farm animal welfare is a field of interest in China: A bibliometric analysis based on CiteSpace. *Animals* **2023**, *13*, 3143. [CrossRef] [PubMed]
52. Yuan, X.Y.; Lai, Y. Bibliometric and visualized analysis of elite controllers based on CiteSpace: Landscapes, hotspots, and frontiers. *Front. Cell. Infect. Microbiol.* **2023**, *13*, 1147265. [CrossRef] [PubMed]
53. Mamat, M.; Li, L.; Kang, S.F.; Chen, Y.Y. Emerging trends on the anatomy teaching reforms in the last 10 years: Based on VOSviewer and CiteSpace. *Anat. Sci. Educ.* **2024**, *17*, 722–734. [CrossRef] [PubMed]
54. Li, H.B.; Xiang, Y.R.; Yang, W.J.; Lin, T.; Xiao, Q.K.; Zhang, G.Q. Green roof development knowledge map: A review of visual analysis using CiteSpace and VOSviewer. *Heliyon* **2024**, *10*, e24958. [CrossRef] [PubMed]
55. Che, S.P.; Kamphuis, P.; Zhang, S.; Zhao, X.Y.; Kim, J.H. A visualization analysis of crisis and risk communication research using CiteSpace. *Int. J. Environ. Res. Public Health* **2022**, *19*, 2923. [CrossRef]
56. Yang, X.F.; Liu, Q.F. Research foundation and hotspot analysis of urban road ecology—A bibliometric study based on CiteSpace. *Sustainability* **2024**, *16*, 5135. [CrossRef]
57. Li, X.; Du, J.; Long, H. A comparative study of Chinese and foreign green development from the perspective of mapping knowledge domains. *Sustainability* **2018**, *10*, 4357. [CrossRef]
58. Qiu, Y.R.; Liao, K.H.; Zou, Y.T.; Huang, G.Z. A bibliometric analysis on research regarding residential segregation and health based on CiteSpace. *Int. J. Environ. Res. Public Health* **2022**, *19*, 10069. [CrossRef]
59. Hu, W.; Li, C.; Ye, C.; Wang, J.; Wei, W.; Deng, Y. Research progress on ecological models in the field of water eutrophication: CiteSpace analysis based on data from the ISI web of science database. *Ecol. Model.* **2019**, *410*, 108779. [CrossRef]
60. Liu, C.; Yuan, Y.; Zhou, A.H.; Guo, L.F.; Zhang, H.R.; Liu, X.D. Development trends and research frontiers of preferential flow in soil based on CiteSpace. *Water* **2022**, *14*, 3036. [CrossRef]
61. Wang, B.; Liu, J.; Liu, Q.; Sun, J.B.; Zhao, Y.X.; Liu, J.; Gao, W.S.; Chen, Y.Q.; Sui, P. Knowledge domain and research progress in the field of crop rotation from 2000 to 2020: A scientometric review. *Environ. Sci. Pollut. Res.* **2023**, *30*, 86598–86617. [CrossRef] [PubMed]
62. Wu, L.J.; Miao, H.Y.; Liu, T.Z. Development in agricultural ecosystems' carbon emissions research: A visual analysis using CiteSpace. *Agronomy* **2024**, *14*, 1288. [CrossRef]
63. Chen, B.; Shin, S.; Wu, M.; Liu, Z. Visualizing the knowledge domain in health education: A scientometric analysis based on CiteSpace. *Int. J. Environ. Res. Public Health* **2022**, *19*, 6440. [CrossRef] [PubMed]
64. Pang, X.M.; Peng, Z.Y.; Zheng, X.; Shi, J.J.; Zhou, B.C. Analysis of research hotspots in COVID-19 genomics based on citespace software: Bibliometric analysis. *Front. Cell. Infect. Microbiol.* **2022**, *12*, 1060031. [CrossRef] [PubMed]
65. Liu, Y.J.; Li, Q.; Li, W.L.; Zhang, Y.; Pei, X.W. Progress in urban resilience research and hotspot analysis: A global scientometric visualization analysis using CiteSpace. *Environ. Sci. Pollut. Res.* **2022**, *29*, 63674–63691. [CrossRef]
66. Xu, C.; Yang, T.; Wang, K.; Guo, L.; Li, X.M. Knowledge domain and hotspot trends in coal and gas outburst: A scientometric review based on CiteSpace analysis. *Environ. Sci. Pollut. Res.* **2023**, *30*, 29086–29099. [CrossRef]
67. Li, J.; Ma, W.; Dai, X.; Qi, M.; Liu, B. China's policy environment's development and path from the perspective of policy sustainability: A visual analysis based on CNKI and WOS. *Sustainability* **2022**, *14*, 16435. [CrossRef]
68. Assche, F.V.; Clijsters, H. Effects of metals on enzyme activity in plants. *Plant Cell Environ.* **1990**, *13*, 195–206. [CrossRef]
69. Wang, W. Literature review on higher plants for toxicity testing. *Water Air Soil Pollut.* **1991**, *59*, 381–400. [CrossRef]
70. Li, M.; Wang, Y.; Xue, H.; Wu, L.; Wang, Y.; Wang, C.; Gao, X.; Li, Z.; Zhang, X.; Hasan, M.; et al. Scientometric analysis and scientific trends on microplastics research. *Chemosphere* **2022**, *304*, 135337. [CrossRef]
71. Qin, F.; Zhu, Y.; Ao, T.; Chen, T. The development trend and research frontiers of distributed hydrological models-visual bibliometric analysis based on Citespace. *Water* **2021**, *13*, 174. [CrossRef]
72. Zhang, Y.; Li, C.; Ji, X.; Yun, C.; Wang, M.; Luo, X. The knowledge domain and emerging trends in phytoremediation: A scientometric analysis with CiteSpace. *Environ. Sci. Pollut. Res.* **2020**, *27*, 15515–15536. [CrossRef] [PubMed]
73. Chen, H.Y.; Tian, Y.X.; Cai, Y.X.; Liu, Q.Y.; Ma, J.; Wei, Y.; Yang, A.F. A 50-year systemic review of bioavailability application in soil environmental criteria and risk assessment. *Environ. Pollut.* **2023**, *335*, 122272. [CrossRef] [PubMed]
74. CoKim, J.; Perez, C. Co-Authorship Network Analysis in Industrial Ecology Research Community. *J. Ind. Ecol.* **2015**, *19*, 222–235.
75. Tang, Y.; Wang, L.M.; Xie, Y.D.; Yu, X.N.; Lin, L.J.; Li, H.X.; Liao, M.G.; Wang, Z.H.; Sun, G.C.; Liang, D.; et al. Effects of intercropping accumulator plants and applying their straw on the growth and cadmium accumulation of *Brassica chinensis* L. *Environ. Sci. Pollut. Res.* **2020**, *27*, 39094–39104. [CrossRef]
76. Liu, L.; Li, X.F.; Huang, K.W.; Zhu, Y.; Li, A.H.; Ao, Q.M.; Liao, M.A.; Lin, L.J. Salicylic acid promotes growth and affects cadmium accumulation of *Cyphomandra betacea* seedlings. *Environ. Prog. Sustain. Energy* **2022**, *41*, e13787.

77. Tang, W.; Liang, L.; Xie, Y.D.; Li, X.M.; Lin, L.J.; Huang, Z.; Sun, B.; Sun, G.C.; Tu, L.H.; Li, H.X.; et al. Foliar application of salicylic acid inhibits the cadmium uptake and accumulation in lettuce (*Lactuca sativa* L.). *Front. Plant Sci.* **2023**, *14*, 1200106. [CrossRef]
78. Liu, Y.J.; Deng, L.L.; Wang, Y.X.; Xia, X.M.; Cui, T.H.; Lin, L.J.; Deng, Q.X.; Zhang, H.F. Effects of *Solanum photeinocarpum* and its post grafting generations straw on cadmium accumulation of loquat seedlings. *Int. J. Environ. Anal. Chem.* **2023**, *103*, 3604–3615. [CrossRef]
79. Yan, T.T.; Xue, J.H.; Zhou, Z.D.; Wu, Y.B. The trends in research on the effects of biochar on soil. *Sustainability* **2020**, *12*, 7810. [CrossRef]
80. Aleixandre-Benavent, R.; Aleixandre-Tudó, J.L.; Castelló-Cogollos, L.; Aleixandre, J.L. Trends in scientific research on climate change in agriculture and forestry subject areas (2005–2014). *J. Clean. Prod.* **2017**, *147*, 406–418. [CrossRef]
81. Zhang, Z. Visualization analysis of the development trajectory of knowledge sharing in virtual communities based on CiteSpace. Multimed. *Tools Appl.* **2019**, *78*, 29643–29657. [CrossRef]
82. Genchi, G.; Sinicropi, M.S.; Lauria, G.; Carocci, A.; Catalano, A. The effects of cadmium toxicity. *Int. J. Environ. Res. Public Health* **2020**, *17*, 3782. [CrossRef] [PubMed]
83. Gallego, S.M.; Pena, L.B.; Barcia, R.A.; Azpilicueta, C.E.; Iannone, M.F.; Rosales, E.P.; Zawoznik, M.S.; Groppa, M.D.; Benavides, M.P. Unravelling cadmium toxicity and tolerance in plants: Insight into regulatory mechanisms. *Environ. Exp. Bot.* **2012**, *83*, 33–46. [CrossRef]
84. Kubier, A.; Wilkin, R.T.; Pichler, T. Cadmium in soils and groundwater: A review. *Appl. Geochem.* **2019**, *108*, 104388. [CrossRef]
85. Haider, F.U.; Liqun, C.; Coulter, J.A.; Cheema, S.A.; Wu, J.; Zhang, R.; Ma, W.; Farooq, M. Cadmium toxicity in plants: Impacts and remediation strategies. *Ecotoxicol. Environ. Saf.* **2021**, *211*, 111887. [CrossRef]
86. Ismael, M.A.; Elyamine, A.M.; Moussa, M.G.; Cai, M.; Zhao, X.; Hu, C. Cadmium in plants: Uptake, toxicity, and its interactions with selenium fertilizers. *Metallomics* **2019**, *11*, 255–277. [CrossRef]
87. Shahid, M.; Dumat, C.; Khalid, S.; Niazi, N.K.; Antunes, P.M. Cadmium bioavailability, uptake, toxicity and detoxification in soil-plant system. *Rev. Environ. Contam. Toxicol.* **2017**, *241*, 73–137.
88. Rizwan, M.; Ali, S.; Adrees, M.; Ibrahim, M.; Tsang, D.C.; Zia-ur-Rehman, M.; Zahir, Z.A.; Rinklebe, J.; Tack, F.M.G.; Ok, Y.S. A critical review on effects, tolerance mechanisms and management of cadmium in vegetables. *Chemosphere* **2017**, *182*, 90–105. [CrossRef]
89. El Rasafi, T.; Oukarroum, A.; Haddioui, A.; Song, H.; Kwon, E.E.; Bolan, N.; Tack, F.M.G.; Sebastian, A.; Prasad, M.N.V.; Rinklebe, J. Cadmium stress in plants: A critical review of the effects, mechanisms, and tolerance strategies. *Crit. Rev. Environ. Sci. Technol.* **2020**, *52*, 675–726. [CrossRef]
90. He, S.Y.; Yang, X.E.; He, Z.L.; Baligar, V.C. Morphological and physiological responses of plants to cadmium toxicity: A review. *Pedosphere* **2017**, *27*, 421–438. [CrossRef]
91. Qin, S.Y.; Liu, H.E.; Nie, Z.J.; Rengel, Z.; Gao, W.; Li, C.; Zhao, P. Toxicity of cadmium and its competition with mineral nutrients for uptake by plants: A review. *Pedosphere* **2020**, *30*, 168–180. [CrossRef]
92. Zhao, X.Y.; Nan, D.Y.; Chen, C.M.; Zhang, S.A.; Che, S.P.; Kim, J.H. Bibliometric study on environmental, social, and governance research using CiteSpace. *Front. Environ. Sci.* **2023**, *10*, 1087493. [CrossRef]
93. Shen, Q.; Sun, Q.; Zhao, A. Bibliometric analysis of research on China's rural environmental governance in CNKI and WOS. *Front. Environ. Sci.* **2024**, *12*, 1429595. [CrossRef]
94. Tian, Y.; Ding, J.; Zhu, D.; Morris, N. The effect of the urban wastewater treatment ratio on agricultural water productivity: Based on provincial data of China in 2004–2010. *Appl. Water Sci.* **2018**, *8*, 144–210. [CrossRef]
95. Bahmani, R.; Modareszadeh, M.; Bihamta, M.R. Genotypic variation for cadmium tolerance in common bean (*Phaseolus vulgaris* L.). *Ecotoxicol. Environ. Saf.* **2020**, *190*, 110178. [CrossRef]
96. Belimov, A.A.; Safronova, V.I.; Demchinskaya, S.V.; Dzyuba, O.O. Intraspecific variability of cadmium tolerance in hydroponically grown Indian mustard (*Brassica juncea* (L.) *Czern.*) seedlings. *Acta Physiol. Plant.* **2007**, *29*, 473–478. [CrossRef]
97. Zhao, J.X.; Bai, Y.B.; Yang, Y.J.; Li, X.L. The impact of aerobics on mental health and stress levels: A visualization analysis of the CiteSpace map. *PLoS ONE* **2024**, *19*, e0300677. [CrossRef]
98. Wang, W.G.; Wang, H.; Yao, T.; Li, Y.D.; Yi, L.Z.; Gao, Y.; Lian, J.; Feng, Y.L.; Wang, S.P. The top 100 most cited articles on COVID-19 vaccine: A bibliometric analysis. *Clin. Exp. Med.* **2023**, *23*, 2287–2299. [CrossRef]
99. Zhang, Y.S.; Lu, J.; Chang, T.Y.; Tang, X.L.; Wang, Q.; Pan, D.; Wang, J.; Nan, H.M.; Zhang, W.; Liu, L.; et al. A bibliometric review of *Glycyrrhizae Radix et Rhizoma* (licorice) research: Insights and future directions. *J. Ethnopharmacol.* **2024**, *321*, 117409. [CrossRef]
100. Abbas, N.N.; Ahmed, T.; Shah, S.H.U.; Omar, M.; Park, H.W. Investigating the applications of artificial intelligence in cyber security. *Scientometrics* **2019**, *121*, 1189–1211. [CrossRef]

101. Dhital, S.; Rupakheti, D.; Rupakheti, M.; Yin, X.F.; Liu, Y.L.; Mafiana, J.J.; Alareqi, M.M.; Mohamednour, H.; Zhang, B.Z. A scientometric analysis of indoor air pollution research during 1990–2019. *J. Environ. Manag.* **2022**, *320*, 115736. [CrossRef] [PubMed]
102. Abouzid, M.; Karazniewicz-Lada, M.; Abdelazeem, B.; Brasic, J.R. Research trends of vitamin D metabolism gene polymorphisms based on a bibliometric investigation. *Genes* **2023**, *14*, 215. [CrossRef] [PubMed]
103. Zhou, X.D.; Zhao, G.H. Global liposome research in the period of 1995–2014: A bibliometric analysis. *Scientometrics* **2015**, *105*, 231–248. [CrossRef]
104. Chen, C.M. CiteSpace II: Detecting and visualizing emerging trends and transient patterns in scientific literature. *J. Am. Soc. Inf. Sci.* **2006**, *57*, 359–377. [CrossRef]
105. Liu, Z.L.; Chen, W.; He, X.Y. Influence of $Cd^{2+}$ on growth and chlorophyll fluorescence in a hyperaccumulator—*Lonicera japonica* Thunb. *J. Plant Growth Regul.* **2015**, *34*, 672–676. [CrossRef]
106. Liu, Z.L.; Tian, L.; Chen, M.D.; Zhang, L.H.; Lu, Q.X.; Wei, J.B.; Duan, X.B. Hormesis responses of growth and photosynthetic characteristics in *Lonicera japonica* Thunb. to cadmium stress: Whether electric field can improve or not? *Plants* **2023**, *12*, 933. [CrossRef]
107. Haider, M.S.; Sheikh, T.M.M.; Jiu, S.T.; Aleem, M.; Shafqat, W.; Shoukat, K.; Khan, N.; Jaskani, M.J.; Naqvi, S.A.; Ercisli, S.; et al. Genome-wide identification of strawberry metal tolerance proteins and their expression under cadmium toxicity. *Horticulturae* **2022**, *8*, 477. [CrossRef]
108. Wang, J.W.; Li, H.Y.; Zou, D.D.; Zhao, J.F.; Fan, L.X.; Wu, T. Transcriptome profile analysis of cadmium tolerance in Chinese flowering cabbage. *Hortic. Environ. Biotechnol.* **2017**, *58*, 56–65. [CrossRef]

MDPI AG
Grosspeteranlage 5
4052 Basel
Switzerland
Tel.: +41 61 683 77 34

*Horticulturae* Editorial Office
E-mail: horticulturae@mdpi.com
www.mdpi.com/journal/horticulturae

www.ingramcontent.com/pod-product-compliance
Lightning Source LLC
Chambersburg PA
CBHW061556210926
51850CB00009B/2116
*9783725853434*